IRON-MAKING SOCIETIES

Iron-Making Societies

Early Industrial Development in Sweden and Russia, 1600–1900

Edited by

Maria Ågren

Berghahn Books
NEW YORK • OXFORD

Published in 1998 by

Berghahn Books

Library of Congress Cataloging-in-Publication Data

Iron-making societies : early industrial development in Sweden and Russia, 1600–1900 / edited by Maria Ågren.

p. cm.

Includes bibliographical references and index.

ISBN 1-57181-955-X (alk. paper)

1. Iron industry and trade—Sweden—History. 2. Iron industry and trade—Russia—History. 3. Iron and Steel workers—Sweden—History. 4. Iron and Steel workers—Russia—History. I. Ågren, Maria.

HD9525.S852I75 1997

338.4'76691'09485—dc21 97-31843

CIP

British Library Cataloguing in Publication Data

A catalogue record for this book is available from the British Library.

Printed in the United States on acid-free paper

CONTENTS

Part III: The Institutional Environment and How It Changed

List of Illustrations

Figures

Table

PREFACE

In the summer of 1989, after hearing a couple of papers dealing with traditional iron-making in Sweden at a small conference of Russian and Swedish historians, Academician Ivan Kovalchenko remarked that this field would be very suitable for a comparative study encompassing Sweden and Russia. His observation was taken seriously and a year later a joint project was taking shape under the leadership of Professor Veniamin Alekseiev in Ekaterinburg and Professor Rolf Torstendahl in Uppsala.

The researchers, ten from Russia and six from Sweden, were recruited from the Institute for Archaeology and History in Ekaterinburg and the Departments of History and Economic History in Uppsala, with the subsequent addition of one researcher from Stockholm. The joint groups met for the first time in Ekaterinburg (then still Sverdlovsk) in early 1991 and then once a year over the next three years, in Sweden and Russia in turn. Four meetings, at which presented papers were discussed in depth have formed the basis for this book.

The discussions were made possible by our interpreters, for our common command of English turned out to be insufficient. Irina Rybina, Elena Alekseieva, Helene Carlbäck Isotalo and Vadim Azbel showed great interest and devotion to the task of making questions and answers intelligible. Elisabeth Marklund Sharapova provided invaluable assistance to the Swedish project members and the editor by translating Russian texts and standardising the use of Russian terms and their spelling. Transliterations are in accordance with Library of Congress principles. Chapters 2 and 5 have been translated by David Grist. All other chapters were written by the Swedish members of the project, and Martin Naylor revised the language. The obstacles of language, distance and deficient communications technology have been a challenge to our scholarly curiosity.

This book has been edited in Uppsala on the basis of the final papers given by the participants and has been made possible by their painstaking labours in front of the computer in the summer of 1994 to produce basic chapter texts. Maria Ågren has organised and supervised the efforts of the others, but each of the Swedish researchers concerned is responsible for the editing of the chapters to which they contribute. However, the book as a whole is a collective product of all the members of the two research groups.

This joint project would not have been possible without generous grants from the fund for Swedish-Russian research cooperation (administered by the Royal Academy of Sciences in Stockholm) and the Swedish Council for Research in the Humanities and Social Sciences.

Uppsala, May 1997 Rolf Torstendahl

Acknowledgements

Figures 3.1 to 9.2 are from *Jernkontoret*, Stockholm, Sweden. Figures 9.3 and 9.4 are from the private collection of Göran Rydén, Uppsala, Sweden. The maps are from Uppsala University Library, Sweden.

The publication of this book was sponsored by the Swedish Council for Research in the Humanities and Social Sciences.

List of Abbreviations

F	(Russian) fond [archival record group]
GAPO	Gosudarstvennyi Arkhiv Permskoi Oblasti [State Archive of the Perm' Region]
GASO	Gosudarstvennyi Arkhiv Sverdlovskoi Oblasti [State Archive of the Sverdlovsk (now Ekaterinburg) Region]
JKA	(Swedish) Jernkontorets Annaler [Journal of the Ironmasters' Association]
KSFPR	(Swedish) Kongliga Stadgar, Förordningar, Privilegier och Resolutioner angående Justitien och Hushållningen wid Bergwerken och Bruken
L	(Russian) list [folio, leaf, sheet]
Op	(Russian) opis' [inventory]
PSZ	(Russian) Polnoe sobranie zakonov Rossiiskoi imperii [Code of laws]
RA	(Swedish) Riksarkivet [National Archives]
TsGADA	Tsentral'nyi Gosudarstvennyi Arkhiv Drevnikh Aktov [Central State Archive of Ancient Documents]
TsGIA	Tsentral'nyi Gosudarstvennyi Istoricheskii Arkhiv [Central State Historical Archive]

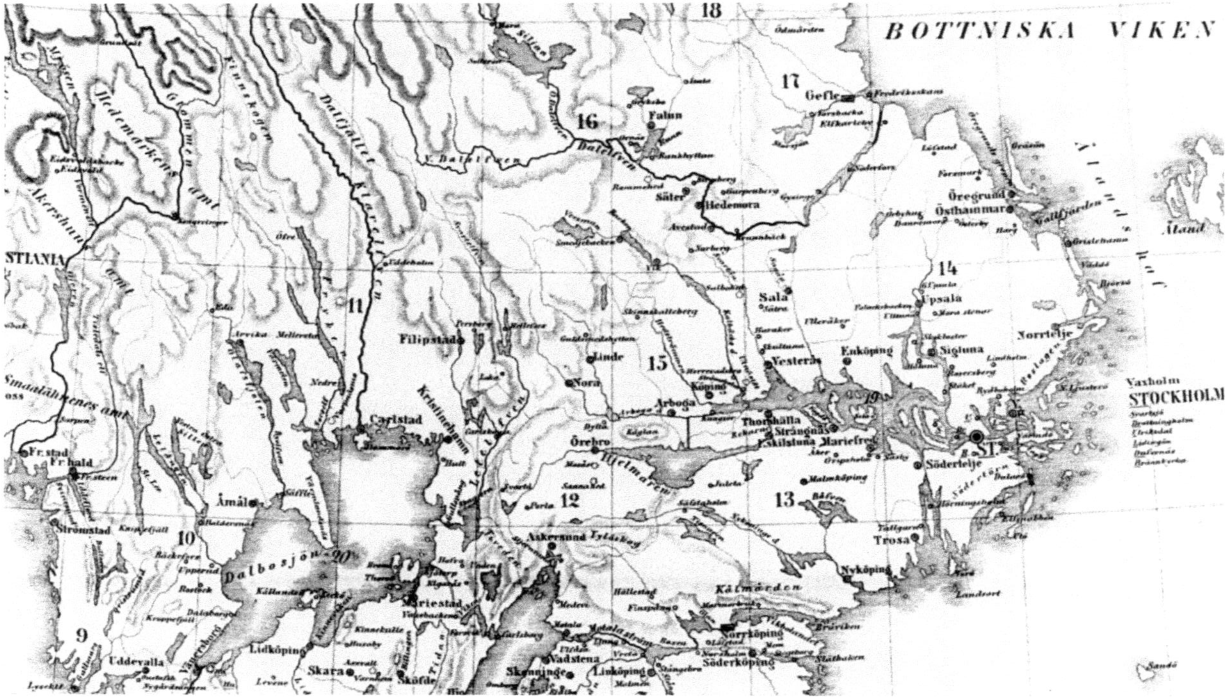

Central Sweden with the Bergslagen region north of the central Swedish lakes

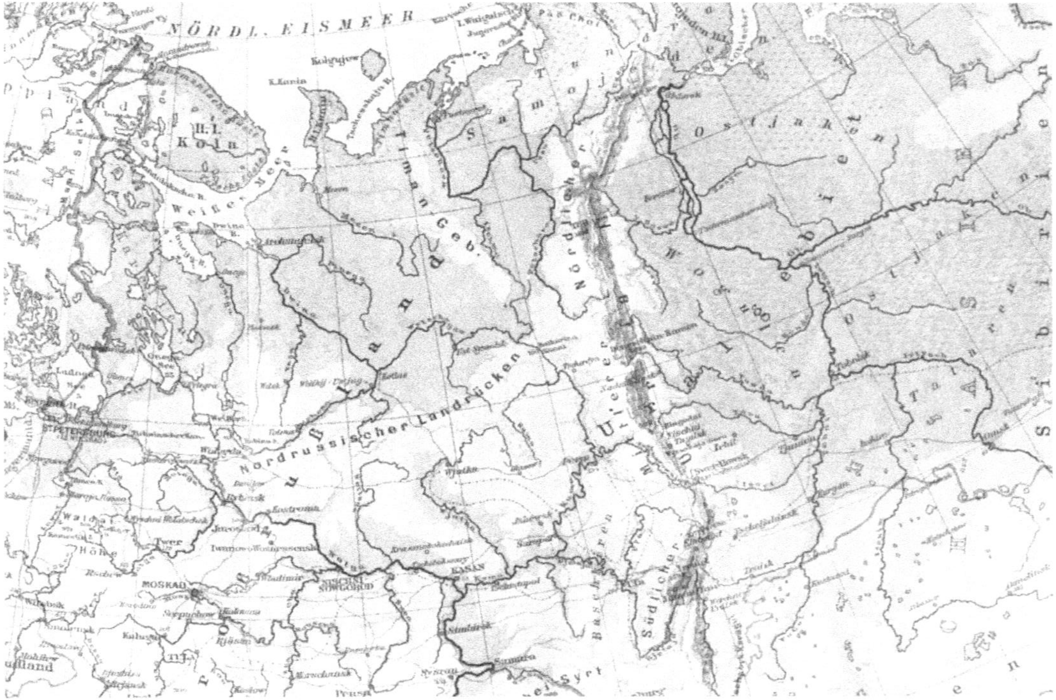

The Ural region of Russia with the towns Perm′ and Sverdlovsk (Ekaterinburg) in the middle

PART I

Chapter One

INTRODUCTION

Swedish and Russian Iron-Making As Forms of Early Industry

Maria Ågren

In 1738, Daniel Tilas, a member of the Swedish Board of Mines, was sent to Russia to investigate the state of the Russian metal industry. Travelling first through the province of Ingria, close to St Petersburg and until quite recently part of the Swedish realm, Tilas visited the deserted copper works at Krasna selo, with its adjacent brass works, and then proceeded to the royal ironworks at Systerbäck. Here, he and his fellow travellers were greeted by Wilhelm De Hennin, who at the time was in charge of Systerbäck and of the rifle factory at Tula, but who had previous experience from the great ironworks in Siberia. Showing his guests round, De Hennin confidently assured them that Systerbäck was a highly profitable production plant, but Tilas was not convinced. In his report to the Swedish Board of Mines, he frankly stated that Systerbäck was nothing but theatre, a place to which it was handy to take foreigners in order to impress them. Continuing his journey into the Olonets region, beyond Lake Onega, he was equally sceptical about what he was shown.[1]

A contemporary of Linnaeus, Tilas paid close attention to the country's endowment of natural resources. His report contains meticulous descriptions of the geology and geography of the places he saw. And, like Linnaeus, he was deeply interested in the possible uses of various resources. In particular, he (and the Board of Mines) wanted to know whether or not the Russian metallurgical

1. D. Tilas, 'Kort Berättelse om en innom Ryska Gräntsen giord Resa, ifrån den 4. Februarii til den 6. April Åhr 1738', Bergskollegii Arkiv. RA (National Archives, Stockholm).

enterprises were economically viable. Tilas therefore tried to calculate the production costs of the different metalworks, comparing these costs with the prices obtainable on the market. To make these calculations, he compiled information about woods, wages, and charcoal prices, making special comment when he was denied access to the account books and hence unable to be more precise. Often, he concluded that the surrounding countryside was sparsely wooded, and that charcoal was expensive. However, Tilas remarked that the difference between production costs and market prices would have been even smaller, and Russian industry even less profitable, had the owners been forced to pay for the work performed by neighbouring peasants.[2]

Tilas's report was not the only one submitted to the Board. During the eighteenth century, several Swedes travelled through Russia, observing the iron industry and providing the Swedish Crown with informed judgements which were important to decision-making in Sweden. Similar information was provided by Swedish envoys. The Russians and their doings thus attracted the keen attention of the Swedes in the 1730s and 1740s. In itself, this was nothing new. There was an old tradition in Sweden of observing the country's eastern neighbour, with whom several wars had been fought.[3] What is more, the attention was mutual. On the Russian side, Sweden was also the subject of interest and curiosity, awe and, sometimes, admiration. For instance, in the days of Peter the Great Russia had systematically mapped out the Swedish administrative system and faithfully copied it.[4] Russians had also travelled through Sweden and studied Swedish ironworks. What might be seen as new is the fact that Sweden now regarded Russia as a dangerous competitor in the economic field and, more specifically, as an intruder in the mercantile sphere that had been dominated by Sweden since the late seventeenth century: the international iron trade.

From the beginning of the seventeenth century, Swedish iron exports had started to rise dramatically. It has been estimated that about 3,000 tons were shipped abroad annually over the period 1627 to 1632. By 1680, the figure had risen to about 27,000 tons, and by the 1730s, even further, to a little more than 40,000 tons a

2. Ibid., 78.

3. K. Tarkiainen, *'Vår gamble Arffiende Ryssen'. Synen på Ryssland i Sverige 1595–1621 och andra studier kring den svenska Rysslandsbilden från tidigare stormaktstid*, Uppsala 1974.

4. C. Peterson, *Peter the Great's Administrative and Judicial Reforms: Swedish Antecedents and the Process of Reception*, Stockholm 1979.

year. The increasing volume of exports was mainly directed towards Western Europe, and from the end of the 1660s Britain was the most important market for Swedish iron. In the first half of the eighteenth century, the majority (75 per cent) of all the iron imported to Britain came from Sweden.[5] Swedish iron thus played a prominent role in the British economy, and Britain was regarded as an extremely valuable commercial partner among Swedish iron producers.

By the mid-eighteenth century, however, Sweden's position was seriously threatened by the appearance of Russian iron on the British market. In 1734, the Swedish envoy in London, Baron Sparre, reported to the Board of Mines that there were loud complaints among British merchants importing Swedish iron. Russian iron was sold at a considerably lower price, although its quality was, they said, equal to that of the Swedish product.[6] It was widely feared that the Russians would make further inroads on the British market, and this was indeed what happened. By the 1780s, Russia had succeeded in obtaining 63 per cent of the British market for bar iron, whereas Sweden retained a mere 35 per cent.

It is against this backdrop that the intense Swedish interest in the Russian iron industry should be understood. Tilas was just one representative, and not the first one either, of a general Swedish eagerness to fathom the potential of Russian iron, to evaluate the quality of Russian ores and to assess the possible cost benefits inherent in the social structure of Russia. For instance, Tilas and many other observers emphasised the subjugated position of the Russian peasantry, or the fact that Russian workmen were serfs.

However, the iron produced in Ingria and in the Olonets region was not that which outrivalled Swedish iron on the British market. The derisory remarks by Tilas concerning west Russian iron may not have been totally unwarranted. Rather, the danger came from the newly founded works in the Ural mountains in Siberia. According to Baron Sparre, this iron was equal in quality to the much-famed Swedish Öregrund iron.[7] It is hardly surprising that the Swedish Board of Mines should request information on these

5. E. F. Heckscher, 'Un grand chapitre de l'histoire du fer: le monopole suédois', *Annales ESC* 1932, nos 14 and 15, 234; K.-G. Hildebrand, *Fagerstabrukens historia. Del I. Sexton- och sjuttonhundratalen*, Uppsala 1957, 35ff., 49, 106.

6. Minister Baron Sparre, 'Underdånig Berättelse den 26 November 1734', Bergskollegii Arkiv. RA.

7. See for example the reports of the Swedish envoy von Ditmer (1726) and of the mining official Simon Lindhielm (1726), Bergskollegii Arkiv, RA.

plants as well. The Ural works were visited by the Swedish prisoner of war Petter Schönström as early as 1721, and by the Swede Nils Psilanderhielm, who held the position of Russian *Berghauptman* from 1737 to 1740.[8]

Schönström had travelled throughout the entire Siberian mining area. He had seen examples of small-scale peasant production of iron near the towns of Kungur and Tomsk, but he had also seen the still relatively few Crown-owned works and the one privately owned ironworks belonging to Nikita Demidov, a former master craftsman to whom Peter the Great had granted many and ample privileges. At the time when Psilanderhielm visited the Urals, there were twelve ironworks owned by the Crown and around twenty privately owned ironworks.[9] The most renowned among these private works were those belonging to the Stroganov and Demidov families. Psilanderhielm noticed that practically all of the Ural works had sufficient wood, and that many of them used mountain ores with low pyrite contents.[10] Moreover, it was not just iron, but also copper which was produced in the region, sometimes at one and the same production plant.[11]

Although iron and copper had been produced in the Ural region since as early as the seventeenth century,[12] the Ural industry was essentially the creation of Emperor Peter the Great. Unhappy about the fact that Russia had to import iron from its enemy Sweden, Peter initiated a search for minerals in the eastern parts of his realm and invited mining experts from Saxony. V.N. Tatishchev was put in charge of the task of founding Crown works in the Ural mountains. Under the subsequent leadership of Wilhelm De Hennin (from 1722 to 1734), the garrison town of Ekaterinburg was founded and subsequently several huge, state-owned ironworks were set up, as witnessed by Psilanderhielm. Iron production was not confined to state-owned plants, however. From the outset, two families (the Demidovs and the Stroganovs) established themselves as leading producers of iron and copper. When De Hennin had returned to western Russia in 1734, where Daniel

8. P. Schönström, 'Berättelse om the Ryske Bergwärkens tilstånd, upgifwen wid återkomsten utur Ryska fångenskapen'; N. Psilanderhielm, 'Berättelse om Ryska och Sibiriska Jernverken Ingifwen til Kongl. BergsCollegium den 8. Decembr. 1743', Bergskollegii Arkiv, RA.

9. R. Portal, *L'Oural au XVIIIe siècle*, Paris 1950, 126ff. (figures based on contemporary reports by De Hennin and Tatishchev).

10. Psilanderhielm, 'Berättelse om Ryska', 16–26.

11. Portal, *L'Oural*, 52.

12. D. A. Kashintsev, *Istoriia metallurgii Urala*, Moscow, Leningrad 1939, 22f.

Tilas met him, his position was once again given to Tatishchev, who allowed the industry to expand even further.[13]

Seventy-one iron or copper works were erected in the first half of the eighteenth century, and 101 in the second half. Production in 1750 can be estimated at around 23,000 tons of cast iron and a little less than 15,000 tons of bar iron. By 1800, these figures had risen to 123,000 and 87,000 tons, respectively. The massive state presence was gradually modified as many of the previously state-owned works were sold to private owners, mainly in the 1750s.[14] In 1750, almost half the blast furnaces in the Urals belonged to the state, but in 1800 only 12 per cent remained state-owned.[15] At the same time as many of the Ural metalworks went privatised, the ownership structure changed in another way as well. Groups of metalworks belonging to one and the same owner tended to become interdependent, forming a number of vast mining districts that had much in common with large landed estates.

In Sweden, iron was produced in the Uppland region, north of Stockholm, in the vast area of Norrland, and in Småland, close to the Danish border. However, the traditional heartland of iron-making was Bergslagen, where it had a much longer history than in the Urals. Iron ore had been extracted and processed at least since the Middle Ages, and Swedish iron in the form of *osmund* had long been sold on markets in northern German cities. However, a decisive turning-point was reached in the early seventeenth century. As Europe was ravaged by protracted wars, Sweden's iron exports started to rise substantially. Now, bar iron superseded *osmund* iron as the main product. New forges were established on the periphery of Bergslagen and in the province of Uppland, and new owners appeared. Most of the plants that had been state-owned in the previous century were sold into the private sector. Merchants such as Louis De Geer and Wilhelm De Besche brought capital and labour from abroad.

The traditional producers of Swedish iron, peasants known as *bergsmän*, who had previously been in charge of the entire production process, were now confined to the task of producing pig-iron, from which bar iron was subsequently made. Pig-iron production remained comparatively small-scale, though, taking place within the limits of individual *bergsman* households and relying on the

13. Portal, *L'Oural*, 60–88.

14. Some of these works were later taken back by the State Treasury owing to serious mismanagement. See Portal, *L'Oural*, 101ff.

15. *Istoriia Urala s drevneishikh vremën do 1861 g.*, Moscow 1989, 270, 274.

wooden resources of the individual farmsteads. The blast furnace, however, was owned by the *bergsman* community as a whole and was used by each of the members in turn.

The entire Swedish industry was supervised by the Board of Mines (founded in 1649). Through meticulous regulations, the Board intervened in most matters of importance to Swedish iron production, and through a comparatively extensive administrative system it had at its disposal fairly efficient means of enforcing its intentions. The whole country was divided into administrative units, each headed by a *bergmästare* who carried out inspections and reported to his superiors in Stockholm. Likewise, the mining areas had special courts to deal with disputes relating to metal production.[16] Interestingly, parts of this administrative system had been copied by the Russians in the early eighteenth century.[17] A Russian Board of Mines was also established, for example, although after the reign of Peter the Great it was dissolved, not to be re-established until the late eighteenth century. It was not until the early nineteenth century that the ideas of state regulation and supervision seem to have been successfully implemented in Russia. At the same time, the Swedish mining administration was finally dismantled, as more liberal economic ideas made the whole concept of regulation appear outdated and even detrimental to society.

Compared to the Ural ironworks, the Swedish production plants were quite small. The original state plan concerning the Ural region was to concentrate all stages of production within a couple of large units, modelled as self-sufficient landed estates. In Sweden, by contrast, concentration was not a goal in the seventeenth and eighteenth centuries. Mining, smelting and forging were generally not intended to be carried out in the same place. Production units were scattered, spatially, and different social estates were put in charge of different stages of production. As time wore on however, marked tendencies towards concentration also appeared in Sweden. Ironmasters gradually attained part ownership and then complete ownership of the blast furnaces. In 1750, approximately 50 per cent of all blast furnaces belonged to ironmasters. Later on, mines also became the property of private companies. This trend was accompanied by changes in the social organisation

16. An excellent survey of the history of the Swedish iron industry is provided in K.-G. Hildebrand, *Swedish Iron in the Seventeenth and Eighteenth Centuries. Export Industry before the Industrialization*, Stockholm 1992.

17. Peterson, *Peter the Great's*, 368ff.

of work in mines, furnaces and forges, and in the woods, where charcoal was produced.

The Role of Early Industry

Iron from the Ural Mountains and from Sweden was thus competing for the British market in the middle of the eighteenth century. Russian iron won. Sweden retained only a relatively small share for its famous Öregrund iron. The details of this are well known, of course, having already attracted the attention of several scholars. In Swedish historiography, economic historians such as Eli F. Heckscher, Karl-Gustaf Hildebrand, Artur Attman, Rolf Adamson and Staffan Högberg have made careful investigations of the fate of the Swedish iron trade, not only in the eighteenth century but also before and since.[18] In Russian historiography, the Russian iron trade was mainly studied by S.G. Strumilin, who provided a huge body of statistical data upon which researchers still rely.[19]

However, the early Swedish iron industry has also attracted the interest of social geographers and social historians. Often, investigations have taken the shape of local studies, focusing on a specific ironworks or region, but within this fairly limited space important historical problems have been handled. For instance, the effects of ironworks on the local settlement structure, resulting in new crofts, have been discussed, as have forest resources and their possible insufficiency as a source of charcoal. Here, names such as Helge Nelson, Jalmar Furuskog, Gunnar Arpi, Olof Nordström and Gabriel Bladh should be mentioned.[20] In the field of social history, the conditions faced by the industrial labour force and the

18. Heckscher, 'Un grand chapitre'; E. F. Heckscher, *Sveriges ekonomiska historia från Gustav Vasa 1:1–2:2*, Stockholm 1935–1949; *Svensk handelsstatistik 1637–1737*, eds E. F. Heckscher and B. Boëthius, Stockholm 1938; Hildebrand, *Fagerstabruken* and *Swedish Iron*; A. Attman, *Fagerstabrukens historia. Adertonhundratalet*, Uppsala 1958; A. Attman, *Svenskt järn och stål 1800–1914*, Stockholm 1986; R. Adamson, *De svenska järnbrukens storleksutveckling och avsättningsinriktning 1796–1860*, Göteborg 1963; R. Adamson, *Järnavsättning och bruksfinansiering 1800–1860*, Göteborg 1966; S. Högberg, *Utrikeshandel och sjöfart på 1700-talet*, Stockholm 1969.

19. S.G. Strumilin, *Istoriia chërnoi metallurgii v SSSR*, Moscow 1954.

20. H. Nelson, 'En bergslagsbygd, en historisk-geografisk överblick', *Ymer* 1913; J. Furuskog, *De värmländska järnbruken*, Filipstad 1924; G. Arpi, *Den svenska järnhanteringens träkolsförsörjning, 1830–1950*, Stockholm 1951; O. Nordström, *Relationer mellan bruk och omland i östra Småland 1750–1900*, Lund 1952; G. Bladh, *Finnskogens landskap och människor under fyra sekler – en studie av natur och samhälle i förändring*, Göteborg 1995.

phenomenon of peasant and forgeman indebtedness have resulted in investigations by for example, Emil Sommarin, Bertil Boëthius and Sigvard Montelius.[21] It is also within the social-historical field that many of the authors of this book have made their earlier contributions. Comparatively little has been said about the political and institutional context of the early Swedish iron industry, but recently Per-Arne Karlsson has highlighted important aspects of this.[22]

In Russian historiography, one of the first – and definitely one of the most influential – discussions of the Ural metal industry was that provided by Lenin in *The Development of Capitalism in Russia*. Here, Lenin declared that the Ural industry was doomed to stagnation, rather than successful capitalism, because of the extensive use of serfs or otherwise bound labourers.[23] Whether the Ural industry bore the seeds of capitalism or not has continued to put its imprint on Russian historiography, as has the role of bound labour. Much attention has thus been devoted to whether or not the Ural enterprises should be classified as manufactures.[24] Problems of classification have also intrigued historians studying the 'ascribed peasants', the question being whether they were *de facto* serfs or not. Here, important investigations were provided by V.I. Semëvskii, for example, who argued that they must be regarded as free state peasants – an opinion which most, if not all, contemporary Russian historians seem to share.[25] By contrast, fairly limited attention has been paid to the function of the household, or to small-scale peasant production of iron.[26]

21. E. Sommarin, *Bidrag till kännedom om arbetareförhållanden vid svenska bergverk och bruk i äldre tid till omkring 1720*, Lund 1908; B. Boëthius, *Gruvornas, hyttornas och hamrarnas folk. Bergshanteringens arbetare från medeltiden till gustavianska tiden*, Stockholm 1951; S. Montelius, *Fagerstabrukens historia. Arbetare och arbetarförhållanden*, Uppsala 1959; S. Montelius, *Säfsnäsbrukens arbetskraft och försörjning 1600–1865*, Falun 1962; *Iggesunds Bruk 1685–1869*, Iggesund 1985.

22. P.-A. Karlsson, *Järnbruken och ståndssamhället. Institutionell och attitydmässig konflikt under Sveriges tidiga industrialisering 1700–1770*, Stockholm 1990.

23. V.I. Lenin, *The Development of Capitalism in Russia*, Moscow 1974, (first edition 1899), 489ff.

24. F. Ia. Polianskii, *Ekonomicheskii stroi manufaktury v Rossii xviii veka*, Moscow 1956; *Perechod ot feodalizma k kapitalizmu v Rossii: Materialy Vsesoiuznoi diskussii*, Moscow 1969; E. I. Zaozërskaia, *U istokov krupnogo proizvodstva v russkoi promyshlennosti xvi-xvii vekov*, Moscow 1970; I. V. Putilova, *Kazënnye zavody Urala v period perechoda ot krepostnichestva k kapitalizmu*, Krasnoiarsk 1986.

25. V.I. Semëvskii, *Krest'iane v tsarstvovanie imperatritsy Ekateriny II*, St Petersburg 1901.

26. S.I. Smetanin, 'Razlozhenie gornozavodskogo naseleniia i formirovanie rabochego klassa na Urale', *Voprosy istorii*, 1982, no. 2, 29–41; K. N. Serbina, *Krest'ianskaia*

A major concern of Russian historiography has been to analyse where the first labour force was found, how it was attracted to the ironworks, and the general living conditions of workers. Here, important contributions were made by Kashintsev, Pavlenko and Cherkasova,[27] and by Kafengauz, who made a detailed local study of the Demidov enterprises in order to unravel the function of large-scale mining economies.[28] Not surprisingly, special attention has been paid to the eighteenth century and to the initial phase of large-scale production in the Urals, to the endeavours of Peter the Great and those implementing his objectives (De Hennin and Tatishchev), and also to the general role of the state as a guarantor of order and as a major owner of ironworks.[29]

One of the most probing and multifaceted analyses of eighteenth-century Ural industry was achieved by the French historian Roger Portal. Writing in the 1940s, Portal was unable to use primary, unpublished sources and had to rely on many of the Russian historians already mentioned. Despite the difficulties posed by the lack of direct access to sources, Portal was nevertheless able to give a holistic and theoretically considered presentation of the Ural industry, devoting great attention to the social organisation of work within the enterprises of the region.[30]

The Swedish and Russian iron industries continue to fascinate us as early examples of larger-scale commodity production. What were the prerequisites for these early industries? What problems had to be overcome in order to make production come about? What were the long-term repercussions of this production? By comparing two industries geared to the same end product (iron), to be sold on the same market (Britain), it will be easier to answer these questions and to identify not only similarities, but also differences. The need for fuel, manpower, and a certain degree of protection from the state was certainly the same in both countries,

---

zhelezodelatel'naia promyshlennost' Severo-Zapadnoi Rossii xvi – pervoi poloviny xix v., Leningrad 1971; K. N. Serbina, *Krest'ianskaia zhelezodelatel'naia promyshlennost' Tsentral'noi Rossii xvi – pervoi poloviny xix v.*, Leningrad 1978.

27. Kashintsev, *Istoriia*; N. I. Pavlenko, *Razvitie metallurgicheskoi promyshlennosti v Rossii v pervoi polovine xviii v: Promyshlennaia politika i upravlenie*, Moscow 1953; A. S. Cherkasova, *Masterovye i rabotnye liudi Urala v XVIII veke*, Moscow 1985.

28. B. B. Kafengauz, *Istoriia khoziaistva Demidovykh v XVIII-XIX vv.*, Moscow-Leningrad 1949.

29. Pavlenko, *Razvitie*; Z. G. Karpenko, *Gornaia metallurgicheskaia promyshlennost' Zapadnoi Sibiri v 1700–1860 gg.*, Novosibirsk 1963; A. S. Orlov, *Volneniia na Urale v seredine xviii v. K voprosu o formirovanii proletariata v Rossii*, Moscow 1979.

30. Portal, *L'Oural*.

but these needs were met in different ways. Institutional solutions varied, and so did the strategies displayed by the parties involved: works owners, workers, peasantry, state officials, and merchants.

Eighteenth-century Swedish travellers devoted themselves to just that: the task of comparing. They compared the Swedish and Russian iron industries in order to pinpoint the advantages and disadvantages of the Russians, and they were aware of some of the most important ones. The Ural region was remote and iron had to be transported over long distances to reach the ports in St Petersburg and Archangel. It was difficult to control production from the capital. On the other hand, the forests were vast, the ore was easily extracted and labour was believed to be comparatively cheap. The Swedish situation appeared to be the opposite. Bar iron production took place largely in the region of Bergslagen, not far from Stockholm. The Swedish administrative system was comprehensive and exerted efficient control. On the other hand, there was a constant fear of complete destruction of the forests, ore was taken from deep pits that filled with water unless pumps were used, and the peasantry could not simply be forced to produce charcoal for the ironworks.

Eighteenth-century observers can thus give us a number of hints as to possible differences and similarities, hints that are valuable from the comparative point of view. Nevertheless, the aim of this book is to take the comparison a step further, to use the Swedish and Russian examples within a more general discussion about early industrialisation in Europe. Given the present state of research within the field of proto-industrialisation, where much effort has been devoted to Britain, to continental Europe, and to textiles, it is now time to include Sweden, Russia, and large-scale iron production in the debate.

The concept of proto-industry was coined and elaborated in the 1970s, mainly by Franklin Mendels and, later, by Peter Kriedte, Hans Medick and Jürgen Schlumbohm.[31] Definitions vary, but

31. F. Mendels, 'Proto-industrialization: the first phase of the industrialization process', *Journal of Economic History* 1972, no. 32; F. Mendels, *Industrialization and Population Pressure in Eighteenth-Century Flanders*, New York 1981; P. Kriedte, H. Medick and J. Schlumbohm, *Industrialisierung vor der Industrialisierung. Gewerbliche Warenproduktion auf dem Land in der Formationsperiode des Kapitalismus*, Göttingen 1977 (English version *Industrialization before Industrialization. Rural Industry in the Genesis of Capitalism*, Cambridge 1981); P. Kriedte, H. Medick and J. Schlumbohm, 'Sozialgeschichte in der Erweiterung – Proto-Industrialisierung in der Verengung? Demographie, Sozialstruktur, moderne Hausindustrie: ein Zwischenbilanz der Proto-Industrialisierungs-Forschung', *Geschichte und Gesellschaft* 1992, Hefte 1, 2.

proto-industry is generally taken to denote small-scale production of commodities within rural households in the early modern period, the products of which were sold on distant markets, i.e. in other regions or abroad.[32] The concept was consciously construed as excluding commodity production by urban craftsmen, agricultural production by rural households, or large-scale commodity production in factories or proto-factories. It was also assumed that proto-industry appeared in places where pastoral farming and certain inheritance practices prevailed,[33] and that it did not thrive in places where feudal institutions were strong.[34]

Proto-industry, in this sense of the word, was thought to have prepared the ground for industrialisation proper in a number of ways. It was therefore regarded as strategic to investigate this phenomenon in order to understand subsequent economic development. Proto-industry was for example believed to have generated demographic expansion, since more people could make a living where land and agriculture were no longer the sole sources of income. It now became possible to combine the cultivation of very small plots (or ordinary-sized plots on low-yielding soils) with the production of commodities such as textiles or wooden items. In this manner, a labour force other than the traditional peasantry started to grow, ready for the industrialisation to come. But proto-industry also contributed to the accumulation of capital, later to be invested in industry proper, and it stimulated the integration of different markets. Exchange of commodities and foodstuffs was initiated, from which industry proper could eventually profit. At a more abstract level, proto-industrialisation was conceived of as a transitional phase between feudal and capitalist society. It was believed to have engendered a process through which independent producers were gradually subjected to putters-out and then reduced to dependent wage labourers; at the same time, putters-out appropriated larger shares of the profits and even the very means of production. In this process, indebtedness often played an important part.[35]

The concept and the connected hypotheses have been widely discussed and criticised. As research has progressed, the connection between early modern commodity production in rural households

32. See *Protoindustrialisierung in Europa*, Vienna 1994, 9.

33. See J. Thirsk, 'Industries in the Countryside', in *Essays in the Economic and Social History of Tudor and Stuart England in Honour of R.H. Tawney*, Cambridge 1961.

34. Kriedte, Medick and Schlumbohm, *Industrialization before*, 6.

35. Ibid.

and later industrialisation has been seriously questioned. Proto-industrial regions were not always the ones that were most successfully industrialised, nor did such activities appear solely in regions where pastoral farming and partible inheritance predominated.[36] Consequently, to explain why proto-industry and industry proper arose in some regions and not in others, the perspective has to be widened considerably. Furthermore, there was no clear and simple connection between proto-industry on the one hand and demographic expansion and social stratification on the other.[37] Domestic production of commodities could take place in areas where population growth continued to be slow; demographic expansion might just as well be concomitant with seasonal under-employment rather than of small-scale commodity production.[38] Capital could indeed accrue in the hands of merchants and putters-out who involved themselves in proto-industrial activities, but it was not always used to bring about industrialisation. It could also be re-invested in land, strengthening the agricultural sector instead.[39] To sum up, the empirically tested connection between domestic production of commodities and industry proper seems weak. Is there any point at all in stressing such a connection by using the word proto-industry?

Some of the disagreement seems to have derived from the fact that the concept was too closely linked to certain empirical observations made at an early stage. Facts about the consequences of small-scale commodity production in *some* regions were assumed to have general relevance, and these consequences were even seen as parts of the definition itself. As it became obvious that the consequences did not always follow, problems arose. Was it appropriate to label something as proto-industry if it did not involve for example, demographic expansion, when Franklin Mendels had put so much emphasis on this issue and it had tainted the meaning of the word? Was it possible to talk about proto-industry in cases where the commodities were not sold on a distant market? Was it relevant to talk about proto-industry when it did not in fact evolve into industry proper? When isolated parts of the proto-industrial

36. See e.g. M. Berg, *The Age of Manufactures 1700–1820,* London 1985, 102ff.; Kriedte, Medick and Schlumbohm, 'Sozialgeschichte', 239.

37. Kriedte, Medick and Schlumbohm, 'Sozialgeschichte', 73–87.

38. See M. J. Daunton, *Progress and Poverty. An Economic and Social History of Britain 1700–1850,* Oxford 1995, 159.

39. P. Servais, 'Les structures agraires du Limbourg et des pays d'Outre-Meuse du XVIIe au XIXe siècle', *Annales ESC* 1982, no. 37.

model were confronted with empirical results, it seemed difficult to save the model as such.[40]

Another important root of discord was the fact that other types of early commodity production were left out, types in which production took place in factory-like units or in an urban setting. Why should the concept be defined in a way which made it impossible to include these branches in the discussion? The decision to exclude them seemed fortuitous, since such production was undeniably early, industry-like, and sometimes evolved into industry proper. Critics such as Karl Heinrich Kaufhold, Wolfgang von Stromer and Marcus Cerman have put the case for including larger production units in general, including urban ones.[41] Martin Daunton has also emphasised that the concept of proto-industry excludes too much to provide a fruitful point of departure when discussing early forms of commodity production. It excludes the whole of London, for example, 'the largest single industrial district in Britain'.[42] It also leaves aside production that was based on non-domestic organisations. 'In the country, there was a wide variety of mines, mills, forges, and furnaces, and historians have made little effort to explain how these fitted into the agrarian economy.'[43]

Criticism of the concept of proto-industry has thus been frequent and severe, pointing at a number of shortcomings. We, the authors of this book, do however feel that the ideas launched by Kriedte, Medick and Schlumbohm were valuable and that they provide a fruitful point of departure for a study which seeks to investigate and compare Swedish and Russian iron-making in the early modern period. For despite the fact that isolated parts of the model have proved to be over-simplifications or unwarranted generalisations, the proto-industrial model nevertheless offered a holistic picture of the tensions and dynamics arising in rural society as its members were drawn into new forms of production. This

40. Kriedte, Medick and Schlumbohm, 'Sozialgeschichte', 79.

41. K. H. Kaufhold, 'Gewerbelandschaften in der frühen Neuzeit', *Gewerbe- und Industrielandschaften vom Spätmittelalter bis ins 20. Jahrhundert,* Stuttgart 1986; W. von Stromer, 'Gewerbereviere und Protoindustrien in der Spätmittelalter und Frühneuzeit' (the same collection); M. Cerman, 'Proto-industrialization in an urban environment: Vienna, 1750–1857', *Continuity and Change* 1993, no. 8; M. Cerman, 'Proto-industrielle Entwicklung in Österreich', in *Protoindustrialisierung in Europa,* Vienna 1994. Cf. also the discussion by J. Schlumbohm, ' 'Proto-Industrialisierung' als forschungsstrategisches Konzept und als Epochenbegriff – eine Zwischenbilanz', in *Protoindustrialisierung in Europa,* Vienna 1994, 28ff.

42. Daunton, *Progress,* 156.

43. Ibid., 169. See also 206 and 231.

was expressed by as early a writer as Rudolf Braun, albeit more metaphorically: when industry spread into the countryside, it was met by rural inhabitants who 'introduced preconditions which favoured or impeded, and sometimes prevented, reception of industrialisation'.[44] But Kriedte, Medick and Schlumbohm attempted to underpin this general picture of a structural conflict with detailed hypotheses that could be tested. Even if the subsequent testing proved some concrete hypotheses to be wrong, the value of having raised questions concerning the interaction of rural and industrial production should not be denied.

Most of all, Kriedte, Medick and Schlumbohm brought the households of small-scale producers into the focus of historians' attention, pointing to the kind of rationality typical of such production units. By formulating questions about the way rural households balanced domestic commodity production against agriculture, and how they sought to adapt demographically to new situations, the theory highlighted the role of household strategies in early industrial development. This was of value to Swedish and Russian historians dealing with early iron production, because previous historiography in this field had often seen early industrial development solely from the point of view of the large-scale enterprises. It is true, of course, that research had acknowledged the quantitative importance of e.g. peasant households, as producers of charcoal and, in Sweden, of pig-iron. With the proto-industrial concept, however, it became easier to perceive the crucial role of the strategies, rationalities and conflicts of peasant and artisan households. By extension, the strategies adopted by ironmasters to adapt available households to industrial work directed attention to *the social organisation of work*. The social organisation of work can be seen as the outcome of a precarious balance of power between ironmasters on the one hand and households on the other. The social organisation of work is the main theme of this volume.

The present book is thus indebted to the work of Kriedte, Medick and Schlumbohm, but it is also indebted to their critics. From the debate on proto-industrialisation we, the authors, have drawn four major conclusions which set their imprint on this volume. The first concerns the way early industry should be conceptualised. The second concerns the wider context of early industry. The third concerns the spatial setting of early industry. The fourth,

44. R. Braun, *Industrialisation and Everyday Life*, Cambridge 1979 (first edition 1960), 10.

finally, concerns the character of the market on which early industrial products were sold.

(1) A focus on domestic production of commodities appears to be far too narrow an approach. There are no convincing arguments as to why factory-like or urban commodity production, for example, should be left aside. Indeed, the current state of research provides a good argument for including them. It has been shown that the actual borders between domestic commodity production, larger production units, and urban commodity production were often blurred. For a long time, these different types of production units seem to have interacted quite well, being interlinked through various modes of exchange and dependence. Early modern commodity production was a mixture of features which it would be difficult, and inadvisable, to disentangle. Even at the micro-level of single households, this mixture can be seen. One and the same household could be supported partly by agriculture, partly by home-based production of wooden spoons, for example, and partly by participation in larger-scale industry. The flexibility and tenacity of such households can often be explained by this diversified economy.[45]

Even if we were to maintain an interest in domestic commodity production, we would still have to admit that this phenomenon cannot be properly understood unless neighbouring forms of production are taken into consideration. To grasp this interdependence, the notion of a *network* is crucial. It was as a result of networking that various forms of commodity production came about, and it is only if we can fathom the extent and functioning of these networks that individual production units can be analysed in a satisfactory manner. The contributors to this volume have thus chosen to conceive of their topic as early industry, a formulation which is not so preladen with a certain meaning and allows for the aim of including many different types of production: domestic production (of commodities or raw materials), co-operative production, and large-scale production in factories or proto-factories.

This can be seen as a return to the original intentions of Kriedte, Medick and Schlumbohm, who gave their book the broad title 'industrialisation before industrialisation', even if their concrete discussions devoted little attention to metal production, for example. The aim of this book is to take up the discussion where they left off, and to engage in the task identified by Daunton: to analyse

45. Kriedte, Medick and Schlumbohm, 'Sozialgeschichte', 84ff.

how the Swedish and Russian iron industries 'fitted into the agrarian economy', how such industries interacted with domestic and co-operative production, and what kinds of social organisations and social conflicts arose in and around mines, furnaces and forges.

(2) There may be many reasons why the hypotheses originally connected with proto-industrialisation have not all been verified. We have already hinted at the danger of drawing too general conclusions from studies of a certain region. Regional results can hardly be generalised without due attention being paid to the *wider context* of that region, and of the regions with which we are comparing it. Paying attention to a wider context means, among other things, analysing the importance of *social institutions*, as pointed out by e.g. Joan Thirsk, Rudolf Braun, Pat Hudson, Martin Daunton,[46] and Shelagh Ogilvie. Ogilvie has even argued that a majority of the discrepancies and contradictions which beset the discourse on early industry and blur the general picture of the phenomenon can be explained if social institutions are fully taken into account. Here, social institutions means 'established rules and practices, through which people have organised their economic, social, demographic, political and cultural activities'.[47] This would encompass a great deal (property rights, community systems, inheritance practices, the new institutions of the early modern state, etc.) and the whole scope of the inquiry then becomes very broad. However, Ogilvie is no doubt right in advocating such a holistic approach, and Kriedte, Medick and Schlumbohm have also put greater emphasis on the wider context of proto-industrialisation in their later writings.[48]

Kriedte, Medick and Schlumbohm did not, however, totally neglect the role of social institutions in their earlier writings. Their work of 1977 (English version in 1981) contained an important hypothesis concerning the relationship between early industry and the politico-institutional environment. It was assumed that such production was, by its very nature, difficult to reconcile with

46. Thirsk, 'Industries'; Braun, *Industrialization*; P. Hudson, 'Landholding and the organization of textile manufacture in Yorkshire rural townships c. 1660–1810', in *Markets and Manufacture in Early Industrial Europe*, London, New York 1991; Daunton, *Progress*, e.g. 153f.

47. S. Ogilvie, 'Proto-industrialization in Europe', *Continuity and Change*, 1993, no. 8; S. Ogilvie, 'Soziale Institutionen und Proto-Industrialisierung', in *Protoindustrialisierung in Europa*, Vienna 1994, 36.

48. See Kriedte, Medick and Schlumbohm, 'Sozialgeschichte', 73, for a severe criticism of Wolfgang Mager for wanting to exclude cultural and politico-institutional aspects. See also Schlumbohm, 'Proto-industrialisierung', 23f., 27.

feudal institutions, and that proto-industry even contributed to the dissolution of feudalism. This was an over-simplification. Domestic commodity production could interact perfectly well with a feudal system. Good examples of this are provided by studies of eastern Europe. In Russia, the *kustar'* industry (cottage industry) was widespread and flourishing, many peasants being at the same time serfs and well-to-do entrepreneurs.[49] The same is true if we take into account larger, centralised production units, such as metalworks. Such forms of early industry were also present within the framework of feudal and/or absolutist regimes. Often, they were actually created and actively promoted by landlords or by the state. It is thus hardly fruitful to postulate hostility between early industry and feudal institutions. The contributors to this book have chosen to ask: what was the actual relationship between social institutions and early industry, and between the Russian and Swedish states and different parts of the production networks?

(3) Despite all the controversies in the field of proto-industrialisation, there is general agreement in one respect. Early industry should be thought of as a *regional phenomenon*. In the early modern period, there were no unified national markets for labour, resources or commodities. Commodity production often took place within a limited space where special prerequisites prevailed, either in terms of natural resources and labour supply or in terms of institutional arrangements and infrastructure. A good example of this is provided by banking and early credit arrangements. As has been shown for example by Pat Hudson, industry often developed in regions where local or regional capital was available.[50] It was only within the 'relatively narrow and comprehensible confines of the region that entrepreneurs dared to change ... the economy', to quote Sidney Pollard.[51] But early industry may, on the other hand, have contributed to the subsequent, gradual creation of more integrated markets, national or international. Specialisation of regions was often concomitant with increased contact between and integration of the very same regions. The ability of one region to produce a surplus of grain and sell it to other regions, where soils were poorer, could trigger early industrial growth in the latter. Such a

49. J. Blum, *Lord and Peasant in Russia,* Princeton 1961, 301ff.; R. Pipes, *Russia under the Old Regime,* Harmondsworth 1977, 192ff.

50. P. Hudson, *The Genesis of Industrial Capital,* Cambridge 1986.

51. S. Pollard, 'Regional and inter-regional economic development in Europe in the eighteenth and nineteenth centuries', in *Debates and Controversies in Economic History. Eleventh International Economic History Congress,* Milan 1994, 78.

spatial division of labour allowed people to move into new and hitherto uncolonised lands.[52]

Obviously, regions can be conceived of in different ways. We can use various criteria to delimit them (such as production, population density, income),[53] but they can also be defined in terms of their internal cohesion, of the relations that tie people and places together.[54] Their delimitations can be seen as naturally given (such as ore deposits), but regions can also be looked upon as social constructs, created by the uses that people make of certain resources to be found within a geographical 'space'. The authors of this book maintain that at least when it comes to early modern iron production, the region of production was the result of the way work was organised socially. It is true to say that Swedish iron production was localised in the Bergslagen region and Russian iron in the Ural region. It is scientifically more fruitful to say that, having been localised in a certain geographical area, iron production shaped the area into a region. This means that a region is a space within which production takes place, but the implications are not as trivial as that.[55]

First, the region may change as production is reorganised. For instance, as the Swedish state demanded that pig-iron be produced solely by the peasantry living in the core region of Bergslagen and that bar iron should be produced closer to the ports, this led to a spatial extension of iron production. The demands of the state affected the social organisation of work, but can also be said to have influenced the regional borders in the mining area of central Sweden, the result being that production was scattered and that several units had to interact by networking.[56]

Second, there is an important aspect of power. If regions are social constructs, who makes these constructs? Who has power enough to enforce a decision about what kind of production should take place where? State officials may have one opinion, ironmasters another, and the local peasantry yet another. In the case of Swedish and Russian iron-making, the creation of production regions and

52. See M. Prak, 'Regions in Early Modern Europe', in *Debates and Controversies in Economic History. Eleventh International Economic History Congress*, Milan 1994, 27, 47.

53. See the discussion in Pollard, 'Regional and inter-regional', 59ff.

54. Pollard, 'Regional and inter-regional', 62; H. Pohl, 'Einführung', *Gewerbe- und Industrielandschaften vom Spätmittelalter bis ins 20. Jahrhundert*, Stuttgart 1986, 8, 10, on the distinction between homogeneous 'spaces' and functional 'spaces'.

55. A. Florén and G. Rydén, *Arbete, hushåll och region. Tankar om industrialiseringsprocesser och den svenska järnhanteringen*, Uppsala 1992, 88ff.

56. Ibid., 95.

production networks provides a crucial key to how the state sought to encourage and govern the economy. Emphasising the regional perspective entails, among other things, an emphasis on the role of the state. However, the existence of a conscious state policy to make people within a certain area co-operate, also accounts for some of the social tensions and conflicts that beset early iron production.

(4) To discuss the early iron industry in relation to market-oriented questions has a long historiographic tradition in Sweden and in Russia. As has been pointed out already, the competition for the British market has often been taken as a point of departure for discussions about Swedish and Russian iron. In the discourse on proto-industry however, the role and structure of the market long remained absent, as Maxine Berg has pointed out. Attention focused mainly on the factors that made it possible to produce the commodities, like the supply of labour, not the ones that made it possible to sell them.[57] Recently, however, greater efforts have been made to introduce market issues into the debate on early industries, with inspiration and impetus from current research on changing consumption patterns.[58]

Thus, demand and consumers have come into focus – once again.[59] The effects of new types of consumer groups (such as the middle class), growing numbers of consumers (due to an increasing population), and new tastes and fashions, are now being evaluated with respect to the successes or failures of early industry. The fact that a certain region managed to transcend the proto-industrial stage and become fully industrialised cannot always be conclusively explained with reference to know-how, labour and access to capital. Some importance at least should also be attributed to the fact that the region may have been the only one to produce a certain type of commodity for which there was growing demand.[60]

The primary aim of this volume is to investigate the social organisation of work, not the character of the market on which iron commodities were sold. Even so, this does not make market-related issues irrelevant, since a better understanding of the iron

57. M. Berg, 'Markets, Trade and European Manufacture', in *Markets and Manufacture in Early Industrial Europe*, London, New York 1991, 3ff.

58. See e.g. *The Birth of a Consumer Society*, eds N. McKendrick, J. Brewer and J. H. Plumb, London 1982.

59. See the clear and probing discussion about different kinds of demand in Hildebrand, *Swedish Iron*, 30–32.

60. Berg, 'Markets, Trade', 10.

market may explain some of the decisions and strategies adopted by ironmasters. For instance, it cannot be denied that Swedish and Russian ironmasters were affected by the fact that the British iron industry became increasingly capable of producing coke-based iron in the late eighteenth century. In the Urals, the slackening demand for Russian iron on the British market made ironmasters reconsider their choice of products. Henceforth, production was redirected to a growing domestic market, in which the local peasantry were important consumers. They bought scythes, spades, padlocks, smith's tools, horseshoes and nails. They bought frying pans, roof iron, saws, wire, oven-doors and pots. The kind of market on which these items were sold was highly diversified. Different items came from different production units (the so-called mining districts), each of them specialising in a certain type of iron commodity. There was thus little direct competition between the different mining districts and the market became compartmentalised. A similar argument about the highly diversified character of the iron market is put forward in chapter 3, where the position of Swedish iron on the British market is discussed.

Two Regions: Bergslagen and the Urals

As was pointed out earlier, the primary concern of this volume is with the 'space' within which the whole production process took place. The focus is on the network that tied different places and social groups together. Consequently, it is somewhat difficult to designate the exact limits of the areas studied. However, it does seem necessary to give those readers who are unacquainted with Swedish and Russian geography a rough idea of the position and size of the regions discussed.

For Sweden, we mainly focus on the region of Bergslagen, situated north of the central Swedish lakes and stretching from the eastern coastline to the Norwegian border. It comprises parts of the provinces of Västmanland, Närke, Kopparberg/Dalarna, Värmland and Gästrikland. Traditionally, Bergslagen denoted the region within which state policy decreed that only pig-iron was to be produced.[61] However, the use of the name to also denote the peripheral areas to which the forges were moved from

61. U. Jonsson, 'Bruksdöden i Örebro, Västmanland och Kopparbergs län', *Bebyggelsehistorisk tidskrift* 1983, no. 5, 105.

the beginning of the seventeenth century is a fairly widely accepted usage.[62] Obviously, this is how Bergslagen will be understood in this book, in order not to exclude the important forges. The area thus comprises an estimated 24,400 square kilometres. In 1805, its total population was approximately 181,300.[63] Regardless of where the exact borders are drawn, there can be no doubt that the Swedish iron-producing region was much smaller than its Russian counterpart.

In Bergslagen, ore had been extracted at least as early as in the Middle Ages, at places such as Norberg, Falun and Bispberg. The ore deposits most frequently exploited were copper, silver and iron. There is reason to believe that large parts of the region were colonised because of the opportunities offered by the combination of agriculture and small-scale metallurgy. To rely solely on agriculture would hardly have been viable.[64] In the early modern period, which is the point of reference for our discussion, the centres of metallurgical production were still highly dependent upon grain grown on the central Swedish plains and delivered by peasants from the peripheral parishes of Bergslagen. Likewise, oxen were brought to the area from southern parts of Sweden.[65] It was not until *circa* 1820 that a county like Kopparberg managed to produce all of its grain within its borders, but this enhanced capacity

62. S. Berger, M. Bohlin and G. Forsberg, *Bergslagsprojektet. ERU-rapport 6*, Stockholm 1981, 18. The authors present and discuss nine different definitions of Bergslagen.

63. These numbers have been calculated on the basis of statistical data provided in G. Sundbärg, *Emigrationsutredningen V*, Stockholm 1910. The following are included: (Örebro county) Karlskoga bergslag, Nora and Hjulsjö bergslag, Grythytte and Hällefors bergslag, Nya Kopparbergs bergslag, Linde and Ramsbergs bergslag, parishes of Näsby, Ervalla, Knista, Hidinge, Kvistbro and Skagershult; (Västmanland county) Skinnskattebergs bergslag, Gamla Norbergs bergslag, parishes of Skultuna, Kolbäck, Svedvi, Säby, Ramnäs, Sura, Arboga lands, Säterbo, Köpings lands, Bro, Kung Karls and Malma; (Kopparberg county) parishes of Söderbärke (excl. Malingsbo), Norrbärke, Grangärde, Säfsnäs, By, Garpenberg, Husby, Stora Tuna (excl. Gustafs), Torsång, Vika, Sundborn, Svärdsjö, Falu lands, Avesta köping; (Värmland county) parishes of Nyed, Norra Ullerud. Ransäter, Järnskog, Färnebo härad, Älvdals nedre härad, Fryksdals nedre härad, Jösse härad; (Gävleborg county) parishes of Hille and Valbo, Österfärnebo, Ovansjö, Högbo, Torsåker, Hamrånge. This area is smaller than the definition at which S. Berger, M. Bohlin and G. Forsberg arrive (*Bergslagsprojektet* map 2), excluding the northern parts of the provinces of Uppland and Södermanland and not including as much of the province of Närke. It should be pointed out, though, that their aim was to define the contemporary Bergslagen region

64. M. Sjöberg, 'Järn bryter bygd – en sanning med variation', *Scandia* 1992, no. 1.

65. J. Myrdal and J. Söderberg, *Kontinuitetens dynamik. Agrar ekonomi i 1500-talets Sverige*, Stockholm 1991.

was mainly to be attributed to its southern-most parts.[66] The upper areas were still dependent on outside supplies, and many people supplemented the meagre incomes they derived from agriculture by engaging in various handicrafts.

In an area where trade and exchange is of vital importance, the infrastructure becomes crucial. In Bergslagen, the necessary conditions were created by waterways. Iron was transported on lakes like Barken and Väsman and on rivers and canals like the Kolbäcksån and Hedströmmen, mainly in the winter season, when heavily loaded sledges could easily be taken on the frozen waterways down to port towns like Arboga, Köping, on Lake Mälaren, or to Karlstad and Kristinehamn, on Lake Vänern. These places lacked the right to engage in international trade, and the iron was therefore shipped on to Stockholm or (from the western parts of Bergslagen) to Gothenburg, from where it was exported.

Compared with southern Sweden, the Bergslagen region displayed a more homogeneous social structure. Land was generally taxable and used by freeholding peasants. There were few landlords, little land exempt from regular taxation, and few tenants. From the seventeenth century on though, larger-scale enterprises were set up. These ironworks were often situated on a kernel of privileged land, which had been freed from the ordinary land taxes. As a rule, the ironmasters were ennobled merchants.[67] Around the ironworks lived a population who were highly dependent on the ironmasters and had to perform various feudal duties in return for their use of the land. Some of these were tenants holding farmsteads situated on the land of the ironmaster, but others were merely crofters, holding smaller units of land. However, practically none of the ironworks were endowed with so much privileged land that they were self-sufficient in terms of woods and labour. Most ironworks had to buy additional, non-privileged land to become more independent. These lands were often purchased from the local peasantry and either converted into new crofts (where a larger workforce could be given a living) or kept as woods, where charcoal could be produced.[68] Thus, the

66. M. Ågren, *Jord och gäld. Social skiktning och rättslig konflikt i södra Dalarna ca 1650–1850,* Uppsala 1992.

67. See R. Bredefeldt, *Tidigmoderna företagarstrategier. Järnbrukens ägar- och finansieringsförhållanden under 1600-talet,* Stockholm 1994, 168, 172. Bredefeldt sees a difference between the first and second halves of the century, with more noblemen in the latter period.

68. See Ågren, *Jord och gäld.*

social structure of Bergslagen was altered somewhat by the introduction of ironworks. However, it was still true that taxable land used by peasants predominated, and that privileged land, landlords and tenants were rare.

As a rule, the production plants established in Bergslagen were quite small. Before 1850, production levels at ironworks (where forging took place) were strictly regulated by the state through the Board of Mines. For instance, a traditional small ironworks in Bergslagen could have a permit to produce 250 tons of bar iron in 1800. As late as 1861, the mean annual production of a blast furnace in the province of Närke was approximately 800 tons. Moreover, only one out of the province's fifty-eight blast furnaces was used continuously throughout the year.[69]

The Ural mountains are situated approximately 1,400 kilometres east of Moscow,[70] reaching heights of 1,600 to 1,700 metres above sea level. As the geographer Johan Philip Strahlenberg proposed in the 1730s,[71] the mountains can be seen as a kind of natural boundary line between the European and Asian parts of Russia. Located in this area was the region of iron and copper production discussed in this book. Like Bergslagen, this region cannot be strictly defined, but the three provinces (*guberniia*) of Perm', Viatka and Orenburg formed its central area. These three provinces together comprised approximately 795,000 square kilometres. In 1719, the population of this region was 858,000, a figure which had risen to just over 6 million in 1858. But this substantial increase did not affect all parts of the Urals, with the northern districts remaining very sparsely populated in the middle of the nineteenth century. The Russian area to be discussed was thus considerably wider and had a much more substantial population than the Swedish area. Iron production was, roughly speaking, confined to an area stretching from the towns of Verkhotur'e and Solikamsk in the north, to the Bashkir lands in the south, and from the towns of Kungur (on the River Kama) and Perm' eastwards to Irbit in Siberia proper.

Like Bergslagen, the colonisation of the Urals was a process connected with the onset of iron production. There is a distinct difference in time and in the type of connection, however. In the

69. Jonsson, 'Bruksdöden', 108.

70. This is the distance between Moscow and Ekaterinburg. Between Moscow and Perm', the distance is 1,200 km.

71. Das Nord- und Ostliche Theil von Europa und Asia (map, Stockholm 1730). University Library of Uppsala.

Swedish case, immigration and small-scale iron-making started as early as in the Middle Ages, with iron production being a prerequisite for settlement. In the Urals, iron-making became part of the rural economy in the seventeenth century, stimulating immigration from the west, but substantial cultivation of the land had already begun before that, independently of iron-making. In the eighteenth century, the population continued to grow, partly as a result of forced migration and partly as a result of cultivation of the land of the southern Urals.

Some parts of the region were dependent on grain deliveries from other areas, those endowed with better soils. In the first half of the eighteenth century, peasants from places such as Shadrinsk, Irbit and Kamyshlov started to penetrate the market of the Middle Urals, travelling to the metalworks in Nizhnii Tagil' and Nev'iansk and selling not only grain but also beef and pork. In the second half of the century, such visits were common.[72] In the middle of the nineteenth century, peasants in the Perm' province supplied all of the ironworks within that province with grain.[73] In the province of Viatka, about 500 peasants were involved in this trade in the 1830s. The different parts of the Ural area were thus gradually interlinked through trade, making it into an economic region in its own right. Quite soon, the trade in foodstuffs was taken over by well-to-do buyers-up of peasant origin, entrepreneurs who could, in some cases, make a fortune from their trade.[74]

For a far-off region like the Urals, infrastructure was vital. Russian roads were notoriously bad,[75] and long-distance transport had to rely on waterways. Iron was thus taken to market by barge, starting on the important Tchesovaia river. Then it was carried, according to the Swedish traveller Samuel von Stockenström (in

72. A. S. Cherkasova, 'Sotsial'no-ekonomicheskie sviazi gornozavodskikh tsentrov i dereven' Urala v seredine XVIII v.', in *Derevnia i gorod Urala v epokhu feodalizma: problemy vzaimodeistviia*, Sverdlovsk 1986, 30.

73. A. Tret'iakov, 'Shadrinskii uezd Permskoi gubernii v sel'skokhoziaistvennom otnoshenii', *Zhurnal ministerstva gos. imushchestv.* 1852, part 45, 183.

74. S. M. Tomsinskii, 'Rassloenie ural'skoi derevni v sviazi s razvitiem manufaktury (pervaia polovina XVIII v.)', in *Ezhegodnik po agrarnoi istorii Vostochnoi Evropy*, Leningrad 1972, 143f.; B. G. Pliushchevskii, 'Imushchestvennoe i sotsial'noe rassloenie krest'ianstva gubernii Bol'shogo Urala v 1-i polovine XIX veka', in *Problemy sotsial'no-ekonomicheskogo razvitiia derevni Srednego Povolzh'ia v period feodalizma*, Kazan' 1986, 94; B. G. Pliushchevskii, 'Gosudarstvennye krest'iane Viatskoi gubernii v 1-i polovine i seredine XIX v.', *Sbornik Udmurtskogo NII.* no. 17, 93.

75. Blum, *Lord and Peasant*, 282ff.; Portal, *L'Oural*, 259.

the 1780s), down the Kama and Ufa rivers to the Volga, then upstream to Lake Ladoga and finally by canal to St Petersburg.[76] An alternative route was described by another Swede, Petter Schönström. According to him, iron was transported part of the way by river, and then reloaded onto sledges.[77] In Petersburg, some of the output was sold to the military forces, and some exported. Some of the iron never left the Urals, and was sold directly from the ironworks.[78]

Socially, the Ural region differed from western Russia, serfdom not being as common here as elsewhere. There was a predominance of state peasants, who were free, possessed land (legally owned by the Crown) and paid taxes.[79] In the seventeenth century, they made up 86 to 89 per cent of the entire peasant population of the Urals. In the middle of the nineteenth century the state peasants of the Perm' province still comprised 74 per cent of its total peasant population.[80] An important group within the population were the so-called Old Believers, who had fled from western Russia to escape religious persecution. Characterised by frugal manners and intense self-discipline, they were perfect industrial workers, providing the Ural metalworks with steady and reliable labour.

The Comparison and the Subsequent Chapters

The purpose of this volume is to compare the social organisation of work in the early iron industry in two regions: the Swedish region of Bergslagen and the Russian Urals. In both regions, bar iron was produced and then offered for sale on the British market in the eighteenth century. In both regions, ironmasters were highly dependent on the local peasantry, and in both cases the state was heavily involved in making production possible. So much for the similarities. There were also considerable differences, which can be

76. S. von Stockenström, 'Anmärkningar rörande ryska järntillverkningen och järnhandeln', section 6, Bergskollegii Arkiv, RA.

77. Schönström, 'Berättelse om', 38.

78. For instance, at the ironworks of the Empress Anna, iron worth 37,474 roubles was produced in 1735. Of this, iron worth 429 roubles was sold directly at the works. See 'Bericht von denen Umkosten, welche Anno 1735', Bergskollegii Arkiv, RA.

79. R. Bartlett, 'The Russian peasantry on the eve of the French revolution', *History of European Ideas*, 1990, vol. 12:3, 401; Portal, *L'Oural*, 13f.

80. *Istoriia Urala*, 198, 299–301; Pliushchevskii, 'Imushchestvennoe', 90.

summed up in two words: size and social structure. The Ural region was much vaster and had many more inhabitants than its Swedish counterpart. Likewise, each Ural production unit was much bigger than those of Sweden, in terms of both output and workforce. Moreover, the social structure was very different, with a high degree of legal force and compulsion in relation to the Russian peasantry (in the form of enserfment or as forced labour, ascription). Swedish peasants, on the other hand, had a considerably freer situation. These differences and similarities are sufficiently interesting to justify an attempt to compare the two systems.

We can however, carry the argument a step further, placing Sweden and Russia within a wider international setting and including, for example, Britain in the discussion.[81] As regards the available workforce, Sweden, Russia and Britain can be seen as three ideal-type cases. In Britain, there was a substantial landless labour force as early as the seventeenth century, available for industry to attract by means of free hire. In Russia, by contrast, there were virtually no free proletarian workers available as the early iron industry started to expand, but the peasantry were gradually subjected to a number of legal restrictions, making it possible to force them to perform auxiliary work for the industry. The Swedish situation can be described as somewhere in between the British and the Russian ones. Unlike Russia, Sweden had a peasant population which was free and could not simply be forced to aid the iron industry with charcoal-burning and transportation. Unlike Britain, Sweden still lacked a substantial landless labour force in the eighteenth century, a labour force which could be offered the job of burning charcoal according to capitalist principles.[82] Consequently, Swedish ironmasters and the Swedish state had to solve a rather complex organisational problem when they attempted to establish Sweden as a leading iron exporter on the basis of a free peasantry. Their solutions and strategies reflect the social structure of their country, just as those of the Russian and British ironmasters do. When juxtaposed like this, Sweden, Russia and Britain stand out as three typical cases, on the basis of which it will be possible to formulate more general hypotheses about the

81. Instead of Britain, a comparison could also be made with Belgium for instance. See A. Florén, 'Social Organization of Work and Labour Conflicts in Proto-Industrial Iron Production in Sweden, Belgium and Russia', *International Review of Social History*, 1994, 39, Supplement, 83–113.

82. See *Ironmaking in Sweden and Russia. A Survey of the Social Organisation of Iron Production before 1900*, eds G. Rydén and M. Ågren, Uppsala 1993, 32f.

prerequisites of early industry and about the way such industries 'fitted into the agrarian economy'.

The kind of comparison carried out here is complex, the complexity arising from the fact that the authors are not simply comparing Sweden and Russia at two distinct points in time. Sweden and Russia are studied continuously over a period of roughly three hundred years. Two spatial levels are also applied in the comparison: the national and, for the most part, the regional. Thus, Bergslagen in the year 1750 is compared with the Urals in the same year, but it is also compared with the previous and the subsequent situation in Bergslagen, and it is constantly related to the general Swedish picture. All of the chapters bear the imprint of the tension arising from this complexity, albeit to a varying degree. No chapter can be said to be an exhaustive comparative analysis of its theme. Sometimes, the emphasis is mainly on differences and similarities between Bergslagen and the Urals, sometimes on chronological change, and sometimes on differences between the regional and the national levels.

It might be argued that the division of this book into chapters is fortuitous, especially with regard to chapters 2, 3 and 4. The iron-producing peasants of the Urals and their Swedish counterparts, discussed in chapter 2, turn up as skilled labour or suppliers of pig-iron in chapter 3, and in chapter 4 they are once again discussed, this time as recalcitrant peasants. Thus, the chapters partially overlap. The borders between them cannot be described in terms of the empirical reality treated, but derive from the different angles, from which the same reality is viewed.

However, this is really because early industry appeared in a rural setting and remained closely knit to peasant life for a very long time. Historical reality itself accounts for the difficulties which arise whenever one attempts to divide early industry into a strictly industrial element and a strictly rural element. Any attempt to disentangle industry from rural life is in danger of becoming bogged down in futile questions of classification: should Swedish peasants selling pig-iron be discussed in the chapter dealing with peasants or in the one dealing with the refining of pig-iron into bar iron? Are we to emphasise that many Ural metalworkers retained a footing in the rural village community, or should we stress the industrial skills they possessed and the tasks they performed? The authors have decided not to choose, but to have both: to discuss the same, complex reality in several chapters, but to do so from somewhat differing standpoints.

As was pointed out in the travel report compiled by Petter Schönström, small-scale production units were still to be found in the hands of Russian peasants in the 1720s. Schönström had seen them with his own eyes near the Ural town of Kungur. Similar observations were made by Tilas on his way to the Olonets region. Tilas, who was otherwise highly critical of the Russian enterprises, maintained that these small plants made excellent iron because of their more careful sifting. Apparently, some of this small-scale production managed to linger on in the Urals, albeit precariously and under constant threat from large-scale works and from the state administration. But on the whole, small-scale peasant production of iron was confined to the task of manufacturing tools within peasant villages. By contrast, Swedish *bergsmän* continued to produce pig-iron in co-operatively owned blast furnaces into the nineteenth century, selling their products to the adjacent ironworks and forming an integral part of the Swedish production network. It seems highly pertinent to begin this volume with a comparative study of these two peasant groups, since they were the starting-points for large-scale industry in both regions and since the differing treatment of the groups provides valuable insights into Swedish and Russian mining policy. Maria Sjöberg, in collaboration with Anton Tomilov, carries out this comparison in chapter 2.

In chapter 3, the focus shifts to the mines, the furnaces and the forges, the places where ore was actually transformed into bar iron. To use their own words, Anders Florén and Göran Rydén, in collaboration with Sergei Ustiantsev, Dmitri Gavrilov and Ludmila Dashkevich, attempt here to 'lift off the roofs' of the forges and show in detail how work there was organised. Against the backdrop of applied technology and the character of the markets on which Swedish and Russian iron was sold, the authors analyse the relationship between the ironmasters and their workforces and the sources of conflict that were built into this.

The subsequent two chapters further analyse the integration of iron production and agrarian society. Here, the focus is on the production of charcoal and the procurement of various services for the ironworks. These were tasks that demanded enormous amounts of labour and, in both countries, they were imposed on the peasantry by the state. Even though the level of compulsion differed, both systems nevertheless generated resistance and conflict, as peasants tried to evade the unwelcome duty of assisting early industry. In chapter 4, Maria Ågren, in collaboration with Nina Minenko and

Igor Poberezhnikov, analyses and compares methods of forcing and/or enticing the peasantry into burning charcoal. In chapter 5, the focus is on the work itself. Maths Isacson, in collaboration with Igor Poberezhnikov, discusses how charcoal was actually produced and how the work process created opportunities for cheating and sabotage.

Proto-industry was made a highly vital field of research by historians with an interest in demography. Many of the hypotheses originally connected with the concept itself and with the name of Franklin Mendels dealt with the demographic consequences of early industry. In chapter 6, Göran Rydén, in collaboration with Svetlana Golikova, discusses differences between Russian and Swedish households affected by the iron industry. As is obvious from this essay, the demographic scope has become much broader. Here a more holistic approach to households and the House is presented, recalling Otto Brunner's famous phrase 'das ganze Haus'.

In the subsequent two chapters, the authors look at the early iron industry from the point of view of institutional arrangements. Both chapters should be seen as sequels to themes launched in chapter 3. There, Swedish *bergsman* households were shown to have been the original labour units in the mines. In chapter 7, Maria Ågren, in collaboration with Vladimir Zhelezkin and Vladimir Shkerin, discusses customary conceptions of property surrounding this traditional kind of work, how such conceptions could interact with absolutist regimes, but how they finally withered away towards 1800. The attitude expressed by the absolutist states towards civil society and its corporations is also shown to carry weight when it comes to comparing the social organisation of Swedish and Russian iron-making.

The issue of knowledge was also discussed in chapter 3. There, it was seen as something which imparted power and, hence, as something which was necessary if ironmasters were to obtain full control of iron-making. Swedish households were shown to have been able to guard their knowledge better than their Russian counterparts. In chapter 8, Rolf Torstendahl, in collaboration with Sergei Ustiantsev and Ludmila Dashkevich, focuses on how the issues of knowledge and education developed further in the nineteenth century, when the position of the household had changed, generally and profoundly. What knowledge was then needed in the iron industry and who was responsible for formal schooling? Contrary to what might be assumed, the role of the state is here ascribed greater importance in the Swedish case.

Together, the following chapters contribute to a more profound understanding of iron-making societies. Focusing on the often contradictory interests of ironmasters and rural households (composed of peasants or forgemen) this book unravels the strategies of both sides, the underlying argument being that these contradictions and strategies are crucial in explaining the form which the iron industry assumed in each country.

Chapter Two

Iron-Making in Peasant Communities

Maria Sjöberg with Anton Tomilov

The art of iron-making has long been known and widely practised in both Russia and Sweden. Long before the seventeenth and eighteenth centuries, when the state in both countries energetically encouraged the iron producers to establish themselves – and even expand – in the international market, iron-making was a natural part of peasant production. What happened to the peasants' iron production when the state and financially strong groups became involved in the iron industry? And what happened to the peasants? The following discussion focuses on both of these questions.

The Initial Stages

At first the iron was extracted from lake and bog ore. This so-called low-technology iron production was common in Sweden as early as the eleventh century, mainly in the regions of Gästrikland, Hälsingland and north-west Dalarna.[1] Iron production in Russia was first set up in the western parts of the country, but later – during the seventeenth century – it also became established in the Urals, as a result of the migration from the western areas.[2] In Sweden the primitive production of iron was an important supplementary source of income for the peasants. It was primarily in those areas where farming conditions were poor that iron-making developed.

1. A. Florén, M. Isacson, G. Rydén, M. Ågren, 'Swedish Iron Before 1900', in *Ironmaking in Sweden and Russia*, Uppsala 1993, eds G. Rydén and M. Ågren, 7–16.
2. N. A. Minenko et al., 'Ural Iron Before the Industrial Revolution', in *Ironmaking in Sweden and Russia*, eds G. Rydén and M. Ågren, 43–44.

The peasants in Lima and Transtrand in north-west Dalarna, for example, worked the molten ore and made tools, amongst other things scythes, which they then sold with the intention of obtaining agricultural produce. The market that was of interest for the peasants in Lima and Transtrand was therefore those areas where agriculture was an important part of the peasants' economy.

Local and regional trade were thus necessary to ensure the survival of the Swedish iron-making peasants. In the Urals, however, there was an abundant supply of fertile soil, and this was also an important reason why a large number of people migrated to this particular area from the west. Just like the iron-making peasants in Sweden the peasants in the Urals also made tools that were intended mainly for agriculture. In contrast to the Swedish iron-producing peasants, the iron-making peasants in the Urals primarily made the tools they needed in their own farming. Another important difference between the toolmaking peasants in the Urals and in Sweden was that iron-making in the Urals developed much later than it did in Sweden. The development of toolmaking in the Urals was linked to the migration from the west, a very different picture to what we find in Sweden. The Russian migrants brought with them knowledge about how to make iron and the improvements that iron products could effect in farming. Just as the peasants who made iron tools in Sweden depended on the earnings that this generated, the development of the primitive production of iron in the Urals was linked with the development of settlements in certain areas.

The primitive production of iron in the Urals can be regarded as a prelude to later developments; it was in this region that iron-making expanded and it was also the iron from this region that competed with Swedish iron on the international market. However, the area of Sweden whose output was shipped out into international waters had no direct connection with the primitive iron-making that the peasants had developed in other regions.[3] Developments in Bergslagen in Sweden were intimately linked with the possibilities of extracting iron from rock ore; in other words iron production based on lake and bog ore was not a direct precursor of what was to occur in Bergslagen.

In medieval Sweden it was well known and established that it was possible to extract iron from rock ore. There is much evidence

3. Å. Hyenstrand, 'Iron and Iron Economy in Sweden', in *Iron and Man in Prehistoric Sweden*, ed. H. Clarke, Stockholm 1979.

that blast furnaces were first used in the thirteenth century, in the area around Norberg. This area seems to have led developments in Swedish iron production, which then spread west from this area towards Värmland.[4] The forest and mining area in central Sweden that came to be called Bergslagen was surrounded by old farming settlements, and in the beginning iron production was carried out in harmony with the annual rhythm of agrarian production. Iron production was largely a seasonal activity and the settlements around the early furnaces were probably not even permanent initially. Wide-ranging sources dating from before the middle of the fourteenth century give us to understand that both spiritual and secular dignitaries – priests and nobles – as well as German immigrants had interests in mining. Subsequently iron production seems to have mainly become the concern of an entrepreneurial peasant group, the *bergsmän*.[5] What then, characterised this peasant group at the beginning of the seventeenth century? Just as was the case with the peasants who extracted iron from lake and bog ore, the *bergsmän's* iron production supplemented the insufficient produce of the land. The *bergsmän's* iron-making was thus from the very beginning intended as a means of obtaining agricultural products in exchange. However, in contrast to the peasants who forged and sold farming tools, the *bergsman* products constituted a less refined item. The *bergsmän* produced – and sold – iron in lumps weighing a few ounces each, so-called '*osmunds*', which the purchasers had to refine themselves. Initially the Swedish *bergsmän* were involved in regional trade where – just like the peasants of north-west Dalarna who smelted bog ore – they received agricultural products in exchange for their *osmund* iron.[6] In contrast to the peasants who made iron tools, both in the Urals and in north-west Dalarna, trading by the *bergsmän* was not just regional; *osmund* iron also constituted a part of Swedish exports up until the beginning of the seventeenth century.

The *bergsmän's* production was, from the very beginning, part of a network that had both regional and international subdivisions.

4. B. Boëthius, *Gruvornas, hyttornas och hamrarnas folk. Bergshanteringens arbetare från medeltiden till gustavianska tiden*, Stockholm 1951, 46–49; G. Magnusson, 'Iron-making in a Long Time Perspective. Some Important Questions', in *The Importance of Ironmaking. Technical Innovation and Social Change.* Papers presented at the Norberg Conference on 8–13 May 1995.

5. Boëthius, *Gruvornas*, 38–39.

6. J. Myrdal and J. Söderberg, *Kontinuitetens dynamik. Agrar ekonomi i 1500-talets Sverige*, Stockholm 1991, 483–487.

There was also an important difference in how production itself was carried out, compared to the peasants who sold products made from lake and bog ore. The medieval blast furnace brought with it the establishment of a basic organisation of the work. This organisation was built, to a greater degree than previously, on cooperation between the different *bergsman* households. The peasants who produced iron from lake and bog ore in simple furnaces were able to carry out their production and also manage the production facilities with the means that were at the disposal of each individual household. In principle two people were needed in order to carry out the smelting of the bog ore: one to work the bellows and one to take care of the furnace.[7] It was possible to manage this work within the framework of one household, as was also the case with the refining, that is the forging, of the molten ore. As with all other peasant production the *bergsmän's* work was organised along household lines, but in addition to the individual household work – which took up most of people's time – the furnace itself was a facility for which several households had responsibility. This cooperation reduced the maintenance costs for each of the separate households, and there were also operational advantages. In contrast to the simple furnaces that were used in the smelting of lake and bog ore, smelting in the blast furnaces could be carried out continuously, without interruption between each refill of charcoal and ore.[8] The heating-up costs were thus reduced for each of the households. According to the royal decrees the *bergsmän* were also forced to engage a furnace-master, whose job it was to supervise the work process in the furnace. The joint smelting periods had the advantage that they only took place once or twice per year and labour costs could also be kept at a reasonable level as this expense was shared by many.

The primary cooperation between the *bergsmän*, however, was with regard to the management of the joint facilities, not the smelting work and the acquisition of raw materials.[9] All the preparation work – with charcoal-making being particularly time consuming –

7. T. J.-E. Pettersson, 'Blästan och blåsningen', in *Lima and Transtrand. Ur två socknars historia. Del 1. Myrjärn och smide,* Stockholm 1982, 92–94, nevertheless points out that it was not unusual for two households to share a furnace.

8. K.-G. Hildebrand, *Swedish Iron in the Seventeenth and Eighteenth Centuries. Export Industry before the Industrialization,* Stockholm 1992, 45, points out that the blast furnace's technical improvements should not be exaggerated. The blasting periods were short and the heating-up costs were high and production was often defective.

9. Hildebrand, *Swedish Iron,* 133.

was carried out separately by each individual household and the smelting itself was also carried out mainly by the labour that each household could mobilise. The joint smelting period was divided up between the different households, each of which had at their disposal a certain number of smelting days. The number of days was linked to how much forest, that is charcoal-making potential, each household had. When the smelting had started the different *bergsman* households followed a previously drawn up rota; when one household had finished smelting, the next household took over, and so on. The *bergsmän's* form of production was therefore – in spite of the cooperative element – largely based on what the individual households could accomplish. Nevertheless, we find here a form of cooperation that was perhaps crucial to the ability of the individual households to be able to carry out production at all. In any case, it should be stressed that this form of cooperation constituted an important difference between the iron-making peasants in the Urals and the *bergsmän* in Bergslagen, and that cooperation among the *bergsmän* continued throughout the eighteenth century.

At the beginning of the seventeenth century the *bergsmän* in principle controlled the whole production process – from the mining of the ore to the smelting in the furnace, the forging process and finally to the selling of the end product. From the middle of the sixteenth century the government had tried to promote through the idea that bar iron would be the best product for export – and here they pointed to the fact that it was not practical to further refine *osmund* iron – but the crown's intentions, however, did not change the balance of power in iron-making – yet.[10] The small number of iron-making peasants who were established in the Urals also had control of the whole production process and here production was even more firmly based on what each household was capable of producing. The forge and the furnace were even often built adjacent to the farms.[11] In spite of the differences that characterised the peasants' production of iron in the regions of Bergslagen and the Urals there was an important similarity: the peasants carried out, and had control of, the whole production process. Around the middle of the sixteenth century the Swedish state intervened quite resolutely in iron-making. It did this both by

10. Ibid., 44.

11. *Istoriia Urala s drevneishikh vremën do 1861 g.*, Moscow 1989, 187; Tsentral'nyi Gosudarstvennyi Arkhiv Drevnikh Aktov(TsGADA), F (Fund) 111 Op (Inventory) 1 D 11 L 23–24. An example of this is reported from the year 1624 in Cherdyn´ *uezd*.

setting up its own production facilities and by directing political measures towards steering iron production in the direction of export sales. Eli F. Heckscher described the change as 'perhaps the first step that Sweden took in the direction of more capitalistic production'.[12] In the sixteenth century this 'step' was still not particularly large; a few decades of the seventeenth century had passed by the time the mining industry changed more radically.

Some way into the seventeenth century the Swedish state withdrew from direct production. Instead the ironworks owned by the state were sold to a new category of owners – financially strong people of rank – who were interested in mining because of the access it gave to capital and business contacts abroad. The new stakeholders not only took over the works that the state had previously established but also set up a large number of ironworks, so-called hammers, themselves. The arrival on the scene of this new category of owners and their eagerness to expand resulted in a dramatic increase in production as early as the middle of the seventeenth century.[13] The number of works also grew considerably and towards the end of the seventeenth century there were, in round figures, no less than 320 ironworks producing bar iron in Sweden, most of which were situated in Bergslagen.[14] The expansion of iron production had a number of effects. The population increased sharply in the whole region and new settlements spread rapidly.[15]

At almost the same time as the Swedish state sold off its ironworks the Russian state first set up a few works in the Urals. However, state involvement in the Urals had, just as in Bergslagen, been preceded by peasant production, and this production came to be increasingly intensified in parallel with the newer and larger works owned by the state. This was seen not least in the growth of the population. Up to the end of the sixteenth century the Urals were quite sparsely populated. However, in connection with the widespread political unrest and several bloody peasant uprisings in Russia, the pressure on the peasants in the central parts of the country increased and many peasants therefore chose

12. E. F. Heckscher, *Svenskt arbete och liv. Från medeltiden till nutid,* Stockholm 1941, 1968, 81.

13. Hildebrand, *Swedish Iron*, 18.

14. Boëthius, *Gruvornas*, 122.

15. N. Friberg, 'The growth of population and its economic-geographic background in a mining district in central Sweden 1650–1750', *Geografiska Annaler,* Stockholm 1956.

– both legally and illegally – to leave these areas. The abundant fertile soil in the central Ural region attracted many colonisers, but there was also a good supply of iron here. Verkhotur´e, Tobolsk and Kungur *uezdy* (districts) were the areas where the migrants first established themselves and there were good deposits of iron ore in all three areas. A couple of figures can illustrate developments. Kungur *uezd* came into being in 1648 and at the beginning of the eighteenth century the original number of inhabitants had increased twelve-fold. There was a similar course of events in Verkhotur´e *uezd*. In 1624 there were 278 households, but just under fifty years later this figure had risen to 1,980 and a further fifty years on there were 3,000 peasant households. The rapid growth of the population and the widespread reclamation of land was considered something positive: the Tsar's envoy, Izbrand Ides, was mightily impressed by this when he travelled through the Urals in the 1690s.[16]

A sharp increase in the population thus took place in both Bergslagen and the Urals. In Bergslagen it was the opportunity to make a living out of iron that was the main attraction, whereas in the Urals it was primarily the fertile soil. But the dramatic developments also created problems in both regions: what role would the peasants play in iron production in the future?

The Peasants and the Production of Iron

When the Swedish state sold off its ironworks to the financially strong groups who then came to constitute a special ironworks proprietor class, it took great care to politically regulate the production of iron. Round about the middle of the seventeenth century the newly established Bergskollegium (Board of Mines) therefore formulated a number of concrete regulations with regard to how iron production should be carried out in practical terms and it was also the officials on this Board – and they were recognised as highly competent – who gave expression to the overall view of how iron production should best be managed.[17] In the regulations of 1649 which prescribed the guidelines for the organisation of the mining industry it was maintained that forging, that is to say bar

16. *Istoriia Urala*, 56, 70–73, 106, 174.

17. B. Ericsson, 'Privilegiegivning till bergverk, järnbruk och exportsågar i komparativt perspektiv', in *Industri og bjergværksdrift. Privilegering i Norden i det 18. århundrede*, Oslo 1985.

iron production, was a bourgeois industry (*borgerlig näring*) that the *bergsmän* should not concern themselves with.[18] Thus, the view that each social group should devote itself to the industry that was appropriate to this group's rank – without interfering in that of other groups – was formalised. Otherwise there was a risk that the social order would be disturbed. However, economic reasons were also given. In a report from 1652 two of the leading officials on the Board of Mines, Christer Bonde and Erik Fleming, emphasised the merits of the small-scale production of *bergsmän*. They claimed that the *bergsmän's* smelting works had a particular stability because they were so obviously combined with the *bergsman's* owning of 'hearth and home, where he, from cattle and various [other] expedients, is greatly helped to [obtain] a living; inasmuch as he works together with his wife and children, whom he in any case must feed, and who have but few needs'.[19] According to Bonde and Fleming the *bergsmän's* solicitude for their families and farms guaranteed that they would not abandon iron production, even if it was not always profitable for the individual *bergsman*. Neither was there any need to fear that the *bergsmän* would be wasteful of nature's resources: the *bergsmän's* production was – per household – far too modest for that.

The state's political regulations therefore aimed to keep the *bergsmän* in iron production, but their work was limited to responsibility for the first phase of production, the mining and the smelting of the ore – so-called pig-iron smelting. The traditional *osmund* forge was energetically opposed; the finished product should instead be bar iron, and the responsibility for this element in the long chain of iron production lay with a completely different group, the bourgeoisie. In the state regulations the aim was for the *bergsmän* to continuously supply the works owned by people of rank with molten ore and the peasant who was not involved in the smelting of pig-iron was encouraged to instead supply the fuel-thirsty forges with charcoal. The overall aim of this policy was to keep the ironworks' production costs at a low level, which the peasant – who was of course forced to produce in order to survive – could contribute to by his integration through certain tasks into the total production process. The low production costs meant that Swedish iron could be sold cheaply on the foreign

18. M. Sjöberg, *Järn och jord. Bergsmän på 1700-talet*, Stockholm 1993, 21–22.
19. C. Bonde & E. Fleming, 'Ett principbetänkande om bergshantering från mitten på 1600-talet. Christer Bonde och Erik Flemings relation till drottning Kristina år 1652', *Jernkontorets Annaler*, Stockholm 1901, 20.

market, which at this time was constituted primarily by Holland, and later mainly by Britain.[20]

The notion that the forests were not an unlimited resource also came from Britain. According to the Board of Mines the devastated forests there spoke for themselves and the mismanagement of the forests was considered to be an important reason behind England's increasing import of iron. In actual fact the English ironworks were being restructured and under these circumstances it was cheaper for the English owners to import iron from Sweden, which was then refined in the English ironworks.[21] However, the constantly recurring reason given by the Board of Mines in all the regulations concerning the country's charcoal resources was the fear that the Swedish forests would be completely cut down. It was considered that this risk was particularly great if all the phases in iron production were obliged to use the same forest. Furthermore, there was the risk of competition for the charcoal, which could lead to higher prices and thus make iron production more expensive overall. It was therefore decided that those units which did not have to have their production in the immediate vicinity of the deposits would be moved to areas outside Bergslagen itself. This was the lot of those who ran bar iron forges, the ironworks owners.[22]

Swedish iron production was thus organised along hierarchical lines and in this rank order it was self-evident that the most work-intensive tasks such as charcoal-making and the acquisition of ore should still be done by the peasants; these tasks had of course been a part of the peasants' work from the very beginning. It was equally self-evident to the state that the phase that required a relatively large capital investment – the bar iron forge – should be transferred to the bourgeoisie. This hierarchical division of the areas of responsibility in the process of producing iron was combined with a geographical division: each group and task was allocated different charcoal purchasing areas. However, as several ironworks had been established in Bergslagen at a very early stage, this division could not be fully complied with in practice,

20. Hildebrand, *Swedish Iron*, 25–26.

21. Ibid., 34–37

22. P.-A. Karlsson, *Järnbruken och ståndssamhället. Institutionell och attitydmässig konflikt under Sveriges tidiga industrialisering 1700–1770*, Stockholm 1990, 50. During the course of the eighteenth century the conflict between the peasants and the industrial ironworks intensified, which underlined the authorities' difficulties in mobilising the peasants for charcoal-burning, 146–147.

though this did not change the goal of the Board of Mines.[23] It was not only the geographical division that was difficult to implement in practical terms. The dividing up of the different groups' obligations also met with problems. As early as the middle of the seventeenth century regulations were issued which absolutely forbade *bergsmän* to sell their charcoal to the forges of persons of rank – which apparently was a common occurrence which should be prevented – and the owners of the ironworks were forbidden to buy charcoal from other areas than those which belonged to each ironworks' charcoal purchasing area.[24]

The royal decrees initially aimed at ensuring that *bergsmän* continued to be producers of pig-iron. But the state was not entirely convinced that the *bergsmän's* small-scale production was of the quality desired. As early as the seventeenth century the *bergsmän* were therefore obliged to hire a furnace-master, who would make sure that the raw materials were of good quality and that the smelting was done in an acceptable way.[25] During the course of the eighteenth century a change could be seen in this respect and it was specially noticeable in the decree of 1766 which regulated work in the blast furnaces.[26] In this decree the furnace-master would – just as before – be responsible for the smelting work and was himself compelled to have certain qualifications in order to be able to work as a furnace-master. In contrast to what had been the case in previous decrees it was now a part of the furnace-master's duties to hire some of the workforce. In the decree of 1766 the *bergsmän* should have as little to do with the actual production of pig-iron as possible; their task was principally to supply raw materials – charcoal and ore – of good quality. It was quite obvious with the issuing of this decree that the state was seeking to develop the division of labour within iron production even further and that it was intended that the *bergsmän's* areas of responsibility should be further reduced.

The political intentions of the state, however, were one thing; how things turned out in practice could be another. As early as the middle of the seventeenth century the state noticed that the owners

23. E. F. Heckscher, *Sveriges ekonomiska historia från Gustav Vasa,* Del 1:2, Stockholm 1936, 486–487.

24. Sjöberg, *Järn och jord,* 64.

25. *Kongliga Stadgar, förordningar, privilegier och resolutioner angående justitien och hushållningen vid bergverken och bruken, (KSFPR),* vol. I, Stockholm 1736–1797, 1638, 104–108, 1649, 154–172.

26. *KSFPR,* vol. III, 1766, 224–243.

of the ironworks were buying up the property of the *bergsmän* in order to acquire the charcoal-making potential of the various farms. As the owners of the ironworks were not interested in producing pig-iron themselves, several furnaces were destroyed and the Board of Mines feared that the total production of pig-iron would fall. In order to prevent such a development the possibilities of acquiring a *bergsman's* land were restricted for people not belonging to the *bergsman* group.[27] The dividing up of production among different groups, where the *bergsmän* were responsible for the production of pig-iron and the owners of the ironworks for the subsequent refinement, had put the owners of the ironworks in a position where they were competing against each other: what could they do to get the *bergsmän* to deliver their pig-iron to their particular ironworks? The solution was credit. A system very rapidly developed, therefore, where the owners of the ironworks paid the *bergsmän* in advance for forthcoming deliveries. But it was not always the case that the *bergsmän's* deliveries covered the credit that they had received and many *bergsmän* therefore fell deeply into debt. This falling into debt contributed greatly towards the *bergsmän's* property continuing to go over to the owners of the ironworks, in spite of the restrictive legislation.[28]

In the course of the eighteenth century there thus took place in practice a displacement of the various parts of production. This displacement meant that the owners of the ironworks came to have at their disposal units for the production of both pig-iron and bar iron, a development which was even more true of the nineteenth century – the century which saw the definitive demise of the *bergsmän's* small-scale production.[29] In the seventeenth century the different stages of production were integrated with each other, but at the same time production as a whole was decentralised. During the eighteenth century iron production was still characterised by integration and decentralisation, but a centralisation process became more and more noticeable. Despite the fact that the position of the *bergsmän* deteriorated in relation to those who had invested in the ironworks, their total production of pig-iron

27. *KSFPR*, vol. I, 1664, 224–227.

28. Sjöberg, *Järn och jord*, 102–103.

29. A. Weinhagen, *Norbergs Bergslag samt Gunillbo och Ramnäs till omkring 1820. Studier i områdets närings- och bebyggelsegeografi*, Lund 1947, 138–139. For the nineteenth century and another area, T. Omberg, *Bergsmän i hyttelag. Bergsmansnäringens utveckling i Linde och Ramsberg under en 100–årsperiod från mitten av 1700-talet*, Uppsala 1992, 60–64.

was still far greater than it had ever been before.[30] The *bergsman's* forge, which really belonged to the bourgeoisie's area of responsibility, did not cease to exist either. In absolute figures the *bergsmän's* bar iron production remained at a constant level throughout the eighteenth century, but in comparison with the ironworks it was of course a question of decline.[31]

A characteristic of Swedish iron production during the seventeenth and eighteenth centuries was that it was based, both in the aims of iron production policy and in practical production, on what small-scale *bergsman* production could achieve. Compared to the time prior to people of rank appearing on the iron production scene, the *bergsmän* had admittedly lost their direct contact with the international market and their main field of operations had been limited, but that part of production which remained within the *bergsmän* group was still an important precondition for the ironworks' production, and it was around this insight that iron production policy was formed. This policy therefore advocated by and large an integration of *bergsman* production and ironworks production, but that this integration should be indirect.

A completely different line set the tone in the Urals. During the course of the seventeenth century trading in iron products expanded in several parts of the region. It was primarily lake ore deposits that were processed. Thus, here the process of iron-making was similar in nature to that of the peasants in north-west Dalarna. The good supply of lake ore made it easy for peasants to make products out of iron without any advanced technology and many peasants therefore became involved in the small-scale production of iron, as was the case in Kungur *uezd*.[32] However, things developed at greatly different speeds in different places in the Urals. For example, iron production expanded early in Tobolsk, particularly in Bagariak and Aramil', whereas the peasants in Verkhotur´e were not at all involved in small-scale iron production even as late as in the 1670s. However, only a few years later the situation in Verkhotur´e was quite different: several furnaces intended for iron production had then been installed, and production subsequently increased rapidly.[33] In contrast to the case of production by the Swedish *bergsmän* prior to the establishment of

30. Hildebrand, *Swedish Iron*, 130.

31. Boëthius, *Gruvornas*, 122.

32. V. Hennin, *Opisanie ural'skikh i sibirskikh zavodov 1735 g.*, Moscow 1937, 34; *Istoriia Urala*, 29, 106, 187, 268, L 120–122.

33. TsGADA, F 111 Op 1 Stb 11 L 23–24, 32–95.

the ironworks in the seventeenth century, the production of iron by the peasants in the Urals was intended only for local and regional trading. The increase in production was therefore, at least initially, the result of the steadily increasing population. Just as in Sweden, Russian iron production policy also aimed to support the expansion of iron production. The state set up works of its own with this purpose in mind – primarily from the time of Peter the Great – but at the same time it encouraged financially strong groups to establish ironworks. We have already seen that Swedish mining policy aimed to divide up the different phases of production between different groups; neither group – *bergsmän* or ironworks owners – should drive the other out of business. In this respect the Russian state followed another line. At the beginning of the seventeenth century the state did not show much interest in small-scale peasant production, but during the second half of the seventeenth century this production was taxed, and this hardly encouraged increased production. At the same time, as conditions for the peasants' small-scale production became tougher, political support for large-scale production in ironworks became more pronounced. The largest ore deposits were reserved for the large ironworks and the peasants were prevented from mining ore for production of their own. At the beginning of the eighteenth century the peasants were forbidden to mine ore at all. This legislation attempted to monopolise the supplies of raw materials for the large ironworks, which in principle would have ruled out the peasants producing iron independently. In practice the policy was not always so easy to implement: conflicts over control of ore deposits between ironworks owners, peasants and state administrators were common and sometimes also led to violence.[34]

Iron production policy in Russia was thus based on peasants being integrated with – and even subordinated to – large-scale export-oriented production. Despite the authorities' stubborn restrictive attitude small-scale production still continued to expand during the first half of the eighteenth century, which indicates – not least of all – that the ironworks owners were not entirely successful in appropriating the iron-ore deposits for their own use. Just as in Bergslagen the ironworks owners in the Urals were dependent on the peasants' labour. The peasants in the villages around the ironworks in Uktus, for example, supplied on a contractual basis a large part of the ore that was used in production at the ironworks

34. TsGADA, F 214 Op 2 Stb 611 L 202–203.

and it is not unlikely that these supply contracts also allowed the peasants to set aside some of the ore that they had mined for production of their own.[35]

Russian iron production policy, which amongst other things aimed to prevent the peasants from producing metal products that they could sell themselves, was difficult to implement since production as a whole – just as in Sweden – suffered from a shortage of labour. This was felt not least by the state ironworks which in the beginning – in the seventeenth century – had great difficulty in keeping production going continuously. Even the famous Nitsa ironworks was forced to temporarily stop production early on owing to a shortage of labour.[36] A problem that was connected with insufficient labour supply was the small amount of ore that was at the disposal of production, and this also brought about interruptions in operations. For example, the Krasnoborsk works existed for only a short time in the seventeenth century before it was definitively closed down. It was closed down precisely because of the limited supply of ore available for production at the ironworks.

Unlike small-scale peasant production, which was established at an early stage in the Urals, large-scale ironworks production was never intended for local and regional trading purposes. Instead it was the state's needs that steered the orientation of production. However, the early ironworks had difficulties in procuring both ore and grain in sufficient quantities and therefore still contributed through their import needs to a certain amount of trade between the regions. Grain, for example, was transported from Pomor'e to the Urals, which also meant that production costs increased.[37]

In both the Urals and in Bergslagen the large-scale establishment of ironworks was preceded by small-scale peasant production. Swedish iron production policy solved the problem that was common to both of the countries – the shortage of labour – by further building upon and preserving the production, that of the *bergsmän*, which was already established. On the other hand, Russian policy went against peasant home-production and encouraged fuller integration of the peasants with the large-scale iron industry. Although information about the production of iron by the Russian peasants is lacking in many respects the evidence

35. Minenko et al., 'Ural Iron', 80.

36. E. V. Vershinin, 'Istochniki po istorii organizatsii pervogo zhelezodelatel'nogo predpriiatiia na Urale', in *Istochniki po sotsial'no-ekonomicheskoi istorii Urala dooktiabr'skogo perioda*, Moscow 1992, 20–21.

37. TsGADA, F 214 Op 5 Stb 457 L 42.

already suggests that this iron-making, despite energetic attempts, did not disappear all together when the ironworks entered upon the scene. Who, then, were these iron-producing peasants and how was their iron-making organised? And in what ways did their conditions differ from those of the Swedish *bergsmän* ?

The Iron-Making Peasants

Just like the Swedish *bergsmän* the iron-making peasants in the Urals controlled the whole production process. They were therefore ore-miners, ore-smelters and forge-workers; and moreover, they were also farmers. The agrarian part of their lives – measured in terms of the size of the landholding – played a more important role in the overall economy of the iron-making peasants than agrarian production was able to in the economy of the Swedish *bergsmän* in the seventeenth century. Even so, the size of the landholding was also a measure of how successful the iron-making peasants in the Urals were when trading with what they had produced. The conditions under which the iron-making peasants in the Urals lived have admittedly not yet been fully elucidated, but there are signs that at the beginning of the eighteenth century they belonged to that category of peasants who had more land at their disposal than was usual for peasants.[38] There are signs of a differentiation of the resource which carried most weight in all peasants' economies – landholding – at the same time as a certain specialisation developed among the peasants in the Urals. In registers of taxpayers and similar records from the early eighteenth century occupational designations were recorded that suggest this. The authorities received taxes from peasants who sold iron products and recorded these taxes under the occupation of forgeman in the fiscal registers. The iron-making peasants in the Urals thus increasingly oriented their economic activities towards iron production, but the beginnings of specialisation were also noticeable within the production of iron, where some worked mainly with forging, while others primarily focused on obtaining ore for the forges.[39] The number of peasants specialising in forging grew,

38. A. Tomilov, 'Metallurgical trades of the Ural peasantry in the mining industry at the beginning of the 18th century', in *Metallurgical Works and Peasantry. Problems of Social Organization of Industry in Russia and Sweden in Early-Industrial Period*, Ekaterinburg 1992, 296.

39. A. A. Kuzin, *Istoriia otkrytiia rudnykh mestorozhdenii v Rossii*, 1961, 40.

which was of course linked to the fact that the peasantry as a whole rapidly increased. But the increasing number of peasants as a whole would not necessarily have resulted in greater numbers with forging as their main source of income if it had not been for the fact that the great demand for iron products meant that this switch-over was quite profitable.

A few figures can illustrate this – allowing for the role played by possible errors in the authorities' records. In 1680 there were 1,117 households recorded in the villages of Nev'iansk, Belosludsk, Ust'-Nitsa, Aiat', Krasnopol'skaia and Utkino, and of these twenty-four were designated forging households – that is 2 per cent in round figures. The number of forging households had at that time increased in the space of barely fifteen years from seventeen to twenty-four, and this increase also continued after 1680. In 1720 records were made of the peasantry involved in various crafts and at that time eighty-two forgemen were recorded in Verkhotur´e *uezd*. In Tobol´sk *uezd* there were 146, and this figure constituted approximately 3 per cent of the peasants. This figure was large but could, in certain areas, be even larger. In the *sloboda* Bagariak it was recorded at the same point in time that no less than 6.5 per cent of the peasants there were forgemen.[40] The above figures demonstrate that the iron-making peasants were on their way to being distinguished from the farming peasants and the driving force behind this development was, as previously stated, the constantly increasing demand for the kind of iron products that the forgeman was able to produce in his forge. The fact that the forgemen in certain cases came to be regarded as country craftsmen is confirmed not least in the register of taxpayers that was drawn up in 1695 in Aramil´. One of the peasants was taxed for his forge and he stated on this occasion that he was not a farmer – even if he had a little hay at his disposal – and therefore he was not prepared to pay tax for the ownership of land. A similar state of affairs came to light in the taxation assessment in 1719 in Kolchedan when one of the peasants was recorded as the owner of a forge without having any farming land at his disposal.[41] In both of these examples the specialisation had gone so far that the iron-making peasants

40. O. N. Vilkov, *Ocherki sotsial'no-ekonomicheskogo razvitiia Sibiri kontsa XVI-nachala XVIII v*, 1990, 287; A. Tomilov, 'Metallurgical trades', 297; Gosudarstvennyi Arkhiv Sverdlovskoi Oblasti (GASO), F 24 Op 1 D 4a, 31–34; TsGADA, F 271 Op 1 Stb 634 L 981–988, 994ob-1034ob.

41. Vilkov, *Ocherki sotsial'no-ekonomicheskogo*, 286 (1695); GASO, F 24 Op 1 D 1 L 140 (1719).

had left the peasant sector of their economy. However, this occurred only in exceptional cases. Such extreme specialisation was rare but the examples show that the peasants did not lack the ability to develop new economic strategies as the economic conditions changed.

The twin process of the differentiation of resources among the peasants in the Urals and the specialisation of the fields of activity was tentative and its importance should perhaps not be exaggerated. It is not easy to map out the social stratification among the peasants during the seventeenth and eighteenth centuries, given the records of households that remain available to researchers. Despite the fact that the information in these records can be unreliable it still seems to have been common – although there was great variation – that the peasants who worked with forging had more land than peasants in generally had at their disposal.[42] This state of affairs could be interpreted as not supporting the conclusion that specialisation among the peasants increased. However, we believe that specialisation within the framework that household production allowed first developed to the level that the household's workforce could manage – both with regard to the production of goods which were to be sold and the production of provisions for their own needs. In this category can be included those peasants with a forge who were dependent on having land of their own in order to be able to provide their households with food; the forge meant that they probably had an income which enabled them to acquire a little more land than the average peasant. Greater expansion than this required that the peasant forgeman's total production – both from the land and from the forge – was integrated into a system of trade, and that the recruitment of labour extended beyond his own household, that is to say on to the labour market.

Examples of this occurred when the forging developed to such an extent that the owner appeared to be the owner of an ironworks, albeit on a small scale.[43] In the aforementioned tax assessment in Kolchedan in 1719, the peasant M.O. Kurilov, for example, paid tax for a 'factory', a forge, arable land and also a little hay. Kurilov's economic activities thus extended over several fields and were, furthermore, so comprehensive that they could scarcely be maintained by the labour that the household could muster, which

42. Vershinin shows that the standard of living varied among the iron-making peasants: 'Istochniki po istorii', 188.

43. The information on labour in this paragraph is based on GASO, F 24 Op 1 D 1 L 28, 31–34 Op 2 D 72 D 17 L 116–125, 128–132; *Istoriia Urala*, 187.

is also indirectly confirmed by the fact that Kurilov himself was entered in the register of taxpayers as being the only man in his household. Kurilov must therefore have been dependent on hired labour. Whether this labour was hired temporarily or permanently doubtless varied from case to case. As Kurilov carried out both farming and industrial activities his need for labour was probably quite great all the time. Therefore there is reason to believe that his activities were supported by hired labour all year round. However, in Kurilov's case there is no direct information about how large his workforce was – here we must make do with conjecture. On the other hand, when the peasant Remezov in Nev'iansk was taxed he gave information where he clearly stated that production in the forge was carried out with the help of his son and a temporarily hired '*guliashchii*'. Hiring labour on a temporary basis was perhaps the most common way for the peasants to manage the everyday work as well as particularly work-intensive phases.

Who, then, constituted this temporarily hired labour? In the fiscal registers from Kataisk in 1720 it can be seen that all the forges were owned by peasants, but there were also several peasants who were designated forgemen even though they did not own a forge. Several of the forgemen who did not have their own forge were relatives of those who did; it could be a son or a brother, but they still were not included as members of the owner's household. There is therefore reason to believe that the forge owners' children could act as temporary labour in the different forges – and not just in a relative's forge. In this respect the passing on of knowledge was traditional; the older generation introduced the younger generation to the work, and this could take the form of the latter temporarily helping out in the various forges. As the number of registered forgemen was greater than the number of forges it was not likely that all the forgemen could take over the work that the older generation had built up, and this could lead to social differentiation. It also meant that the large-scale ironworks were able to attract those workers who had a knowledge of how to forge good iron, but who did not have the financial resources to carry on production on their own.

A common feature of the small-scale iron-making that the peasants in the Urals established and the production of the Swedish *bergsmän* was that it was based to a large extent on the household's workforce. The individual forgeman's production in the Urals was on a small-scale and very similar to the work of a craftsman, where the household was the primary source for the recruitment of labour and the possibilities of expansion were thus limited. The demographic conditions in Bergslagen during the seventeenth and eighteenth centuries have not been mapped out to any great extent, but the studies that exist point out that the *bergsmän* in general had large households, though the explanations vary. The *bergsmän* in Skinnskatteberg had many children and consequently their households were large. The *bergsmän* thus met the constant need for labour by having large families.[44] Investigations into a later period, the eighteenth and nineteenth centuries, confirm that the *bergsmän* still had large households, but show that this was because they hired a great deal of labour.[45] Other investigations – which admittedly deal with a smaller geographical area – have pointed out that the number of maids and farm-hands was connected with which part of their life cycle the *bergsmän's* families were in during the eighteenth century. Early on in parenthood the husband and wife were in greater need of servants than during those periods when there were half grown-up children who could help with the work.[46] The aforementioned studies all show that the need for manpower was mainly met by means of the family and temporarily hired labour. Just as was the case for the children of the peasants working in forges in the Urals, it was not possible in eighteenth century Sweden for all the *bergsmän's* children to inherit that which their parents had built up. As far as we know it was usually one of the sons who took over the farm that the household's production was dependent on, and this automatically ruled out all of the women, in addition to a number of sons who had no prospects of inheriting.[47] The need for hired labour enabled the *bergsmän's* children – like the forgemen's children in the Urals – to learn about iron production in a practical way, but there was also a basis here for social differentiation – just as in the case of the peasants in the Urals. It can be seen that in the second half of the eighteenth century resources – above all land – were increasingly differentiated. The landowners became fewer but each and every one of them had more land at his disposal than previous generations of *bergsmän*.[48] This process continued and in the

44. D. Gaunt, 'Familj, hushåll och arbetsintensitet. En tolkning av demografiska variationer i 1600– och 1700– talens Sverige', *Scandia* 1976.

45. M. Isacson, *Ekonomisk tillväxt och social differentiering 1680–1860. Bondeklassen i By socken, Kopparbergs län* Uppsala 1979, 96; Omberg, *Bergsmän i hyttelag*, 90.

46. Sjöberg, *Järn och jord*, 51–55.

47. Ibid., 169.

48. Ibid., 97.

middle of the nineteenth century there is evidence, for example, that the people who owned a lot of land in the parish of By in Bergslagen had stuck to the production of iron.[49] This was similar to the forgemen in the Urals, who also – thanks to market-oriented production – managed to obtain more land than the average peasant. The differentiation also meant that at the same time as some managed to obtain more land there were others who were gradually forced to part with land.

In contrast to the peasantry in the Urals the *bergsmän* were established with the express aim of the state that they would specialise in an activity that would benefit the national economy as a whole, that is to say the production of pig-iron. However, how the *bergsmän* actually carried out their work varied. In some places they stuck to the whole production process and like Kurilov in Kolchedan they could appear to be ironworks owners – albeit on a small-scale. In other places a division of labour developed within the *bergsman* group where some *bergsmän* mainly mined ore, while others carried out the subsequent refinement process. This division of labour was particularly striking in Värmland, where the intentions of the legislation regulating the production of iron were best followed. In some areas the *bergsmän* developed their forging and made products intended for the home market. There was also trade here between the craftsmen of the towns and those involved in iron production. For example, the forgemen who made scythes in the town of Hedemora in Bergslagen bought their iron from the surrounding ironworks and the *bergsmän* in the area. In some areas, for instance in the parish of Söderbärke, the *bergsmän's* economy focused on agriculture instead; the reclamation of land was comprehensive and the stocks of cattle were large.[50] There were therefore tendencies towards specialisation within Bergslagen and among various groups of *bergsmän*. As did the peasants in the Urals, the Swedish *bergsmän* were able to develop new strategies as the economic conditions changed.

The above discussion stresses the inner dynamic of which the peasants in both Bergslagen and the Urals were a part. But this state of affairs was not free from the pressure that the state in both countries put on production, and neither was it independent of the

49. Isacson, *Ekonomisk tillväxt*, 85.

50. Heckscher, *Sveriges ekonomiska historia*, 185–186; Isacson, *Ekonomisk tillväxt*, 82–83; M. Ågren *Jord och gäld. Social skiktning och rättslig konflikt i södra Dalarna ca 1650–1850*, Uppsala 1992, 103–113; Omberg, *Bergsmän i hyttelag*, 104–105.

pressure that the large-scale ironworks constituted for the peasants' way of organising their lives. The ultimate reason why the peasants found themselves on a collision course with the ironworks was the constant problem of the shortage of labour – a shortage that the states attempted to solve by mobilising the peasants.

The Peasants and the Ironworks

The iron-producing peasants in the Urals, who with their forging provided the local population with farming tools, were a coveted group for the state ironworks because of their expert knowledge. The peasants who were estranged from agrarian production in favour of forging therefore obtained work at the state ironworks and came to form a group of state craftsmen with a fixed remuneration from the state.[51] The forgemen who kept their homes in the peasant villages thus took part in locally focused trading where they were in control of the finished product. In parallel with this they placed their manpower at the disposal of a production process where they were completely cut off from the final destination of the products. The local and regional demand for wrought-iron products contributed to their being able to concentrate their economic activities on the production of iron products even further. By earning a living at the state ironworks the peasants were able to strengthen their specialisation in forging, but because the wages were regulated the state ironworks found it difficult to attract permanently employed labour on a voluntary basis. Instead, the peasant forgemen were forced to take temporary posts at the state ironworks. In the 1640s there were several examples of forgemen being deported to ironworks and despite the fact that this transfer of labour aroused protests among the peasantry the process continued.[52]

The ironworks in the Urals not only suffered from a shortage of labour; the supply of iron ore was not as great as the authorities desired. These difficulties were handled in different ways. The Mazuevskii ironworks, which was the largest in Kungur, was founded in 1704 to produce bar iron. The process of smelting ore to obtain pig-iron was not a part of the ironworks' operations. Instead the peasants supplied this ironworks with molten ore. The authorities' ban on the peasants producing iron was circumscribed here

51. *Istoriia Urala*, 187.
52. TsGADA, F 214 Op 5 Stb 17 L 38.

by the peasants' deliveries taking place in the utmost secrecy. Not even the large-scale ironworks – whose activities were meant to be favoured by the rigid legislation – were served by the state's regulations. The peasants and the ironworks' administration thus traded in collusion. This trade was similar to that which the Swedish peasants legally maintained with the ironworks. In contrast to the state of affairs in Sweden the trade that took place in connection with the Mazuevskii works was completely illegal. Furthermore, there is evidence that as late as the 1740s operations in Mazuevskii were based on the ironworks illegally purchasing iron products from the peasants.[53]

In spite of the fact that Russian iron production policy strictly prohibited the iron-making that the peasants themselves had established, there are thus signs that peasant production continued anyway – and then in cooperation with the large-scale ironworks. This cooperation was not only based on the peasants feeding the ironworks with raw materials; the opposite also occurred. The peasants who forged farming equipment for the local market sometimes preferred to work with the iron produced in the works and therefore bought their iron from the works.[54] This trade with the peasantry that the large-scale ironworks maintained meant that some of the iron which – if the state's intentions had been followed – should have reached the export market instead contributed to maintaining regional trade. There was also a similar state of affairs in Sweden. An example of this is perhaps how the previously mentioned iron-producing peasants in north-west Dalarna, whose iron-making was based on bog ore and for that very reason was never a part of the production of iron intended for export, solved the problem of the shortage of ore. They also obtained their raw material from the works that produced bar iron.[55] Another example is the trade between the forgemen of the Jäder ironworks during the seventeenth and eighteenth centuries on the one hand and town-dwellers and neighbouring peasants on the other, where the peasants – unlike the peasant forgemen in the Urals – supplied

53. A. A. Preobrazhenskii, 'Rabotnye liudi na Urale v XVII-nachale XVIII veka. K voprosu o rynke rabochei sily', in *Iz istorii rabochego klassa Urala*, 1961, 29, 55, L 117ob, 119ob; A. A. Preobrazhenskii, *Ural i Zapadnaia Sibir' v kontse XVI–nachale XVIII veka*, Moscow 1972, 286; GASO, F 24 Op 1 D 1 331 L 574ob-575.

54. GASO, F 24 Op 1 D 17, 109–110, 112–113.

55. T. J.-E. Pettersson, 'Myrjärnshanteringens storhetstid och nedgång', in *Lima och Transtrand. Ur två socknars historia. Del 1. Myrjärn och smide*, Stockholm 1982.

raw materials in order to obtain wrought-iron farming equipment in return.[56]

In addition to the export market for which production in both regions was intended there was thus trade between the peasants and ironworks – in both the Urals and Bergslagen – which had primarily developed from the daily requirements of the production of the peasants and the ironworks, in order to be able to continue at all. But in both regions the ironworks' production put pressure on the peasants. When the state works were established in the Urals in the seventeenth century the neighbouring peasant forgemen were forced to pay a part of their production in the form of taxes. As this method was often used it meant that the peasant forgemen were to a large extent indirectly drawn into large-scale production.[57] The Swedish situation was characterised by a steering of small-scale production over to large-scale production. The Swedish *bergsmän* were never directly forced to supply the ironworks, but the economic conditions often severely tested the *bergsmän's* independence. In those cases where the *bergsmän* were forced to give up their independence because of their debts and to become rustics under the thumb of the ironworks, the ironworks owners were at the same time given the opportunity – at least theoretically – to force them to deliver.[58]

In both the Urals and in Bergslagen the ironworks' supply of raw materials and labour was associated with a large measure of uncertainty. Iron-making at the Uktus works, for example, relied on the peasants' taxes at the beginning of the eighteenth century, but the peasants were not particularly reliable; deliveries were insufficient and sometimes there were no deliveries at all. The ironworks received not only the peasant forgemen's taxes as deliveries for production: the villages around the works were also forced to pay tax in the form of labour and of course it was those peasants who were knowledgeable in forging who were primarily considered for this tax liability.[59] The forgemen's skill was particularly useful in connection with production at the Uktus ironworks: this works made horseshoes and other tools which were

56. A. Florén, *Disciplinering och konflikt. Den sociala organiseringen av arbetet. Jäders bruk 1640–1750,* Uppsala 1987, 124–133. The forgers' selling was partly unlawful, 134–135.

57. TsGADA, F 1 111 Op 2 Stb 638 L 2–3; E. I. Zaozërskaia, *U istokov krupnogo proizvodstva v russkoi promyshlennosti XVI–XVII vekov,* Moscow 1970, 35.

58. Sjöberg, *Järn och jord,* 120–122.

59. GASO, F 24 Op 1 D 17 L 116–125, 128–132.

then sold in an increasingly expanding inner market, amongst other places to Siberia.[60] The Swedish ironworks that were obliged to buy pig-iron from the *bergsmän* could not in practice – as with the works in the Urals – rely on regular deliveries. The considerable debts that the *bergsmän* in the parish of Söderbärke in southern Dalarna had during the first half of the eighteenth century suggest this and at the same time show the method that the Swedish ironworks owners were obliged to employ, namely continuing to give unsecured credit.[61] Furthermore, because of the extent of the credit, deliveries were not as cheap for the ironworks owners as the iron production policy had intended. The owners who were obliged to buy charcoal and pig-iron from the peasants also, of course, realised the high costs attached to these deliveries.[62]

For example, production in the Ramnäs ironworks, situated on Hedströmmen in Västmanland, was dependent on purchased charcoal and pig-iron throughout the eighteenth century, which meant high production costs compared with the works which produced their own raw materials.[63] Supplies of pig-iron came from the nearby Västerbergslagen *bergsman* villages – often in the winter when the ice-covered streams could support the heavy iron wagons. One of the villages upon whose *bergsmän* Ramnäs relied was Saxe, situated approximately fifty kilometres from Ramnäs. The uncertainty to which production at Ramnäs must have been subject is shown by the fact that the deliveries from Saxe were both of varying size and irregular. Periodically there were no deliveries at all. Production at Ramnäs was, however, not only dependent on the deliveries from Saxe; in the ironworks' accounts several villages are mentioned and this meant that the works' production – apart from a slump in the 1770s – was kept constantly at a relatively high level. For several years during the second half of the eighteenth century the works bought more pig-iron than was used in annual production, which was probably a way of reducing the uncertainty.[64]

60. GASO, F 24 Stb Op 2 D 72 L 340.
61. Sjöberg, *Järn och jord*, 117.
62. Karlsson, *Järnbruken och ståndssamhället*, 195.
63. K.-G. Hildebrand, *Fagerstabrukens historia. Del 1. Sexton-och sjuttonhundratalen*, Uppsala 1957, 275; B. Boëthius, 'Ramnäs bruk under inflationen på 1700-talet. Studier i Jakob Tersmedens enskilda räkenskapsbok', *Historiska Tidskrift* 1954:1; B. Bursell, *Träskoadel. En etnologisk undersökning av lancashiresmedernas arbets- och levnadsförhållanden på Ramnäs bruk vid tiden kring sekelskiftet 1900*, Stockholm 1974, 25–26.
64. Sjöberg, *Järn och jord*, 130–131.

Contrary to the situation of the state ironworks in the Urals, the Swedish works – which in the eighteenth century were under private ownership – were not able to obtain tax revenue in the form of labour. However, the works received – on the payment of duty – the charcoal that the peasants paid in tax. The Swedish state thus also used taxes as a means of redistributing resources in favour of the ironworks. There was a similar idea with regard to pig-iron. The *bergsmän* were forced to carry out a certain kind of production – pig-iron refining – in order to justify the right of possession of their land. Pig-iron refining was taxed and this tax was paid in kind; the iron was then sold through the state. The pig-iron which was thus made available to the owners of the ironworks when it was sold by the state enabled the ironworks to acquire pig-iron over and above that supplied by the *bergsmän*.[65]

In contrast to the ascription system that applied to the peasants in the Urals, the Swedish taxation levy was based on what the individual *bergsman* farms should achieve. In the Urals it was rather the villages that acted as a kind of taxation unit and they were therefore collectively forced to mobilise skilled labour. This had far-reaching consequences. It was to a large extent the ascribed peasant forgemen who constituted the workforce at several ironworks in the 1720s, a workforce which had by then become permanent.[66] In the long term these labour taxes meant that the peasant forgemen gradually came to lose their links with small-scale peasant production, and instead became the wage-earning resource that the ironworks required to maintain production. The link between the Swedish *bergsmän's* small-scale production and the work for wages at the ironworks during the eighteenth and nineteenth centuries is not as clear cut. Generally speaking *bergsman* production was kept as an independent part of production as a whole, which was not the case in the Urals. During the first half of the eighteenth century several *bergsmän* lost their land and then became tenants under the ironworks; but in contrast to what was the case for both the ironworks' workforce in the Urals and conditions in the Ural villages – where there was now marked specialisation – the Swedish tenants' production like that of the independent *bergsmän*, comprised both an agrarian section and a section which was connected with iron production. This combination was in a

65. Ibid., 64.

66. TsGADA, F 214 Op 5 Stb 814, GASO, F 24 Op 1 D 17 L 116–121, 122–125, 128–132.

very concrete way an obstacle for the ironworks owner who sought to integrate his workforce, and it illustrates – at a more general level – the cross-fire that early industrial iron production was in general forced to cope with. Peasant production in both regions will now be summarised around this theme.

Peasants and Iron: Some Concluding Remarks

The most striking thing about developments in Sweden is of course the *bergsmän's* changed role. Initially they had control of the whole production process and also had direct contact with the market themselves. When production as a whole became focused on bar iron the *bergsmän's* role was reduced to providing the ironworks with some kind of semi-manufactured product. They thus lost control of the finished product and they also lost their direct contact with the market. This is a well-known pattern in all early industry, which – irrespective of whether it was textile manufacturing or other production – in general was fed with semi-manufactured products by dependent suppliers who in tough conditions combined agrarian and industrial production.

It was characteristic of the work that the *bergsmän* did that capitalist interests took over contacts with the market, something which is also well known from other proto-industry. A conflict of interests arose here, which at the concrete level was about the aim of production.[67] The *bergsmän's* production was primarily devoted to ensuring the survival of the present and future generations, while the ironworks owners and the business men aimed to achieve profits on the international market. Subsistence production was integrated – and even collided – with production for the international market and this opposition coloured not least of all Swedish iron production policy. The political division of the various stages of iron production was justified by the *bergsmän's* very need to constantly produce. But the small-scale *bergsman* production was limited in its capacity to produce the amount of iron that the ironworks owners desired. This is clear not least from the considerable granting of credit, which exposed the conflict between production for subsistence and production for international market interests,

67. P. Kriedte, 'The Origins, the Agrarian Context and the Conditions in the World Market', in *Industrialization before Industrialization*, eds P. Kriedte, H. Medick and J. Schlumbohm, Cambridge 1981.

which came to characterise *bergsman* production throughout the eighteenth century.

In practice, however, the *bergsmän* developed economic strategies which made them less dependent on the international market and instead favoured their own regional trade. There were opportunities for this in the diversified production that characterised peasant production in general during the seventeenth and eighteenth centuries, but it came to be developed even further by the *bergsmän*.[68] In some places agrarian production was strengthened, in others the forge was developed, but irrespective of which strategy was employed it was combined with continued pig-iron refining, and it was precisely this diversification which was the strength of the economy of those *bergsmän* who continued as *bergsmän* after the eighteenth century. There would have been little chance of maintaining diversified production if it had only been based on what the household could achieve. But with the form of cooperation that the *bergsmän* practised with the furnace household production could extend over several fields of activity – and at the same time maintain and develop small-scale trading.[69] The form of production that the *bergsmän* practised is – together with the support of the authorities – an important explanation for the fact that the *bergsmän* kept their share of iron production throughout the eighteenth century.

If the term indirect integration characterised the Swedish *bergsmän's* position in the production of iron, the term fusion can be applied to the peasant forgemen in the Urals. Their skill was needed in the rapidly expanding state ironworks and their household-based production was therefore opposed in iron production policy. However, just as was the case in Bergslagen, developments in the Urals were not completely clear cut; the peasants' own production was far more capable of surviving than the authorities would have wished. As early peasant production in the Urals was primarily aimed at meeting the local demand for farming equipment it seems unlikely that this production totally ceased to exist; the need remained and was hardly met by the large-scale and export-oriented ironworks. Even if the authorities opposed the peasants' forging obvious difficulties arose when they tried to enforce these regulations. It must have been particularly difficult to

68. J. de Vries, *The Economy of Europe in an Age of Crisis, 1600–1750*, Cambridge 1976, 30–36.

69. A. Florén and G. Rydén, *Arbete, hushåll och region. Tankar om industrialiseringsprocesser och den svenska järnhanteringen*, Uppsala 1992, 53.

monitor the deposits of lake and bog-ore, which suited the technically simple and flexible household production, but which was less suitable for large-scale ironworks. As far as we know the smelting of bog ore lived on in Siberia in the 1720s, when Petter Schönström, who was mentioned in chapter one, travelled around Russia. The smelting of bog ore also continued in Sweden throughout the seventeenth century and in both cases the vitality can be explained by the fact that technically simple iron production satisfied a different market to the one in which the large-scale ironworks were active. Another explanation is that this primitive production of iron did not require forms of cooperation outside of the individual household. The form of cooperation that the peasants in the Urals maintained was primarily a question of the tax levy and it was a form of cooperation which in the long run meant that peasant production lost ground: the peasant villages were primarily bled of the skilful forgemen. At the same time the ironworks and the peasants – as was the case in Sweden – produced partly for different markets, which must have contributed to the vitality of peasant production. Unlike the Swedish *bergsmän*, however, the peasant group quite soon specialised. Of course the increasing demand was a factor here, as in the long run it made a combination of several fields of activity difficult. The specialisation meant that it was the forgemen who were primarily sent off to fulfil the taxation duties of the village, but also – if they remained – that they eventually became more craftsmen and wage-earners than peasants.

The terms fusion and integration underline the different ways the two systems had for maintaining production intended for the international market at the same time as subsistence production continued. Let us stress that this was true at a general level. In actual fact there were examples of parallel activity in both regions: where peasant production lived on side-by-side with the large-scale ironworks – albeit illegally. This was associated with the fact that there was not only an international demand but also a domestic one, the extent of which has yet to be determined.

Chapter Three

The Social Organisation of Work at Mines, Furnaces and Forges

Anders Florén and Göran Rydén
with Ludmila Dashkevich, D.V. Gavrilov
and Sergei Ustiantsev

Introduction

In the introductory chapter the phases in the development of the iron industry during the early industrial periods were discussed. Broadly speaking, Ural iron-making has been categorised as a large-scale industry based on the social foundation of the feudal estate, while the Swedish industry was smaller in scale and showed a social structure in which peasants, *bergsmän*, merchant capitalists and ironmasters played important roles. In the following discussion this will be further elaborated to examine how social change can be seen in the organisation of work in different parts of the chain of production in the two regions. Quite literally, our ambition is to go down the mines and to lift the roofs off the furnaces and forges to get glimpses of the everyday life of work and production there. Such a study will also embrace the conflicts and discords built into these organisations.

It is almost a truism to state that work in the mine, the blast furnace and the forge was of the utmost importance in the process of making iron. However, this is an important starting point in a study devoted to the social organisation of the industry, viewed in a long historical perspective. As has been indicated in earlier chapters, and will be further developed below, the iron industry was an undertaking that was much broader than 'pure' iron-making.

The industry was firmly located in an agrarian setting where, for instance, charcoal-burning and transport played decisive roles. Nevertheless, it was the actual making of iron that governed the other activities; this was the core of the process. Even more specifically, it should be said that in Sweden and Russia it was the making of bar iron in the forges that was the centre of the organisation; the forges were the heart of the iron-making communities, governing the other activities. This was quite natural, as the design of the social network of production in both countries was directed towards the international market in bar iron.[1]

Iron production at furnaces, forges and partly also mines in Sweden and Russia can be seen as islands of industrial activity in a huge sea of non-industrial activities. This metaphor is not altogether unproblematic, however. The agrarian surroundings of iron-making cannot be seen as totally separate from the industrial activities themselves; there was always an interdependence of relations between the world of peasants and that of iron workers (Figs. 3.1 and 3.2). It is, furthermore, difficult to make a clear-cut distinction between industrial and 'non-industrial' activities in a branch of industry during the early industrial period. In this book, however, we have decided that the production carried on at the furnaces and forges should be seen as more or less industrial. Iron-making should also be taken to include the further manufacturing of the product, such as nails, wire, plate, arms etc. We have also included the work done in the mines in this 'category', although as the following will show this is perhaps more problematic.

Mines

As was indicated in the opening chapter of this book and further developed in chapter 2, iron-making in both regions was originally dominated by work done by the peasant household. A simple production organisation was based on ores from lakes and bogs, which demanded neither costly devices nor a particularly

1. A structure which distinguishes the industry in these two regions from many other iron-producing areas in pre-industrial Europe, e.g. in England, the southern Netherlands or the German lands. For a comparison of the Russian, Swedish and Belgian industries see, A. Florén, 'Social Organization of Work and Labour Conflicts in Proto-industrial Iron Production in Sweden, Belgium and Russia', *International Review of Social History* 39 (1994), Supplement, 83–113.

FIGURE 3.1 Small Swedish Ironworks Integrated into an Agrarian Community, Dalsland, 1758

FIGURE 3.2 Small Ironworks Surrounded by Arable Fields, Närke, Sweden, 1785

complex organisation of work.[2] It is worth underlining that the making of iron from lake and bog ore existed in both countries as a parallel and alternative to the larger-scale industry during the early modern period.[3] During the eighteenth century the Swedish crown showed a keener interest in this traditional method and a few large establishments were built up in eastern Finland.[4] The Swedish traveller Daniel Tilas witnessed such a peasant-based industry when in 1738 he visited the Olonets region in western Russia. Tilas described how groups of peasants melted soil rich in iron into primitive blooms. These were then sold to forges owned and directed by merchants.[5] In a report from Russian and Siberian ironworks five years later the Swedish engineer Nils Psilanderhielm summed up by saying that this peasant industry produced and sold between fifty and sixty-five tons of iron each year. Both travellers also visited the blast furnaces at what they called Konzoser, which was operated with lake and bog ore bought from the local peasantry.[6] The Swedish former prisoner of war Petter Schönström, in the 1720s, and the English traveller Daniel Edward Clarke, at the end of the eighteenth century, both mentioned the same sort of ore digging in the Tula region, south

2. The existence of such a peasant-based industry in pre-industrial Russia has been discussed by several researchers, predominantly concerning the textile industry. See O. Crisp, 'Labour and Industrialization in Russia', in *The Cambridge Economic History of Europe,* vii: 2, 336; R. Portal, 'Manufactures et classes sociales en Russie au xviiie siècle, *Revue Historique,* 1949; R.L. Rudolph, 'Agricultural structure and proto-industrialisation in Russia: Economic development with unfree labor', in *VIIIe Congrès international d´ histoire Economique, Budapest 16–22 août 1982, section A2: La protoindustrialisation: théorie et réalité,* 4; R.L. Rudolph, 'Family structure and proto-industrialization in Russia', *Journal of Economic History,* 1980, 111; S.H. Baron, 'The Transition from Feudalism to Capitalism in Russia: A Major Soviet Historical Controversy', *American Historical Review,* 1972, 719.

3. The proto-industrial scythe-making industry which is discussed in M. Isacson and L. Magnusson, *Protoindustrialisation in Scandinavia,* ch. V, was for example mainly based on ore from bogs and lakes.

4. For the investigations in Sweden see, S. Rinman, *Allmogesmidet i Dalarna 1764 års undersökningar,* ed. Carl Sahlin, Stockholm 1936.

5. D. Tilas, 'Kort Berättelse om en innom Ryska Gräntsen giord Resa, ifrån den 4. Februarii til den 6. April Åhr 1738', Bergskollegii Arkiv, RA, (National Archives) Stockholm. Parts of this manuscript have been published in D. Tilas, 'Curriculum vitae I-II 1712–1757', in *Historiska Handlingar* 38:1, Stockholm 1966, 87–105.

6. N. Psilanderhielm, 'Berättelse om Ryska och Sibiriska Jernverken, Ingifwen til Kongl. BergsCollegium den 8. Decembr. 1743', Bergskollegii Arkiv, RA, Stockholm; Tilas, 'Kort Berättelse'. Lake ore was used in Swedish blast furnaces well into the nineteenth century, mostly in the region of Småland. See *Bergs-Collegii Underdåniga Berättelser* during the 1830s and 1840s, Bergkollegii Arkiv, RA, Stockholm.

of Moscow.[7] Schönström adds that ore from bogs and lakes was also used outside Tomsk, in Siberia, by peasants who made their own iron.[8]

The technological development of the blast furnace in Europe during the later Middle Ages was related to a change from lake and bog ore to ironstone and was thus an important factor behind the development of mining. Generally, the higher productivity and the more extended campaigns of the blast furnace raised the demand for ore.[9] In Sweden the relationship between these two developments was of an even more complex nature. Swedish ironstone, in general, had a low phosphorus content, giving an iron of high quality, but it also required a more highly developed furnace technology. Recent research has thus argued that the use of Swedish ironstone demanded the blast furnace. This line of reasoning has served to explain the use of the blast furnace as early as the twelfth and thirteenth centuries in Sweden.[10]

Historical research on the European iron-mining industry has, if compared with studies of the later phases of the production process in the furnace and the forge, or with the mining of silver or copper or even with the coal industry, been rather scarce. Gregorius Agricola's work *De Re Metallica,* published in 1556, has been frequently used in more popular descriptions of the industry in order to give the impression that a large-scale and technically advanced industry developed as early as the later Middle Ages.[11] However, as Agricola shows, underground mining was far from being the only mode of ore extraction in late medieval and early modern times. The extraction of ore in the southern Netherlands, an important area for iron production, especially during the sixteenth century, was for example carried out by small groups of workers. Each crew consisted of two or three people, often from the same family. The mines they worked are most accurately described as no more than a hole in the ground, three metres in

7. P. Schönström, 'Berättelse om the Ryske Bergswärkens tilstånd, upgifwen wid återkomsten utur Ryska fångenskapen', Bergkollegii Arkiv, RA, Stockholm; E.D. Clarke, *Travels in Various Countries of Europe, Asia and Africa. Part the Fourth,* London 1810, 193.

8. Schönström, 'Berättelse om'.

9. R.F. Tylecote, *The Early History of Metallurgy in Europe,* New York 1987, 330.

10. N. Björkenstam, *Västeuropeisk järnframställning under medeltiden,* Stockholm 1991, 72.

11. G. Agricola, *De Re Metallica,* translated by H.C. Hoover and L.H. Hoover, London 1912, book 4–5; P. Wolf and F.L. Mauro 'L'âge de l'artisanat', in *Histoire Générale du Travail,* ed. L-H. Parias, Paris 1977, 308.

diameter and twenty metres deep.[12] In Bergslagen, such a simple technology was impossible to use. As the ore was enclosed by sections of solid rock more developed methods than just a spade and pick-axe were needed to extract it. Explosives were used to break the rock apart. Gunpowder was used from the end of the seventeenth century, but the medieval technique of fire setting, heating up the rock with a fire and then pouring water on it, was a low-cost alternative to the expensive powder and seems to have been the dominant mode of working during the early modern period.[13]

Geological conditions thus placed certain demands on the technology used, but they also gave the mines their shape. The French traveller G. Jars, who visited Sweden in the latter half of the eighteenth century, noted that in the mines:

> La méthode d'exploiter est toute différente de ce qui est en usage en Allemagne et en France, elle n'est même praticable que pour des filons de la nature de ceux de Suède qui ont de la largeur et la solidité. On les travaille en général comme on creuse une carrière, c'est à dire en faisant une ouverture aussi grande que la longeur et solidité du filon peuvent le permettre.[14]

Unfortunately Jars never visited Russia, and the Siberian mines, which makes it more uncertain whether his conclusion about the exclusiveness of the Swedish mining methods actually holds true. Open-shaft mining, as described by Jars, was also used in the Urals. But it is possible that the geology made mining easier and less capital-intensive than in Sweden. In the first part of the eighteenth century, sand or earth which was rich in iron also seems to have been used by the large-scale establishment in the Urals. In a report from 1728 De Hennin, for example, mentioned iron sand at Kungur. He also noted that in general the amount of ore did not increase with the depth of the shaft and that iron ore as well as silver and copper lay just beneath the surface.[15] In this connection, an account written in 1729 by the Swedish envoy Joachim von Ditmer

12. Florén, 'Social Organization of Work', 92. The same could be said for the mine in Somorostro outside Bilbao in the Basque Country. The ore was only extracted close to the surface in small pits worked by small crews of miners. See R. Angerstein, 'Resan ifrån Madrid genom Biscayen, Frankriket och Lothringen', Jernkontoret, Stockholm.

13. K.-G. Hildebrand, *Swedish Iron in the Seventeenth and Eighteenth Centuries: Export Industry before the Industrialization*, Stockholm 1992, 78.

14. G. Jars, *Voyages métallurgiques*, Paris 1769, 108.

15. 'Generalliutenanten Hennings relation om ryske bergwerkens upkomst och början i september 1728, utaf Tyskan öfwersatt', Bergskollegii Arkiv, RA, Stockholm.

seems relevant: 'At these mines no gunpowder blasts are needed in order to unveil the ore as most of it is taken in the soft soil or from slate stone.'[16]

It is possible that the situation changed during the eighteenth century to a heavier dependence on ironstone. In his account of the Siberian industry, prepared in St Petersburg in the 1780s, Salomon von Stockenström, for example, discussed the iron mountain outside Nizhnii Tagil', Vysokogorsk, where no actual shafts were needed. As the entire mountain was so rich in iron, pieces of it could be used directly.[17] As in Sweden, underground mining was rare, a situation that remained unchanged until the nineteenth century. Of thirty-two ironworks in the Urals, examined in 1807 to 1809 by P. E. Tomilov, a Perm' *berg-inspector*, only seven used ironstone from underground mines.[18] In a recent archaeological investigation of the Norberg region, one of the central production areas in Bergslagen, nearly a thousand open pits have been traced and located. Some of them are no more than a slight depression in the ground while others are of greater depth, although rarely more than thirty metres deep.[19] In other parts of Bergslagen, such as the sites visited by Jars, in Värmland in western Bergslagen, and in Dannemora towards the Baltic coast, as well as in the Nora and Linde region north of Lake Mälaren, mines were considerably deeper. Several shafts with a depth of more than one hundred metres were worked.[20]

In his impressive work on the roots of the capitalist system, Fernand Braudel discusses the development of the central European mining industry during the sixteenth century. He assumes that when production became technically more demanding, its social setting changed from collective ownership by small proprietors to administration by the state or local aristocrats. The worker-proprietor could not cope with the higher costs and,

16. Joachim von Ditmer's letter 21 Aug. 1729 to the Board of Mines, Bergskollegii arkiv, RA, Stockholm.

17. S. von Stockenström, 'Anmärkningar rörande så Ryska Jerntillverkningen som äfwen Jernhandeln', 1787, Bergskollegii Arkiv, RA, Stockholm.

18. *Gornozavodskaia promyshlennost' Urala na rubezhe XVIII-XIX* vv., Sverdlovsk 1956, 154–223.

19. I.-M. Pettersson, *Norbergs Bergslag,* Stockholm 1994, 309; A. Weinhagen, *Norbergs Bergslag samt Gunnilbo och Ramnäs, till omkring 1820,* Lund 1947, 129.

20. H. Nelson, 'En bergslagsbygd', *Ymer* 1913, 314; Hildebrand, *Swedish Iron,* 81; M. Nisser, 'Forsmark – ett av vallonbruken kring Dannemora gruvor', in *Forsmark och vallonjärnet,* Forsmarks kraftgrupp 1987; J. Furuskog, *De värmländska järnbruken,* Filipstad 1924, 112.

according to Braudel, merchant capitalists were equally reluctant to make heavy monetary investments. Consequently the initiative was taken over by princes or the state.[21]

Braudel's ideas perhaps hold true for the region which he is primarily discussing but this view seems too narrow as an overall picture of the complex development of early industrial mining as a whole. This is not to deny the role played by the state in the mining industry in both Sweden and Russia. The royal prerogative to the mines was indicated in the first Swedish decrees concerning the industry during the fourteenth century and explicitly stated under the proto-absolutist regime in the sixteenth century. The legal aspects of this problem will be discussed further below, in chapter 7. What we want to stress here, rather is that there were alternative ways of coping with the new costs, alongside the large-scale state-owned establishments that Braudel discusses, not least the various types of cooperative arrangements.[22]

First, however, it should be stressed that there were also ways of avoiding some of the heaviest costs of mining, i.e. the costs of hauling and drainage. In the Norberg region open-air mines were usually planned so that one of the walls had a gentle slope making it possible to enter the shaft with a cart or sledge in order to transport the ore from the mine.[23] The problem of drainage was often avoided by simply abandoning the pit once it had filled with water. This strategy probably partly explains the large number of pits in the Norberg area. We also find several complaints about such action, in both Sweden and Russia. In the general charter for the Swedish iron industry, from 1649, miners were criticised for deserting older and richer mines for smaller ones.[24] In Russia the academic I. I. Lepëkhin, exploring the Urals in 1769 to 1771, wrote that miners there, like moles, were only digging the surface.[25] The promotion of more centralised mines in Sweden stemmed from a desire on the part of the Crown to organise the industry in a more efficient manner in order to maximise tax revenue. It seems to

21. F. Braudel, *Marknadernas Spel. Civilisationer och kapitalism 1400–1800*, vol. 2, Stockholm 1986, 297.

22. See also the discussion in C. Sabel and J. Zeitlin, 'Historical Alternatives to Mass Production', *Past & Present* 1985, 140.

23. Hildebrand, *Swedish Iron*, 78.

24. Järnbergsordning 1649 §§ 5, 11, 12, published in *Kongl. Stadgar, Förordningar, Privilegier och Resolutioner angående Justitien och Hushållningen wid Bergwerken och Bruken* (KSFPR), vol. I, Stockholm 1736.

25. I.I. Lepëkhin, *Dnevnye zapiski puteshestviia po raznym provintsiiam Rossiiskogo gosudarstva*, St Petersburg 1802, chapter 2.

have been successful in this regards, and a concentration of mining activities can be observed in the central mining areas during the seventeenth and eighteenth centuries.[26]

In the seventeenth and eighteenth centuries the supervision of mines was one of the key responsibilities of the Board of Mines. The local representative, the *bergmästare*, was required to report each year on the depth and shape of the mines as well as the standard of their technical equipment. A central laboratory to examine the quality of the ore existed as early as the mid-seventeenth century. It expanded during the last decades of that century, and during the same period a special office was established for making drawings of the mines and a director of mining mechanics was appointed.[27] However, even before the establishment of a technical staff at the central administrative level, the Swedish state had been actively involved in investments in the mines. In the Nora and Linde regions, the establishment of new water-driven devices in the iron mines during the latter part of the 1630s, for example, was partly financed by the state.[28]

In Russia, during the first half of the eighteenth century, attempts were also made to centralise the supervision of the mines. In instructions drawn up by De Hennin, the same German terminology as in Sweden was used for the various positions. Each mine was to be supervised by a *Bergmeister* who had a *Steiger* and a *Bergsgesvoren* to help him in his work, and as in Sweden the *Markschneider* made drawings and plans of the mines. The administration seems to have met with serious problems, however. What was lacking was primarily an experienced and educated administrative staff. De Hennin himself realised this and wrote that it was necessary to employ foreigners for these demanding posts.[29] As with other parts of the Russian legislation, these instructions mainly concerned the state-owned ironworks.

However important the similarities in administrative superstructure might have been, there were nevertheless substantial differences between the mining industries of the two countries. In a

26. Weinhagen, *Norbergs Bergslag*, 31; Nelson, 'En bergslagsbygd', 315.

27. Joh. A. Almquist, *Bergskollegium och Bergslagsstaterna 1637–1857*, Stockholm 1909. See also S. Lindqvist, *Technology on Trial. The Introduction of Steam Power Technology into Sweden 1715–1736*, Uppsala 1984, 95–107.

28. Letters to Bergmästare Sten Andersson from the Board of Mines 26/10, 27/10 1637, 14/7 1638, Bergmästaren i Nora och Linde med flera bergslagers arkiv E1:1, Uppsala Landsarkiv, Uppsala; Nelson, 'En bergslagsbygd', 315.

29. R. Portal, *L'Oural au XVIIIe siècle. Étude d'Histoire Économique et Sociale*, Paris 1950, 181.

report to Catherine I after a two-year visit to Sweden, V. N. Tatishchev wrote in 1726 that the Swedish industry was far better organised, not the least on the technical side. He was impressed by the depth of the Swedish mines and by the high standards of the water-driven pumps and hauling machines he had seen.[30]

The more active policy of the Swedish state might explain the difference mentioned by Tatishchev. This brings us back to the arguments put forward by Braudel, that the investments demanded by the industry led to royal or princely involvement. It is important to stress, however, that this fact did not change the overall picture of the social organisation of the Swedish industry. As will be discussed in chapter 7, up to the first part of the nineteenth century the mines were – with the important exception of the Dannemora mine in Uppland and the Utö mine in the Stockholm archipelago – mainly the property of the *bergsmän*. Even though the state helped to finance some of the technical equipment, a large proportion of the work was done by the petty-proprietors. The mining courts organised the different contributions and made sure that each part-owner made his contribution to the work to be done. In 1651 the local court at the Dalkarls mine in Nora, for example, checked that each *bergsman* had delivered his share of the logs needed for work on the dams and water-wheels.[31] As the *bergsmän* obtained the logs from their own forests and used their own means of transportation and labour force, the monetary costs were kept down.

Even if state control of the mining industry was firmer in Sweden than in Russia, it nevertheless did not go beyond the gates of the mines. Compared with the careful regulation of the skills and duties of furnacemen and forgemen in Sweden, which we will come back to later, the regulation of miners was rather pendulous. The same thing could be said of the Urals. The detailed prescriptions that De Hennin laid down for work in the furnaces and forges have no counterpart when we turn to the mines.[32] An important reason for this lack of regulations was surely that until mid-nineteenth century mining was not as a rule a full-time occupation. The mining season also differed from region to region. The Norberg mines were predominantly worked in winter when conditions for

30. Portal, *L'Oural*, 99; P. Alefirenko, 'Ekonomicheskie vzgliady V.N. Tatishcheva', *Voprosy Istorii* 1948, vol. 12, 91.

31. Record of the mining court 12/2 1651, Bergskollegii Arkiv, Advokatfiskalämbetet EIIh:1, RA, Stockholm.

32. Portal, *L'Oural*, 179.

transport were more favourable.[33] In Nora and Linde, and also in the Urals, summers were preferred.[34]

It seems logical to assume that, to be profitable, the investments made in technical devices would have called for more continuous production. This type of economic rationality, however, does not hold good for the regions and periods studied. As neither of the countries exported iron ore, production was reared towards internal consumption. As mines or parts of mines were usually allotted to certain furnaces or ironworks, the amount of ore sold on the market must have been fairly negligible. The mines were worked only as long as was necessary to meet the demand created by the number of furnaces actually in the hands of the proprietors.

Another important factor behind the seasonal pattern of work was the lack of a labour force or to be more precise, the fact that the workers also had to be used to perform other tasks. D.V. Gavrilov has described the situation in the Urals as follows: '... today a worker works with an axe; tomorrow he is sent to an iron foundry; in a month or two to a furnace; then to ore extraction and washing, etc. So the artisan will try everything during his lifetime, but will master nothing.'[35]

The mining engineer A.I. Antipov, studying the Ural mines as late as in 1857, levelled the same sort of criticism against:

> ... the most false and erroneous view of works owners that the mining economy could do without a skilled labour force. A mine worker has the same skill as a metalworker, a blacksmith or any other workers; it is versatile work and quite often demands from a worker sharpness, knowledge and skill, and hence it should enjoy even more respect than the other occupations.[36]

The Ural ironmasters' desire to make full use of their labour force of serfs or ascribed peasants thus seems to have been an obstacle to further specialisation. In Sweden, it was the integration of workers in the household of the *bergsman* which restrained the process. The members of his household alternated between tasks

33. Hildebrand, *Swedish Iron*, 78.

34. E. Bergensköld, 'Nora Bergslag vid 1784 års slut', in *Om Noraskog vol. 1*, ed. J. Johansson, Stockholm 1875, 46.

35. D.V. Gavrilov, *Rabochie Urala v period domonopolisticheskogo kapitalizma, 1861–1900*, Sverdlovsk 1985, 116–117.

36. A.I. Antipov, 'Kharakter rudonosnosti i sovremennoe polozhenie gornogo, t.e. rudnogo, dela na Urale', *Gornyi zhurnal* 1860, no. 1, 63.

performed in the furnace, on the land and sometimes also in the forge. From May until St Michael's Day wrote the *bergmästare* for the Nora and Linde areas in 1786: 'Sons and daughters work in the mines, and in the wintertime they work at the furnace. The wife also works at the furnace. She crushes iron ore and also transports pig-iron to the city. Because of this the *bergsman* is not able to work in the mines during the wintertime, even if the snow and ice were to permit him to perform such tasks.'[37]

These two modes of production, the serf/ascribed peasant mode in Russia and the household system in Sweden, formed the overall framework for mining. At a more concrete level, however, the actual organisation varied from one mine to another. In Sweden, during the early modern period, it is possible to single out three different organisational strategies. The first two were firmly based on the household of the *bergsman*. Each household either formed a work crew of its own or, more often, sent individual members to the mine, where a joint crew was formed, working under the command of a paid overseer. In the first case the ore extracted by one household was kept separated from the ore of other households while, in the second case the ore from the mine had to be shared among the proprietors.[38] In both of these slightly different ways of organising production both the female and the juvenile share of the workforce was very high, in certain mines in the Nora and Linde area as high as 40 per cent during the first half of the nineteenth century.[39]

A third way of organising work in the mines gradually emerged. Some households hired workers for the mine work, and a small population of more or less specialised miners thus came into existence around certain mines. In some regions this group grew in such a way during the seventeenth century that it gave rise to a new way of organising work in the mines. This was the case around the Persberg mine in Värmland, for instance, where it was argued that the distance between the furnaces and the mine prevented the *bergsmän* from sending their own workers to dig the ore, and in Norberg where a substantial market for ore existed which probably called for a higher level of production. In both places this group of owners-workers also hired labourers and sold

37. Bergenskiöld, 'Nora Bergslag', 46.

38. J. Granlund, 'Greksåsars bergsmän i hyttelag och gruvlag', *Meddelanden från föreningen Örebro Länsmuseum*, 188; H. Henriksson, 'Kvinnor i Gruvarbete', *Med Hammare och Fackla* xxxiii, 1994, 118–120.

39. Henriksson, 'Kvinnor i Gruvarbete', 132.

the ore to the *bergsmän* or ironmasters.[40] During the eighteenth century these mining entrepreneurs and not the *bergsmän* were generally seen as the owners of the mine.[41] We also find this specialised workforce at the important mine of Dannemora, which supplied the ironworks on the east coast of Sweden with ore for the production of high-quality iron. An important difference was that in Dannemora the mine was owned by the different ironworks. Its operation was thus organised and supervised by the ironmasters and not by a group of *bergsmän*.[42] As the need to improve productivity was probably greater at these mines, it is possible that the proportion of female workers was lower. For example, at Dannemora, the overseer complained during the early eighteenth century that some worker's households were sending their daughters to work in the mine, while the men engaged in more urgent or profitable occupations.[43]

During the first half of the nineteenth century, the household-based organisation was challenged by the change in ownership of the mines. When ironmasters became owners or part-owners of mines, to a greater extent, they tried to organise work in a way appropriate to their interests. It was more convenient for them to use groups of hired hands placed under a single command. An owner-worker who hired miners for a fixed sum of money could thus be engaged. The system was criticised by the authorities, since it was seen as leading to wasteful and dangerous management, primarily seeking short-term profits. As the new order implied higher monetary costs for the remaining *bergsmän*, it gave rise to protests and conflict. A more appealing alternative from the authorities' and owners' point of view was to foster a stable workforce of miners who were hired on a yearly basis. This eliminated the risk which the involvement of an owner-worker always involved but it did nothing to avoid the monetary losses incurred by the *bergsmän*.[44]

40. Plakat och förordning angående järnvräkeriet 15/12 1671, § 4, KSFPR vol. I; A. Weinhagen, 'Gruvor, hyttor och hamrar under 1600-och 1700-tal', in *Norberg genom 600 år. Studier i en gruvbygds historia*, Uppsala 1958, 288; Furuskog, *De värmländska*, 113; Hildebrand, *Swedish Iron*, 73ff.; K.-G. Hildebrand *Fagerstabrukens historia, del I, Sexton och Sjuttonhundratalen*, Uppsala 1957, 258.

41. Hildebrand, *Swedish Iron*, 82.

42. Nisser, 'Forsmark'.

43. Letter from the overseer to the mining court in Dannemora, 6/2 1716. Bergmästaren i Gävleborg, Uppsala och Stockholms arkiv, F1:4, Uppsala Landsarkiv, Uppsala.

44. Henriksson, 'Kvinnor i Gruvarbete', 120–123.

In Nora and Linde the new system became dominant during the first half of the nineteenth century, resulting in the growth of a stable force of miners and the creation of mining communities round the bigger mines.[45] The change also meant a masculinisation of the work, as the *bergsmän* no longer sent their wives, daughters or maids to the mines.[46] The creation of a specialised workforce had been made possible by the rapid increase in population in Sweden from the latter part of the eighteenth century. The increase was especially marked in certain mining areas, not the least because of migratory moves from agricultural regions.[47]

As a rule serfs or dependent peasant workers were used in the Ural mines. As in Sweden, though, there were regional differences. At ironworks in the western Urals, founded in the second half of the eighteenth century on the estates of the Stroganovs, where there were no ascribed peasants, serf peasants were occupied with the provision of ore, while ironworks in the mid Urals had a mixed workforce of serfs and ascribed peasants. As in other spheres of the iron industry, however, ascribed peasants were predominantly used for auxiliary tasks. Works in southern parts of the Urals, established late in the eighteenth century, used serfs together with a substantial number of wage-labourers.[48]

Reflecting the variation in the socio-economic settings of the workforce, there were different ways of organising their daily toil. Like the work on the estate, most work was organised directly by the ironmaster and his staff. At smaller mines, however, operations could also be left to an entrepreneur, who was contracted to organise production. He either used his own family to work the mine or hired workers whom he provided with horses and tools. Peasant families, wrote a contemporary, 'during the time they are free from agricultural work undertake to extract and to deliver to the plant

45. Nelson, 'En bergslagsbygd', 329.

46. Henriksson, 'Kvinnor i Gruvarbete'. Concerning the general trend towards a masculinisation of work in the Swedish iron industry see A. Florén, *Genus och producentroll. Kvinnoarbete inom svensk bergshantering, exemplet Jäders bruk 1640–1840*, Uppsala 1991 (second edition 1995). As regards the mines a similar development has been analysed by Chr.Vanja, 'Mining Women in Early Modern European Society', in *The Workplace before the Factory*, eds T.M. Safley and L.N. Rosenband, Cornell University Press 1993.

47. H. Norman, 'Den utomäktenskapliga fruksamhetens lokala variationer. En fallstudie från Örebro län', in *Historieforskning på nya vägar. Studier tillägnade Sten Carlsson 14.12 1977*, Uppsala 1977.

48. Iu. Essen, *Istoriia gornozalozhnykh SSSR*, Moscow 1926, vol. 1; S.G. Strumilin, *Istoriia chërnoi metallurgii v SSSR*, Moscow 1954, vol. 1.

such and such an amount of ore; at the same time, quite often in the past and, to my mind, even now, they themselves choose the place where it is more convenient for them to extract the ore'.[49]

As the ironworks often relied on a mixture of large and small mines, they often used both types of social organisation. Nevertheless, from the late eighteenth century, more individual ownership of the mines also appeared. During the 1820s, for example, the Brenëvs, merchants from Ekaterinburg, owned twenty mines. During the succeeding decade, however, such a holding of iron mines was prohibited, and private mining did not develop further.[50]

The extent to which wage-labour was used thus varied at a regional level, being correlated to the conditions in which each enterprise was established. Some Russian researchers, however, have also indicated a shift from the use of unfree to free labourers in the mines during the late eighteenth and nineteenth centuries, i.e. a trend towards a system of iron ore extraction organised by contractors and individual ore suppliers. At the Turchaninovs' Sysert' ironworks, for example, 57 per cent of the ore required was obtained in this way from 1816 to 1818. At ironworks in the Viatka province, contractors and individuals supplied most of the iron ore and they were paid for each *pud* of ore extracted.[51]

As had been the case in Sweden, a precondition for this social organisation of mine work was the existence of a surplus agrarian population or at least a strata of peasants who were suitable for wage labour. The population of the Urals increased during the eighteenth century. Whether this was due to a natural increase or migration is difficult to establish. In this context, however, it is not just the size of the population that matters, but also its social composition, that is, the existence of a group of peasants who could not find land or enough land to support their families.

Furthermore, it is of vital importance to try to view the development that occurred from yet another angle and to understand why the ironmasters chose to use free or semi-dependent labour instead of serfs for mining. The process of contracting workers seems to have been governed solely by a desire to secure the biggest possible profits. As in Sweden, the system was criticised for leading to various problems. Excavation work was carried out

49. E.I. Ragozin, *Zhelezo i ugol' na Urale*, St Petersburg 1902, 7.

50. V.Ia. Krivonogov, *Naëmnyi trud v gornozavodskoi promyshlennosti Urala v doreformennyi period 1800–1860, Naëmnyi trud v gornozavodskoi promyshlennosti Urala v doreformennoe vremia*, Sverdlovsk 1964.

51. Ibid.

without timber, ore was left unextracted if a hard ore or water inflow was encountered, and mines were abandoned from time to time for other tasks. The ironmasters were aware of these problems, but the lower cost involved enticed them, and thus the squandering went on.[52]

It is questionable, however, whether the use of free hired labour really was more profitable. When the estate economy has been dealt with in international discussions, it has often been argued that costs which did not originate outside the estate economy were not looked upon as outgoings.[53] In this case the ironmasters seem to have chosen such costs instead of using forced labour. The explanation could of course be that the expense of supervising the workforce was increasing and that the cost in money or in kind was nevertheless lower for the contract mode of organisation than for the serf mode.

It is easy to see that serfs and ascribed workers caused problems. There were frequent complaints that they did not carry out their tasks, that they were absent from work etc. At the same time as several serious uprisings occurred among the peasantry in the latter part of the eighteenth century, ascribed peasants were released in 1769 from their obligation to work in the mines. This decision naturally hastened the spread of free hired labour in the mining branches of production. The choice between one mode of social organisation and another was thus in this context guided not just by considerations of profitability, but rather by the limits of what was socially possible. The protests of the feudally bound workers seem to have advanced the change towards a system based on wage labour.

To return once again to Braudel, he stresses that strikes and uprisings among the miners were common. More recently David Levine and Keith Wrightson, as well as Jan Lucassen have also underlined that the collectivity of miners often put them in a good position to struggle against their patrons.[54] Their conclusions

52. According to von Stockenström the same attitude existed towards the forests, which were exploited in the same way. Stockenström, 'Anmärkningar rörande'.

53. M. Myska, 'Pre-industrial iron-making in the Czech lands: the labour force and production relations circa 1500 – circa 1840, *Past & Present* 1979, 57; W. Kula, *An Economic Theory of the Feudal System. Towards a Model of the Polish Economy 1500–1800*, London 1976, 37.

54. D. Levine and K. Wrightson, *The Making of an Industrial Society, Whickham 1560–1765*, Oxford 1991; J. Lucassen, 'The other proletarians: Seasonal Labourers Mercenaries and Miners', *International Review of Social History* 39 (1994), Supplement, 186–193.

seem very apt as descriptions of the Russian situation, and should be considered when the profitability of the mining industry is discussed. Waves of social unrest among miners also arose in Sweden, but do not seem to have been a significant problem during the seventeenth and eighteenth centuries. They predominantly occurred at mines where the traditional system had made way for larger-scale production based on more or less proletarian workers; as for instance at the silver mine in Sala, the copper mine in Falun and the iron mine in Dannemora.[55] The more common system of mining, centred on the households of the *bergsmän*, did not generate this type of labour conflict.

It thus does not seem valid to assert that problems with the workforce played a vital role in the development of the social organisation of Swedish mines. As has been indicated above, the most important social conflict was instead that between *bergsmän* and ironmasters, which should be understood as a contradiction between a household-based and a capitalist economy, rather than as a class struggle within one and the same economic system. The *bergsmän* were reluctant to accept monetary spending and to break up the traditional symbiosis between agriculture, mining and iron production, which for ages had formed the basis for their household economy. The conflict can thus be regarded neither as a struggle against feudal compulsion, as in the Urals, nor as an industrial conflict designed to raise wages or ameliorate living conditions, as was the case in mines with a more proletarianised workforce; it should be seen rather, as has in fact been discussed concerning the French coal industry or iron production in the Pyrenees, for example,[56] as a proto-industrial conflict against further penetration of the industry by expanding industrial capital.

Mining is usually said to be capital intensive. This was especially true in the cases of Sweden and the Urals, where the iron mines usually required investments in pumps as well as in hauling machines. However, this does not presuppose a social organisation dominated by either the state or industrial capitalists. The

55. A. Florén, 'Klasskamp utan fackförening, om förindustriella arbetskonflikter', *Arkiv för studier i arbetarrörelsens historia*, no. 45, 1990, 15.

56. G. Lewis, 'The constraints of a protoindustrial society on the development of heavy industry: the case of coalmining in the south-east of France, 1773–1791', in *Innovations and Technology in Europe*, eds P. Mathias and J.A. Davis, Oxford 1991; J. Cantelaube 'La société Pyrénéenne, La forge à la Catalane et Le Haut Fourneau (XVIIe – XIXe siècles)' in *The Importance of Iron-Making. Technical Innovation and Social Change*, vol. 1, Stockholm 1995, 194–202.

Swedish example clearly shows that cooperation among the petty-producers could bear the costs of establishment, as long as they could be met from the resources of the part-owners' households. In the Urals the costs were instead met by feudal means, using the resources of the estate and the unpaid labour of serfs and ascribed peasants – a mode of production which was undermined by the perpetual discords between peasants and lords, conflicts which chipped away at the system and led to social change.

However, and in Braudel's favour, it has to be said that the support of the state was crucial in both regions. In Sweden the mines were carefully supervised and the Crown made direct investments in their technical equipment. As was mentioned in the preceding chapter, a legal framework to secure the *bergsman* mode of production was also created during the seventeenth century, and persisted up to the first half of the nineteenth century. In the Urals it was the state that guaranteed the feudal rights of the ironmasters.

To say that mining in the Urals was organised as a feudal estate and its counterpart in Bergslagen was based upon the peasant household would be an over-simplification of what was a complex reality, however. It is indisputable, though, that the estate organisation existed in Russia while it did not in Sweden. During the eighteenth century this large-scale organisation nevertheless began to make way for a more easily controlled and cheaper system of subcontracting peasants. Workers were contracted individually, but there are no signs of a peasant cooperative mode of production as in Bergslagen. In the same period, a more professional workforce of miners developed in Bergslagen, as a consequence of the social erosion of household production.

The new system of professional wage earners working the mines meant a radical departure from both the rationality of the *bergsman's* household in Sweden and with the Russian estate mentality. In Bergslagen the *bergsmän* tried to resist the innovations which meant higher monetary outgoings and preferred for as long as possible to use members of their own households as labourers. The penetration of industrial capital into mining meant, however, that a social carrier of a new rationality became dominant in the trade. In Russia the shift towards wage labour is harder to explain, though it seems to have been simultaneous with the development of the feudal estate economy in the form of the mining districts from the latter part of the eighteenth century. The assumption here has been that the cost of supervising the work done by ascribed or serf peasants was considerably higher than the price for contracted 'free' peasants. The peasants' resistance was thus an important factor behind the ironmasters' choice of strategy.

Blast Furnaces

As was indicated in the introductory chapter, the most striking dissimilarities between iron-making in the Urals and Bergslagen related to the size, centralisation and concentration of the ironworks. In order to achieve a rational distribution of water power, fuel and labour, the Swedish policy was to create and maintain a social as well as a geographical dividing line between pig-iron making and bar iron making, that is, between blast furnaces and forges. In the Urals, furnaces and forges usually formed part of a large industrial complex which also included workshops, mills and smithies for the making of different metal wares. It was not uncommon, however, to have units with forges only. Even if it became more frequent during the latter part of the eighteenth century for Swedish ironmasters to produce their own pig-iron, the pattern of separate furnaces and forges persisted. Industrial complexes of the Ural type were rare in Sweden before the mid-nineteenth century, and they were mainly to be found in Uppland, a region which lacked *bergsmän* and where Walloon forging, which favoured a combined location, was practised.

While the 'physical shape' of the workplace seems to have been fairly similar in the mines of the two regions, which were of the open-shaft type, we find a striking difference when we turn to the furnaces. Russian blast furnaces were much larger constructions than their Swedish counterparts, located within much larger industrial enterprises. In the iron industries of these two countries we have an example illustrating the discussion started by Charles Sabel and Jonathan Zeitlin about alternatives to mass production.[57] Even in a wider European perspective, however, the large-scale Russian ironworks seem rather outstanding, at any rate during the eighteenth century. The scattered Swedish iron-making trade cannot be regarded solely as a consequence of an efficient industrial policy, but rather as a logical outcome of social and economic preconditions for the iron industry in Western Europe. Peter Mathias has underlined that spatial disintegration was typical of the early industrial iron industry, as a consequence of its

57. Sabel and Zeitlin, 'Historical Alternatives'.

need for charcoal and water power. He also mentions the important role played by the transport costs. These were factors which:

> ... all combined to enforce an iron industry of widely scattered, very small units. Expansion with such a matrix of raw materials and technology enforced a greater Diaspora of tiny blast furnaces and forges to more remote woods and streams, in a search for sources of charcoal and power for the tilt hammers. This precluded concentration at sites favourable for the market or the labour force, larger plant and furnaces, economics of scale and external economies leading collectively to low-cost mass output.[58]

This conclusion, which in itself seems most accurate, does not, however, necessarily imply the position held by Mathias, and even more clearly stated by E.A. Wrigley, that a change in the basic ecological metabolism of the industry from charcoal to coal was a necessary precondition for centralised iron-making.[59] It cannot be denied that this was certainly the case both in Britain and Belgium and Germany from the early nineteenth century, but examples from other areas tend to undermine the general value of this conclusion. As Maxine Berg has pointed out, both American and Swedish examples, and we can also add the Russian one, clearly show that centralised industries were also possible on the basis of the traditional fuel/power technology.[60] The Mathias/Wrigley view can also be questioned from another angle. It seems clear that in the metal trades in which coal was used from at least the early seventeenth century, such as in the making of nails, guns and wires in the West Midlands, Walloonia and Westphalia, the small-scale and proto-industrial character of production prevailed well into the nineteenth century.[61]

58. P. Mathias, 'Resources and Technology', in *Innovations and Technology in Europe*, eds Mathias and Davies, 20.

59. E.A. Wrigley, *Continuity, Change and Change. The Character of the Industrial Revolution in England*, Cambridge 1988, chapter 1. See also the discussion in P. Verley, 'La révolution industrielle Anglaise, une révision (note critique)', *Annales ESC* 1991.

60. M. Berg, 'Revisions and revolutions: Technology and productivity change in manufacture in eighteenth-century England', in *Innovations and Technology in Europe*, eds Mathias and Davies, 56.

61. G. Hansotte, *La clouterie liégeoise et la question ouvrière au XVIIIe siècle*, Bruxelles 1972, 9–13; C. Gaier, 'Un artisanat spécialisé: l'armurerie liégeoise', in *La Wallonie le pays et les hommes*, vol. IV, eds R. LeJeune and J. Stiennon, Bruxelles 1981, 200; R. Leboutte, *Reconversions de la main-d'oeuvre et transition démographique. Les bassins industriels en aval de Liège XVIIe–XXe siècles*, Paris 1988, 131; K.-H. Kaufhold, *Das Metallgewerbe der Grafschaft Mark im 18. und frühen 19. Jahrhundert*, Dortmund 1976, 46–47; M. Rowlands, *Master and Men in the West Midland Metalware Trades before the Industrial Revolution*, Manchester 1975.

The use of charcoal was thus not an automatic brake on economic development, no more than the early experience of coal in certain branches of iron-making made them the forerunners of industrial progress. It seems more relevant to focus not on fuel/power technology as such, but on the social relationships of the industry and in this case chiefly on the relationship between agrarian society and industrial production. The social organisation of both coal mining and charcoal-burning was, during the eighteenth century in Walloonia and also in France and Britain, mainly based on work performed by part-time peasants and petty producers.[62] The development of coal mining in most parts of Europe thus seems to show the same pattern as has been discussed above concerning the Swedish and to some extent also the Russian iron mines: the centralisation and proletarianisation of the industries, entailing a break with the agrarian society, a process belonging to the nineteenth century. The same reshaping of social relations is, for a slightly later period, detectable as regards the coal burners in Sweden, a point which will be further developed on below, in chapter 4.[63] In this connection it is also important to take into account the objectives of the state and its policy towards the iron industry.

Technical Organisation

The traditional Swedish timber-clad blast furnace (*mulltimmerhytta*) was a fairly primitive construction. The actual furnace, the 'stack', was a creation of fire-proof sandstone which was covered in a layer of sand and stone kept in place by a timber structure (Fig. 3.3). They were seldom more than eight metres high. Technological development was very slow and changes were introduced in a patchy way. In the seventeenth century, leather bellows were replaced by bellows made of wood, and furnaces made from stone were introduced by immigrating Walloons during the same century.[64] The furnaces were only used for a few

62. H. Hasquin, *Une mutation, le pays de Charleroi aux XVIIe et XVIIIe siècles*, Bruxelles 1971, 288; H. Watelet, *Une industrialisation sans développement. Le bassin de Mons et le charbonnage de Grand-Hornu du milieu de XVIIIe siècle au milieu du XIXe siècle*, Louvain 1980, 100; Levine and Wrightson, *Making*, 221; Lewis, 'Constraints', 1991; R. Trempé, *Les Mineurs de Carmaux 1848–1914*, vol. I, Paris 1971, chapter II:2; D. Woronoff, *Histoire de l'industrie en France du xvie siècle à nos jours*, Paris 1994, 120.

63. Se also M. Ågren, *Jord och gäld. Social skiktning och rättslig konflikt i södra Dalarna, ca 1650–1850*, Uppsala 1992, 254.

64. Hildebrand, *Swedish Iron*, 40f.

FIGURE 3.3 A Blast Furnace, Västmanland, Sweden, 1849

months every year, as pig-iron production was only a part-time occupation for the *bergsmän*. They divided their time between mining, charcoal-burning, agricultural work and iron production, and work at the furnace predominantly took place in early spring. The average production period often lasted less than a hundred days in the early eighteenth century, and it was even shorter in the seventeenth.[65]

The further development of Swedish blast furnaces up to the beginning of the nineteenth century does not show any major leaps, but rather steady and gradual development, with minor alterations to the process. Better dams were constructed, with a better and more regulated flow of water, and the stone-built furnaces also increased in size. Of perhaps even greater importance, however, was the process of concentration which began in the early eighteenth century. Small furnaces, belonging to *bergsmän*, were closed down and production was transferred to larger units with a higher number of part owners. While the average number of *bergsmän* engaged at each furnace in the Norberg area was two at the beginning of the sixteenth century, it had more than tripled

65. Ibid., 35–45.

two centuries later.[66] The causes behind these changes were diverse. In some areas competition over water-power and woodlands led to a closing down of furnaces in the least favourable natural surroundings. As Maria Sjöberg has shown in her study of a *bergsman* community close to Norberg, the relative prices of rye and pig-iron also developed in a direction that was unfavourable to the *bergsmän* during the eighteenth century. The diminishing profitability of iron-making thus led many producers to concentrate on agriculture.[67] However, an alternative must have been to rationalise production and strive towards bigger and more productive establishments. Another important factor in this process was that the policy of the Board of Mines was to promote such a concentration. In 1766 it wanted to force the *bergsmän* to concentrate their work in fewer establishments.[68] This policy was never really implemented, but the decision reveals the board's intentions.[69]

Primarily as a consequence of the process of concentration, working hours were extended and towards the end of the eighteenth century the average production period came close to one hundred and fifty days a year.[70] Both output and productivity increased as a result. It has been indicated that production per furnace rose from about 600 kilograms a day in the 1620s to almost to one and a half tons in the 1680s. One hundred years later just under three tons were produced.[71]

The national restrictions on iron production that were imposed on the Swedish industry from the middle of the eighteenth century quite certainly diminished the producers' eagerness to increase pig-iron production. When this mercantilist framework for the industry withered away during the nineteenth century, it was also followed by a more positive attitude towards technical and organisational changes at the furnaces.[72] Technological development took off at the beginning of the nineteenth century.

66. Ibid., 127; Weinhagen, *Norbergs Bergslag*, 69; Nelson, 'En bergslagsbygd', 326.

67. M. Sjöberg, *Järn och jord. Bergsmän på 1700-talet*, Stockholm 1993, 123.

68. Masmästarordning 1766, § 13, KSFPR vol. II.

69. R. Ringmar, 'Bergsmän i Bergslag. En undersökning av bergsmansbruket i Ovansjö och Torsåkers socknar, Gävleborgs län (Diss., department of Economic History, Uppsala University, 1960) 27–52.

70. T. Omberg, *Bergsmän i hyttelag. Bergsmansnäringens utveckling i Linde och Ramsberg under en 100-årsperiod från mitten av 1700-talet*, Stockholm 1992, 50; Weinhagen, *Norbergs Bergslag*, 67.

71. Weinhagen, *Norbergs Bergslag*, 66.

72. Hildebrand, *Fagerstabrukens historia*, 222.

Related to this was an increase in the influence of the ironmasters over production. In the mid-eighteenth century about half of the nation's pig-iron was produced in establishments owned by ironmasters. To an increasing extent, ironmasters had built blast furnaces of their own and also acquired shares in the *bergmän's* furnaces. This process has been clearly demonstrated in several regional studies, which also point to some dissimilarities in the development that occurred. In the mid-eighteenth century the *bergsmän* still owned about 90 per cent of the furnaces in the Nora and Linde region, a figure which a century later had fallen to 10 per cent.[73] In Norberg the *bergsmän's* share of the furnaces was as low as 59 per cent as early as the middle of the eighteenth century, a figure which then remained stable until the beginning of the nineteenth century.[74] Along with this process many furnaces were closed down during the latter half of the eighteenth and the early years of the nineteenth century. The total number of blast furnaces in the traditional pig-iron region dropped from more than two hundred in the mid-eighteenth century to about one hundred and fifty at the beginning of the following century.[75]

With the change in ownership structure, it may be assumed that the work was performed in a more efficient way, i.e. it imposed a capitalist rationality aimed at maximising production and profitability. More capital also became available for improvements in blast furnace technique. This does not mean, however, that the *bergsmän* who survived as producers in the nineteenth century were totally ignorant of technical improvements. It is on the contrary quite clear that their cooperative arrangements could cope with the demands of the industrial age.[76] However, it can be asked whether the group in itself had not by then undergone a transformation, namely the one that Karl Marx once called the real revolutionary way in the development of a capitalist mode of production, from producer to non-producer. Several studies have underlined that a process of social differentiation took place among the *bergsmän*.[77]

The first feature that was singled out for improvement was the roasting of the ore, and during the 1830s and 1840s new roasting kilns were invented. Other important changes from this time

73. Omberg, *Bergsmän i hyttelag*, 137–141.

74. Weinhagen, *Norbergs Bergslag*, 138.

75. Omberg, *Bergsmän i hyttelag*, 45.

76. Ibid., 82.

77. M. Isacson, *Ekonomisk tillväxt och social differentiering 1680–1860. Bondeklassen i By socken, Kopparbergs län*, Uppsala 1979; Omberg, *Bergsmän i hyttelag*; Sjöberg, *Järn och jord*.

were the advent of hot blast in the 1830s, and the ever increasing size of the furnaces. There was also a continuing extension of the production time during the year. At the end of the Napoleonic wars, total production was about 80,000 tons, an output which had risen by 60 per cent towards mid-century. A smaller number of furnaces thus produced a larger amount of pig-iron. Daily production rose from three tons at the beginning of the century to four tons in the middle. This trend was clearer at the furnaces owned by ironmasters and situated outside the traditional pig-iron region. At this time, these furnaces often formed a part of a larger industrial complex.[78]

Improvements not only occurred in quantitative terms, however, but also in the quality of the pig-iron. Since at least the mid-eighteenth century it had been a well known fact that the quality of pig-iron made by the *bergsmän* left a lot to be desired, and the Swedish Ironmasters' Association devoted considerable resources to finding a remedy for this. In discussions on the future of the Swedish iron industry in the 1820s and 1830s, the poor quality of the pig-iron was singled out as one of the most important problems to be solved.[79] Attempts to find technical solutions to this problem were therefore supported financially both by the Board of Mines and the Ironmasters' Association.

In the first half of the eighteenth century Wilhem De Hennin, moving to the industry in the Urals, wrote a description of the ironworks of that region. The picture he drew could be said to be representative of the basic structure of the industry for the coming century. According to him, each ironwork usually had two blast furnaces. In some cases even larger units were built, with three or four furnaces together. However, in the eighteenth century, these were a rare exception.[80] As a rule, one of the furnaces was in operation while the other served as a spare, taken into production when repairs were carried out on the first one. This was the case at the ironworks at the so-called Konzoser in western Russia which Psilanderhielm visited in 1743. He explained: 'Of these two furnaces, only one was working while the other is kept as a spare one. The furnace which is now operating has been doing so for two years.'[81] During his involuntary stay in Siberia during the first

78. A. Attman, *Fagerstabrukens historia, Adertonhundratalet*, Uppsala 1958, 35.

79. G. Rydén, 'Gustaf Ekman, Jernkontoret och lancashiresmidet – Ett inlägg i synen på teknisk utveckling', *Polhem, Tidskrift för teknikhistoria* 1994/2, 76–79.

80. GASO, F.24. Op.3.D.100; Portal, *L'Oural*, 188.

81. Psilanderhielm, 'Berättelse om Ryska'.

decades of the century, Petter Schönström made similar reflections, for example when discussing the state-owned establishment at Alapaika Zavod.[82]

An undated and anonymous Swedish memorandum on the Russian blast furnace from the end of the eighteenth century stated that the Russian furnace was similar to that which was current in Sweden. However, as the author wrote in the following sentence that the Russian furnace was as a rule built of bricks, his report obviously did not take into account the dominant type among the Swedish *bergsmän*, the timber-clad furnace.[83] The Ural furnace was at that time built more solidly than its Swedish counterpart. This difference had the consequence that, while the Swedish furnace could work continuously for some years before it had to be rebuilt, a technical prerequisite for continual production on a longer-term basis existed in the Urals.[84] To achieve this, dams and canals had to be constructed to secure a constant supply of water. With the severe Siberian climate this caused problems and a good deal of De Hennin's work was concerned with the question of water supply.[85]

As a rule the Siberian furnaces were slightly taller than their Swedish counterparts.[86] A rather extreme example was the construction by Akinfii Demidov in 1743 of a thirteen-metres high furnace. As the other two furnaces at his works were of the more modest size of nine metres this giant edifice should be seen as a form of conspicuous consumption, bringing its owner status and fame, rather than a technical innovation. At state-owned ironworks, furnaces reached an average height of 8.5 metres in the same period.[87] By the end of the century their height had increased by about two metres. In the above-mentioned Swedish

82. Schönström, 'Berättelse om'.

83. 'Anon., manuscript', Bergskollegii Arkiv, FIIIa:1, RA, Stockholm.

84. Hildebrand, *Fagerstabrukens historia,* 224. Hildebrand mentions that it was rare for a furnace to last for more than ten campaigns without major repairs. When the Swedish traveller Reinhold Angerstein visited the Habsburgian Netherlands in the middle of the eighteenth century he was astonished that the solid construction of the Walloon furnaces enabled them to be worked for five or six years, which was quite amazing in the view of the experiences he had from his own country. R. Angerstein, 'Resan genom Nederländerna', Jernkontoret, Stockholm. Se also A. Florén, 'Voyager, c'est apprendre! Le voyage de l´ingénieur suédois Reinhold Angerstein en 1755', *Patrimoine industrielle Wallonie-Bruxelles,* no. 33, 1996, 10ff.

85. Portal, *L´Oural,* 186.

86. 'Anon., manuscript'.

87. Portal, *L´Oural,* 192.

report, the size of the furnace is said to vary between six and fifteen metres, but in general to be between eight and ten metres.[88]

With these differences between the Swedish and Russian furnaces in mind it seems logical to assume a higher level of production at the Russian plants and also a longer period of production. In 1718, for instance, the blast furnaces in the Urals were blown for 185 days on average, and this had been extended to 251 days by 1827.[89] Both the average production per furnace and the average daily production rose during this period. In the first half of the eighteenth century De Hennin calculated that average production at the state-owned ironworks was between four and five tons per day.[90] In the anonymous Swedish report the daily output of the ironworks at Kamensk, even as late as 1789, was said to be no more than five tons. The author adds, however, that production at this works, owing to its moderate size and the poor iron ore used, was comparatively low. At the other works discussed he thus mentions a daily output of around 440 *pud*, or about seven tons. The figures in his report are thus similar to those calculated by Strumilin, and indicate an increase in the daily production of nearly 100 per cent during the century.[91]

Thus even in the first half of the nineteenth century daily production of pig-iron in the Russian furnaces was twice that achieved in Sweden. It must be emphasised, however, that both production and productivity rose considerably in both regions during the eighteenth century. In neither of these regions can this be fully explained by any major technological shift. The key seems instead to have been a concentration of the works and an extension of the production period. In Sweden this was also associated with a change in ownership patterns.

The Organisation of Work

It is not surprising to find that the organisation of work was very different in the two regions. Swedish pig-iron making was based on the collective organisation of the *bergsmän* who divided their

88. 'Anon., manuscript'. Strumilin calculated the average height of the early nineteenth century furnace to be 10.5 metres. S. G. Strumilin, *Istoriia chërnoi metallurgii v SSSR*, Moscow 1954.

89. Figures based on Strumilin, *Istoriia*, 180.; N.A. Minenko et al. 'Ural Iron Before the Industrial Revolution' in *Ironmaking in Sweden and Russia. A Survey of the Social Organisation of Iron Production before 1900*, eds G. Rydén and M. Ågren, Uppsala 1993, 67.

90. Portal, *L´Oural*, 192.

91. Strumilin, *Istoriia*, 425ff; Minenko et al. 'Ural Iron ', 66.

time and resources between various occupations, of which work at the furnaces was only one. Contrary to the Ural ironmasters' endeavour to secure continuous production at the furnaces, the *bergsmän* had to maintain the seasonal character of this work, as they were also responsible for burning the charcoal, working the mine etc. The work done at the furnaces owned by *bergsmän* can, furthermore, be characterised as a mixture of collective and individual undertakings. The maintenance of the furnace as well as the heating before the actual blowing and the closing of the campaign were tasks that were performed collectively by all the part-owners of the furnace. The actual making of pig-iron, however, was an individual venture, in which the *bergsman* and his household used their own ore, charcoal and labour to produce iron. Collective work was planned and controlled by the joint owners at special meetings, at which one of them was elected each year as furnace bailiff, *hyttfogde*, with the task of ensuring that collective decisions were properly executed.[92] From the middle of the eighteenth century this tradition began to be formalised, with written minutes and proceedings supervised by the *bergmästare*.

At the beginning of the seventeenth century, the peasants themselves worked at the furnaces, but in order to improve the quality of the pig-iron and to better control production, it was stated around the middle of the century that specially trained workers responsible for production should be hired and paid for each day they worked at the furnace. Their responsibility also meant that they could be prosecuted and fined at the mining court if anything went wrong during the campaign. This court also ensured that master furnacemen, *masmästare*, had the knowledge and skills needed to perform this task. Furnacemen thus passed a guild-like examination before they were accepted as masters.[93]

In the middle of the eighteenth century, regulations on the training of furnacemen became more specific. The mining courts were also told by the Board of Mines to be more careful with the examination not only of masters, but also of hands and chargemen. A chargeman, or filler, filled the furnace with charcoal and ore. He took orders from the master furnaceman, but had a special responsibility for checking the quality of the raw materials.[94] An alderman was also elected among the masters in a district to ensure that the furnaces in the region were properly built and

92. Granlund, 'Greksåsars bergsmän', 157.
93. Masmästarordningar 6/11 1638, 6/7 1649, KSFPR, vol. I.
94. Masmästarordning 1766, § 14, KSFPR vol. II.

operated.[95] Corresponding to these aldermen at the local or regional level was the office of a senior master furnaceman, *övermasmästare*, at the Board of Mines, who had the job of supervising and improving furnaces at a national level.

The objective of the authorities was thus to professionalise furnace work crews who in the mid-eighteenth century were supposed to be hired on a yearly basis. This aim sprang from the growing concern over the quality of the pig-iron produced by the *bergsmän*, which was also reflected in the desire to promote a concentration of the trade. However, these efforts were only partly successful. It was often said that several furnaces lacked master furnacemen and that masters alternated between different workplaces. Members of the *bergsmän's* households also often played an important role as workers (Fig. 3.4). In 1759 the senior master furnaceman, Sven Rinman, complained that 'at furnaces owned by the *bergsmän* it is the habit that the *bergsman* himself works as a labourer, each one of them on their tiny part of the campaign'.[96]

Even if the *bergsman's* position at the blast furnace had been challenged by the trained workers, his household remained in charge of the auxiliary and simpler tasks at the furnace, as well as undertaking the necessary transport, charcoal-making and ore breaking. This included all the members of these households, with wives and daughters often involved in transport. As was the general pattern in urban trades, widows also succeeded their dead husbands if there was no adult male heir in the family (Fig. 3.5). Women thus played an important part in the industry.[97]

Legally, the *bergsmän* were free to sell the pig-iron to whom ever they wanted, but not to export it. During the second half of the eighteenth century, however, regulation of the market by the state became more rigid. The peasants' freedom was thus rather limited. There were also other restraints on a free market, since many *bergsmän* were tied to ironmasters by debts. As the prospects for food production on the poor soils of the mining districts were very uncertain, ironmasters and merchants often advanced money or goods to the *bergsmän*. This relationship between the producer of

95. S. Montelius, G. Utterström and E. Söderlund, *Fagerstabrukens historia V. Arbetare och arbetarförhållanden*, Uppsala 1959, 146.

96. Quoted from Å. Kromnow, 'Övermasmästarämbetet under 1700-talet', *Med Hammare och Fackla* ix-x, 32.

97. Florén, *Genus och producentroll.*

FIGURE 3.4 A *Bergsman's* Household at Work, Värmland, 1860

FIGURE 3.5 A *Bergsman's* Widow, 1837

pig-iron and the ironmaster often took the form of a *Kaufsystem*, or in some cases even a fully developed putting-out system.[98]

The process of concentration among pig-iron producers and the changing property relations during the latter part of the eighteenth and the first half of the nineteenth century made it possible to establish more stable working conditions, with a master furnaceman and his helpers contracted to work at each furnace on a steadier basis. This was especially the case at the furnaces owned by ironmasters. As production at these works was rarely continuous the workforce remained very small, often consisting of no more than the master furnaceman, the charge-man and one or two hands.[99] During the nineteenth century, as production time was extended, attempts were made to enlarge the crew and to enforce a stricter division of labour. Towards the middle of the nineteenth century the master furnace keeper and his assistants were joined by roasting-kiln workers, slag hauliers, ore crushers etc.[100]

The remaining *bergsmän's* furnaces were not unaffected by the new tendencies. Traditionally, as was stated above, there was a certain reluctance among the *bergsmän* to incur monetary outlays but during the eighteenth century a process of economic differentiation took place among them, which probably also affected their attitudes and mentality. It has thus been asserted that *bergsmän* who went on producing pig-iron adopted a more capitalistic rationality. Their investments in new technology, which entailed costs and required knowledge from outside their households, are one sign of this change. Another is that the importance of wage labour also seems to have been growing at the *bergsman*-owned furnaces. At Löa furnace in the Nora and Linde region the tasks previously carried out collectively were entrusted during the first half of the nineteenth century to individuals, who were paid for their work. The *bergsmän* did not even produce the charcoal, lime or ore needed to heat up the furnace themselves, this being purchased instead on the market by the furnace bailiff (*hyttfogde*).[101] At the end of the nineteenth century a former *bergsman* sketched the causes of what he saw as the decadence of the *bergsmän* and the iron industry in the same region as follows:

98. Sjöberg, *Järn och jord*, chapter 5.
99. Montelius, Utterström and Söderlund, *Fagerstabrukens historia*, 149–151.
100. M. Larsson, *Arbete och lön vid Bredsjö bruk. En studie av olika löneprinciper och lönenivåer för olika yrkeskategorier vid Bredsjö Bruk 1828–1905*, Uppsala 1985, 35–48.
101. G. Rydén, 'Bergsmän, skog och tackjärn', *Historisk Tidskrift* 1994, 656.

> … their forefathers worked. They accompanied their farmhands to the forest and drove their horses and carts themselves. They took part in all the various tasks until they were too old to do so. But how is it today? They hire transporters for all their needs, they have 16 or 18 workers at the furnace and as many miners (or they buy the ore).[102]

Turning to the Urals, we get a different picture of the scale and organisation of furnace work. From the beginning of the eighteenth century, we find specialised categories of workers performing the tasks which in Sweden were carried out by the members of the *bergmän's* households or, at the furnaces owned by ironmasters, by a handful of skilled workers and their apprentices. The organisational framework was codified by De Hennin.[103] He stated that a furnace was to be operated by one head foreman, one assistant foreman, four apprentices, four chargemen, and eighteen unskilled workers. In addition to this there was a team of bricklayers, one senior and three junior workers, who undertook the construction and maintenance of the furnaces. Four workers with horses transported ore, limestone and slag. Tapping of the furnace and mouldings in sand were the responsibilities of special foremen, with their own team of apprentices, hands and founders.[104] At the Konzoser ironworks in western Russia, visited by Psilanderhielm in 1743, the furnace crew consisted of eight skilled workers. With the exclusion of the assistant foreman, this followed the same organisational matrix that had been proposed by De Hennin.[105]

From the rolls of workers drawn up at different ironworks at the beginning of the nineteenth century it is evident that the organisation of work had not changed much since the beginning of the previous century. The prime reason for this stability was that the technology had also remained unchanged.[106] The staff list at the Rastorguev works in Kasli thus comprised one foreman, two apprentices, three workers, two chargemen, twelve ore crushers, four coal carriers, two ore carriers, two slag carriers and two founders.[107] The crews then increased towards the middle of the century. The

102. A. Larsson, 'Nora socken förr och nu', *Från bergslag och bondebygd 1982*, 13. See also chapter 7, below.

103. V. De Hennin, *Opisanie ural'skikh i sibirskikh zavodov 1735 g.*, Moscow 1937.

104. In 1737 such a staff list was proposed to the government by V.N. Tatishchev, and it was legally approved, while the private works used De Hennin's work as a guideline and shaped their organisation according to it.

105. Psilanderhielm, 'Berättelse om Ryska'.

106. GASO, F.24. Op.3.D.100.

107. GASO, F.24. Op.2 D.320.

manager of the Neivo-Rudiansk works, Fëodor Tret'iakov, stated in a report that sixty-one people worked at the furnace.[108]

Overall control of pig-iron production was in the hands of the blast furnace foreman. According to De Hennin he was to supervise the construction of the furnace, be responsible for receiving roasted ore, supervise work at the furnace and monitor the quality of the pig-iron. These duties remained the same until the mid-nineteenth century, and they were all listed in Fëodor Tret'iakov's report in 1860. At the same time, the increased number of workers at the blast furnace made a certain specialisation and division of both labour and supervision necessary. In his report Tret'iakov listed the duties of each member of the crew.

The role of management was crucial to the organisation of the works in the Urals, and administrative control of production even expanded over time. Even at the beginning of the eighteenth century, as De Hennin wrote, experimental melting was carried out using strictly fixed qualities and quantities of ore, fuel and flux to produce a certain amount of iron. At state-owned works, a foreman could not change these proportions, and if he did not achieve the stated amount of pig-iron an expert from the *Oberbergamt* was sent for. He then inspected each stage of melting and the state of the blast furnace. If the foreman was to blame, he had to pay a fine. If the pig-iron was of poor quality, the foreman was reported to the manager. He would then assemble the other foremen at the plant to discuss the reason for the poor quality. If a foreman was at fault, the manager could fine him or inflict corporal punishment, and if the case was repeated, the foreman would be demoted to the position of an apprentice and another person would be given his post.[109]

At this point it is interesting to note the traces of the same corporate structure for organising work that prevailed in Sweden. There was a hierarchy based on training and skills within the furnace work crew and there was also the body of *Oberbergamt* responsible for quality and skills. This structure, however, did not survive in the social climate of the Ural industry, and what we see is only a few remnants that survived at state-owned works. At private works, the structure was different. Akinfii Demidov, for

108. L.A. Dashkevich, S.V. Ustiantsev, Z. F. Tret'iakova, 'Ob obiazannostiakh sluzhashchikh i zavodskikh liudei pri zavodskikh proizvodstvakh', in *Organizatsiia proizvodstva i truda v metallurgicheskoi promyshlennosti Urala XVIII- nachala XIX v.v. Nauchnye doklady*, Sverdlovsk 1990.

109. Hennin, *Opisanie*, 153–154.

instance, imposed very rigid control indeed. His correspondence with the Nizhnii Tagil' works gives instructions on how to deal with the work of foremen.[110] During the nineteenth century the administrative staff grew and their duties expanded. In 1803 a supervisor was included in all crews at blast furnaces at every private ironworks. In 1839, at Nizhnii Tagil', overall administrative control of the whole ironworks was exercised by a manager. An assistant manager was responsible for work at the blast furnace and the finery. He kept a special journal, in which he noted all the information obtained from the supervisors. The latter organised the actual work on the shop-floor in accordance with orders given by the manager and kept records of the work done by each worker. The supervisor of the blast furnace checked that ore brought from the mines was properly burnt, kept records of incoming ore, flux, sand etc., and also oversaw the blowing machines and other technical devices. He monitored the amounts of the different materials, and once a week he examined the volume of ore and flux in stock, as well as that of coal.[111]

Developments in the two regions seem in a sense to have converged towards the middle of the nineteenth century. By then blast furnaces in both regions were fairly large enterprises with well developed technological standards. They were used throughout the year by a large workforce organised according to the principles of specialisation and division of labour. Behind this situation, however, lay two completely different historical developments.

To return to the argument advanced by Peter Mathias and E. A. Wrigley we saw that they analysed the trend towards large-scale industrial establishments in terms of the introduction of a certain fuel/power technology and the availability of labour and a commodity market. A supply of labour was crucial for all early industrial enterprises, not least the iron industry. What was not discussed by Mathias and Wrigley was the different ways and means of enrolling and binding such a labour force. A developed labour market did not exist in either Sweden or the Urals during the seventeenth or eighteenth centuries. The solution in the latter region was to use unfree labour in order to build up a large-scale industry. This policy led to an endeavour among the ironmasters to maximise output by running the furnaces continuously, thereby

110. A.S. Cherkasova, 'Khoziain ili tvorets slavy Starogo sobolia', *Metallurg* 1993, no. 8, 53–56.
111. GASO, F.643. Op.I. D.774.

making full use of the workers bound to them or to their works by legal bonds. The habit of building two furnaces, sometimes even under the same roof, reflects this endeavour as does the increased division of labour.

The supervision of workers was strengthened during the eighteenth century, at the same time as a more complex system of work organisation was developed. At the beginning of the following century, an administrative system of supervisors was built up, in parallel with the increasing specialisation among the workers. The higher productivity of the Ural furnaces, compared with their Swedish counterparts, can be attributed partly to the larger size of the furnaces, but perhaps more clearly to the severe discipline imposed there.

Thus a framework for large-scale industrial production, built with the support of the Tsarist state by means of a feudal power system, was established in the Urals at the beginning of the eighteenth century. In Sweden the 'industrialisation' of the blast furnaces was the result of a more prolonged process of development. During the seventeenth century, pig-iron production was almost completely dominated by the cooperative mode of production of the *bergsmän*. In the perspective of the European world economy of the time, this mode could perhaps be seen as an 'alternative to mass production'. From the standpoints of the Swedish state, the ironmasters or the *bergsmän* of the seventeenth or early eighteenth century, it is perhaps more correct to say that it was difficult to perceive a profitable alternative to this cooperative or at least small-scale and scattered organisational framework for the production of pig-iron. Though it was incorporated in the household-based economy of the *bergsmän* pig-iron production was deemed to be a seasonal occupation. This was not a problem for the ironmasters or for the state, but it was in their interests to ensure that the quality of the iron did not deteriorate and that the price remained low and stable. It is in this perspective that the supervision of the training of furnace workers is to be seen, together with the desire to concentrate the furnaces in order to save forest resources and thus to keep down the prices of both charcoal and pig-iron.

In order to control quality as well as to cut costs, it could be beneficial to the ironmaster to own his own furnace. However, he had to calculate the cost of hiring workers and to keep them busy when the furnace was not in operation. Unlike the Russian ironmaster, he had no feudally bound supply of labour to count on. When furnaces were first taken over by ironmasters, it was probably not with the aim of increasing production. As long as the national output of bar iron was restricted by the state, such an endeavour was not economically sane. Only with the advent of the nineteenth century, and changes in property relations combined with the development of the labour market and a more liberal economic policy, do we find a more rapid process of technological change and the development of a more industrial mode of organisation at the Swedish blast furnaces.

Forges

The most conspicuous contrast between Russian iron-making and its Swedish counterpart lay, as has been noted several times already, in the scale and social composition of the two industries. Bar iron production was no exception to that, and indeed it was the part of the production chain in which these dissimilarities were most striking. The Russian forges were well integrated into the economy of the feudal estates, while bar iron making in Sweden was organised along more capitalist lines. Russia's forges were also bigger than the corresponding production units in Sweden. As with the blast furnaces, we are thus dealing with entities which in one respect exhibit certain similarities – they both make bar iron – but which in other respects show important differences, that is in size and social composition.

It is important as a starting point for this section, however, to dwell briefly on the resemblance between the two regions. It has been stated above that in two related respects the Russian iron industry was a mirror of the Swedish one. The first and also the most obvious of these is that both countries were important exporters of bar iron, and that the overall structure of the two industries showed important similarities. Closely connected with this feature is the way in which bar iron making shaped the rest of the trade. We are dealing with two export-oriented industries in which the making of a commodity destined for export dominated a large trade with extensive roots in so-called 'traditional' society.

The Swedish industry was older than its cousin in Russia and its export orientation had been an important feature of it ever since late medieval times at least. Sweden was an important supplier of *osmund* iron and later bar iron to the international market. At the outset the main destinations had been the Hanseatic towns of Danzig and Lübeck, but from the seventeenth century on a change of destination occurred, with a growing amount of bar iron entering Dutch ports, at the same time as exports rose very significantly. During the next century there was another change of destination and Britain became the most important market for iron. Towards the middle of the century, exports to Britain constituted almost 60 per cent of the total Swedish iron exports. This situation remained very much the same well into the nineteenth century.[112]

As was mentioned in chapter 1, the Russian iron industry was much younger and it was in the 1720s that iron from the Urals began to be looked upon as a serious competitor with Swedish iron on the British market. The Russian share of that market expanded as the quantities coming out of Russian ports rose, and during the 1780s iron from the Urals was sent to Britain in amounts almost double those emanating from Sweden.

The situation on the world market means that, in this section, we are comparing two entities, the Swedish and Russian iron industries, which for more or less a century were subjected to factual comparison. Swedish iron and Russian iron were compared on the international market, in this case basically the British market, for bar iron. Primarily in terms of quality and price, bar iron from Sweden was compared with bar iron from Russia.

This 'factual' comparison or, to put it another way, the competitive situation between Swedish and Russian producers is of course an important element in our comparative venture. Developments on the market affect and change the organisation of production, just as organisational changes might affect the market situation. In the following, we will be playing down this aspect, since very little research has been undertaken on these matters. However, there are two fairly evident features on the British market which ought to be mentioned. First we are dealing with a market that was expanding rapidly from at least the last few decades of the seventeenth century. British producers could not meet this ever increasing demand and imports were thus rising. At first Spanish iron played an important role, but Swedish iron became an increasingly dominant force on the market, followed by iron from the Urals from the mid-eighteenth century. The second feature that should be emphasised is that the British market was not a 'one market' in which all types of iron were substitutable. Iron from different countries and regions

112. Regarding this and the next paragraph, see chapter 1 above. See also Hildeband, *Swedish Iron;* C. Evans and G. Rydén, 'Iron in Sweden and Britain 1600–1800: Interdependence and Difference', in *The Importance of Ironmaking. Technical Innovation and Social Change,* ed. G. Magnusson, Stockholm 1995.

was used for different purposes and therefore did not enter into direct mutual competition.[113]

'Öregrund iron', for instance made by the Walloon method at ironworks around the Dannemora mine, had more or less a monopoly of the market for steel iron in Sheffield. Other Swedish brands had a very strong position in areas where strength and hardness were essential, such as edge tools and naval equipment such as anchors. On the other hand, Swedish iron was seldom used in one of the staple trades in Britain, nail-making. One of the reasons Swedish iron was not used in the making of more 'ordinary' forgings was given to the Swedish traveller Reinhold Angerstein by a British blacksmith: 'Horseshoes of Swedish iron would be more durable, but less profitable for those who make them.' Swedish iron was tough, as opposed to the often cold-short British iron, giving dependable products, but it was also more expensive to work.[114]

Russian iron, on the other hand, shared many of the qualities of British iron, and therefore entered into more direct competition with it. In spite of a few brands, such as Zobel iron from the Demidov works, which competed with Swedish Öregrund iron on the Sheffield market, the majority of Russian iron was destined for the same segments of the market that were covered by British producers. As early as 1737 it was said of an iron that was probably from the Tula region, that 'the Moscow iron was used solely for nail-making, and took the place of ordinary English tough'.[115] Around fifteen years later Angerstein noted that Russian iron was cold-short, like the British product, and that a slight price difference existed between Swedish and Russian iron, with the latter the cheaper of the two.[116]

Angerstein's conclusion does not hold true as a description of all Russian iron, however. As early as in November 1730 an

113. For a more extensive treatment of the topic dealt with in this and the next paragraph see Evans and Rydén, 'Iron in Sweden and Britain'. See also A. Florén and G. Rydén, 'A Journey into the Market Society. A Swedish Preindustrial Spy in the Middle of the 18th Century' in *Societies Made Up of History. Essays in Historiography, Intellectual History, Professionalisation, Historical Social Theory and Proto-industrialisation*, eds R. Björk and K. Molin, Uppsala 1996.

114. R. Angerstein, 'Dagbok över resan genom England åren 1753, 1754 och 1755', Jernkontoret, Stockholm. See also K.-G. Hildebrand, 'Foreign Markets for Swedish Iron in the 18th Century', *The Scandinavian Economic History Review*, Vol. VI, No. 1, 1958, 22.

115. *Journal of the House of Commons*, XXII, April 1737, 854.

116. Angerstein I, LVI. See also C. Hyde, *Technological Change and the British Iron Industry 1700–1870*, Princeton 1977, 104.

experiment involving Russian iron was performed in Stockholm. Naturally it was prompted by the alarming reports about the rapid expansion of the Russian industry. The iron had been sent to Sweden a year earlier by the Swedish diplomat Joachim von Ditmer and was examined in a shop owned by the well-known Swedish engineer Mårten Triewald.[117] A large number of officials from the Board of Mines and the Board of Commerce were present and the results were carefully analysed. Most of the iron was of quite good quality, but the Swedish experts saw it of no use for steel-making. Some Siberian brands were cold-short, but not to the same degree as the samples from the Olonets and Tula regions, which were considered completely worthless. Some of the bars were also regarded as badly worked, whereas for example iron bars from the Demidov works contained good, tough iron at one end, but were cold-short at the other.[118] A similar account was given by the Swedish diplomat Carl Sparre, in a letter from London in 1731. The Russian bars, he wrote, were 'thick and hard to handle, with a loop in the middle because of the long transport from Siberia'.[119]

The difference in quality between Russian and Swedish iron should perhaps not be exaggerated, as substitution was always a possibility if the prices changed in one way or another. It is important, though, to bear in mind that British consumers of iron seem to have preferred Swedish iron for purposes requiring toughness, while British and Russian iron were more often used for ordinary purposes. The fight for a share of the British market was thus more of a competition between British and Russian ironmasters, with their Swedish rivals more as spectators. Nevertheless, the Swedish ironmasters and authorities saw the invasion of Russian iron as a serious treat.

Russian iron was cheaper than its Swedish counterparts. As Sparre saw it, this was a direct outcome of a reduction of customs duties on the commodity. It was also assumed that production costs were lower in Russia than in Sweden. When the Swedish authorities discussed the problem, they thus elaborated different types of solutions. One was to convince the British government to lower the taxes on Swedish iron, another to persuade their own

117. Ditmer´s letter to the Board of Mines 21/8 1729, Bergskollegii Arkiv, RA, Stockholm. In this period Triewald had installed the first Swedish newcomen machine at the Dannemora mines, see Lindqvist, *Technology on Trial*.

118. 'Records of Bergsdeputationen 1734', RA, Stockholm.

119. Carl Sparre, letter to the Board of Mines 12/1 1731, Bergskollegii Arkiv, RA, Stockholm.

government to give economic support to exports. A vital point in these discussions was that a further increase in production costs in Sweden could not be accepted.[120]

The factors behind the more severe regulation of the Swedish iron industry in the middle of the eighteenth century have been much discussed by Swedish historians.[121] Karl-Gustaf Hildebrand has linked the restrictions on iron production in Sweden to the development of the Russian industry. When Swedish iron exports fell, he asserts, it made room for an increase in Russian exports on the British market.[122] Viewed from the perspective of the debate in the 1730s, however, the converse seems to have been the case, i.e. the increase in Russian exports forced Swedish ironmasters and authorities to try to limit production costs. One way of doing this, already in use towards the end of the seventeenth century, was to restrict competition for raw materials on the national market. This in turn meant limiting iron production. When Salomon von Stockenström visited Russia in 1787 he suggested that the Russian ironmasters ought to be persuaded to impose similar restrictions on their production and exports so that they would be more in accordance with market demand. Such a monopolistic policy by the two leading producers on the world market should, he believed, be favourable for the Russian national economy as well as for that of Sweden. No immediate steps in this direction were taken by the Russian government, however.[123]

The Workplace

A starting point for this section is that the overall structure of the industry in both regions was roughly the same: bar iron was to be made and then sold on the international market. At the technical level we could add another similarity: both in Bergslagen and the Urals bar iron was made by the same method, the German forging method. The basic procedure was as follows. Pig-iron was melted on a bed of charcoal in a hearth. In order to raise the temperature, air was supplied through a *tuyère*. The actual refining took place when the iron was in a molten, or semi-molten, state. The refined metal was taken to a hammer for shingling, a process in which slag and other impurities were squeezed out of the iron. The

120. 'Records of Bergsdeputationen 1734', RA, Stockholm.
121. For a summing up of the discussion see Florén and Rydén, 'A Journey into'.
122. Hildebrand, *Fagerstabrukens historia*, 155.
123. Stockenström, 'Anmärkningar rörande'. See also chapter 7, below.

blooms, as the refined iron was called, were taken back to the hearth for a second, and perhaps even a third, round of heating, reheating, or a welding process. The iron was then sharpened into bars under the hammer.[124]

A second method of making bar iron was imported into Sweden during the 1620s, the Walloon process. It was primarily installed at some ironworks around the Dannemora mine in Uppland. During the same period this method was also introduced at the Tula works in central Russia. The basic procedure was the same as with the German method, but there were also some important differences. The most obvious one was that in German forging only one hearth was used for both refining and reheating, while two separate hearths, a finery and a chafery, were used in the Walloon method. There was also a difference in the shape and quality of the pig-iron used. Walloon forging used a pig-iron made from ironstone from the Dannemora mine, and it was cast in long and very heavy pieces called *geuses*, while German forging used smaller ingots of pig-iron. Walloon iron made in Sweden was considered a very high quality product very suitable for steel-making. Consequently, as was noted above, the output of ironworks using this method was sold on the Sheffield market for steel iron.

Apart from the fact that some ironworks in both Sweden and Russia used the Walloon method, it can also be noted that the technical similarities, that is the using of the same forging method, might in fact disguise some important differences. It is perhaps better to talk not of *a* forging method, but rather of a *family* of closely connected methods. In the Urals, for instance, there were about ninety modifications of the German method during the eighteenth century. These differed from each other in minor details regarding the dimensions and proportions of the hearth and the hammer and in the order of laying the charge and the fuel, etc.[125] In Sweden it was common to make a major distinction between *koksmide* and *butsmide,* the latter variant being criticised for resulting in an inferior product.[126]

124. Tylecote, *The Early History*, 233f.

125. A.G. Kozlov, 'K voprosu o predposylkakh tekhnicheskogo perevorota v gornozavodskoi promyshlennosti na Urale', *Voprosy istorii Urala. Sb.10*, Sverdlovsk 1970, 7.

126. C. Rinman, *Handbok uti den gröfre jern- och stålförädlingen*, Falun 1823, 143–159. The German method was, of course used in the German lands, and many different variants existed there. See the editions of *Jernkontorets Annaler* (JKA) in the 1820s; A.G. Tamm, *Försök till besvarandet af den af herrar bruks-ägare i Wermland och Dahl år 1827 framställde prisfråga*, Karlstad 1831.

Bar iron, then, was made by roughly the same methods in both Bergslagen and the Urals, but there the similarity between the two regions ends. In the accounts given by Swedish travellers in the Urals during the first half of the eighteenth century, hand-operated bellows – which had been the normal technique in the Tula and Olonets regions – were said to be used at several ironworks, a peculiarity that had no counterpart in Sweden by that time. However, they also carefully underlined that all the larger works, owned by the state, used water-powered bellows.[127] According to the French historian Roger Portal, hand-driven bellows were still used at Demidov's large ironworks at Nev'iansk as late as in the 1760s.[128]

In his account from Russia, Samuel von Stockenström discussed the construction of the forges in some details. He did not dwell upon the bellows, but concluded that an important difference between Russian and Swedish forges was that in the former, anvils and other heavy equipment were made of cast iron, while in Sweden they were forged. Stockenström made a drawing of a Russian forge, with one hammer, two hearths and two huge, presumably water-powered, bellows.[129] As he never visited the Urals but based his drawing on a model in St Petersburg, however, he overlooked the most stunning difference between Bergslagen and the Urals.

As with the pig-iron making, there was a striking difference in size and scale between Swedish and Russian bar iron making, and this is clearly visible in the forges. From the early eighteenth century, the forges built in the Urals were much bigger than their Swedish counterparts. If the latter contained two hearths at most, forges built in the Urals were often at least twice that size. They were rectangular buildings of varying length, with the water-wheels for the hammers placed in a row along one of the walls, and the hearths along the opposite one. The bellows, with connected water-wheels, were placed along the short walls at either end of the forge. Larger forges were equipped with two pairs of bellows, one by each wall, while small forges had only one pair.

This general forge design remained virtually unchanged for more than a century, and well into the 1830s. It was only in two respects that anything happened during this period. First wooden

127. Schönström, 'Berättelse om'; S. Lindheim, 'Berättelse till Kongl May:ts om Ryske bergverken i gemen den 18/2 1726', Bergskollegii Arkiv, RA, Stockholm; J. von Ditmer, 'Berättelse till Kongl Maj:t om bergverken i gemen den 25/2 1726', Bergskollegii Arkiv, RA, Stockholm.

128. Portal, *L'Oural*, 214, 376.

129. Stockenström, 'Anmärkningar rörande'.

constructions were replaced with stone buildings. In the first decade of the nineteenth century, fifty-one forges out of a total of seventy-six were made of either stone or brick.[130] In connection with this trend forges also increased in size. The older timber-built forges rarely had more than four hearths, while newer forges built from brick and stone sometimes included up to twenty. According to S. G. Strumilin, thirty-six ironworks in the Urals in the mid-eighteenth century were equipped with a total of 178 hearths, giving an average of five hearths per forge. In 1860, 130 works had 1,064 hearths in all, or more than eight on average.[131]

It is important to bear in mind, though, that not all these hearths were in use at the same time. In large and medium-sized forges, a few hearths served as reserves. They were kept in case any of the others were to be damaged. The forge crew then had to change hearth while their old one was repaired. This arrangement also had the advantage that ironmasters with more than one forge could move forgemen and use all the hearths in one forge if another one could not be used, for instance as a result of a lack of water. It thus offered flexibility.

Forges in Sweden did not undergo any major alterations during this period either. A forge built according to the German method during the seventeenth century remained very much the same until at least the 1820s. The common set-up consisted of two hearths and one hammer. This was in fact true of both the German and the Walloon forges. In the former, two similar hearths were used for the entire process, while the latter consisted of one finery and one chafery. The hearths were supported with air from separate bellows until the years around 1800, when the first blowing machines were introduced.

Technological progress in the Swedish iron industry gradually took off at the beginning of the nineteenth century. The first roasting kilns, as was mentioned in the previous section, was one sign of this, the new blowing machines another. A further indication of a growing awareness of the importance of change came in the 1830s and 1840s, when Swedish ironmasters were very quick to introduce 'hot blast' at their forges as well as their furnaces. Hot blast was invented by James Neilson in Scotland in 1828 and already at the beginning of the 1840s it was a fairly common practice in Sweden.

Another new feature introduced at this time was the equipping of forges with an extra hammer. In both Sweden and Russia the

130. N.S. Alfёrov, *Zodchie starogo Urala. I polovina XIX veka*, Sverdlovsk 1960, 21.
131. Ibid.

common pattern had been that two hearths shared one hammer. This caused some inconvenience, since operations at one hearth could disrupt those at the other, and the forges were not utilised efficiently. The delay in the shingling of blooms resulted in both greater losses of pig-iron and higher fuel consumption. Older forges that were rebuilt to incorporate new blowing machines and hot blast were often also provided with an extra water-powered hammer. This process was most rapid in eastern Sweden, where a few totally new forges were also built, some with as many as four hearths and the same number of hammers.[132]

There were technical advances in Russia, too, although they followed a slightly different route. Both new blowing machines, often made in cast iron of British invention, and one hammer to each hearth seem to have been introduced earlier than in Sweden. The former entered the Urals in the last decade of the eighteenth century and spread rapidly. The extra hammer was introduced even earlier, but, as the construction of the Salda works in the 1760s shows, it was probably fairly slow to catch on. When this big works was built it included eighteen hearths and nine hammers in four forges. During the first half of the nineteenth century, however, the 'extra' hammer was practised in most places.[133] It is only when it comes to hot blast that the developments in Sweden seem to have been more rapid, and also more penetrating. The method was tried in the Urals in the 1840s, but did not have an immediate impact.[134]

In spite of some differences in timing, it is possible to see the technological development of the iron industry in both Sweden and the Urals in terms of two more or less distinct phases. During the eighteenth century nothing really happened, and the forges remained very much the same. In Sweden this phase commenced during the late sixteenth century, while it started around a hundred years later in the Urals. The first forge in that region was built in Nev'iansk in 1701. The standard forge in Sweden during this period had two hearths and one hammer, while Russian forges included more than twice as many. Ironworks in Sweden sometimes consisted of two forges, and ironworks with four hearths altogether was considered large. In Russia, most

132. G. Rydén, *Hammarlag och hushåll. Om relationen mellan smidesarbetet och smedshushållen vid Tore Petrés brukskomplex 1830–1850*, Stockholm 1991, 64–75.

133. S.V. Ustiantsev and E.V. Logunov, *British Technological Experience and Ural Mining and Iron-Making Works of the XIX century*, Ekaterinburg 1992, 71.

134. Ibid., 85.

ironworks had at least twice as many hearths, and many of them had even more.

During the second period, incorporating roughly the first half of the nineteenth century, technological development was somewhat faster. Ironmasters in both regions started to erect bigger forges: in the Urals, they sometimes incorporated up to twenty hearths, while Swedish forges doubled their size to four hearths. Influences from Britain were also taken on board, in the form of new blowing machines, hot blast etc.

During this phase the dominance of the German forging method was also gradually withering away. From the 1830s, attempts were made to introduce new methods in both countries. In Sweden there was a debate about whether to use puddling or to develop the Walloon method for ironstones from mines other than Dannemora, while in Russia attempts were made to introduce both puddling and the Comtois refining method. In both countries, once again, developments were very similar in that the transformation seems to have been rather slow and patchy. It was only from the 1850s that puddled iron had any real impact in the Urals, and it was not until 1848 that the most famous mining district, Nizhnii Tagil', changed from German forging to the Comtois method.[135]

In Sweden it was not until the second half of the 1840s that a new forging method was developed. It was an adapted Walloon method, named after its British origin in Lancashire. The Lancashire method spread in Sweden after the ironmaster and metallurgist Gustaf Ekman had developed it for Swedish conditions.[136]

The Workers and Their Work

Up to the seventeenth century, the peasants of Bergslagen constituted the social base for the production of iron. They took care of the whole production chain, from mining the ore to selling the finished commodity, which at this time was *osmund* iron. The rather static nature of this peasant corporation meant, however, that it could by no means respond efficiently to the rise in demand for iron that occurred on the international market during the seventeenth

135. Minenko et al. 'Ural Iron', 67–68 and 92; Ustiantsev and Logunov, *British Technological Experience*, 71–78, and 84–86; T. Esper, 'Industrial Serfdom and Metallurgical Technology in 19th-Century Russia', *Technology and Culture*, vol. 23, 1982, 598–601.

136. A. Attman, *Svenskt järn och stål*, Stockholm 1986; Rydén, 'Gustaf Ekman'.

century. The ten-fold increase in iron exports from Stockholm perceivable during that century was instead due largely to a massive influx of foreign merchant capital and its investment in, above all, forges.[137] The result of this process was, as was described in the introductory chapter, a socially and spatially divided industry, in which the *bergsmän* remained in control of mines and blast furnaces, but lost control of the production of wrought iron.[138] In the Nora and Linde region the proportion of forges that were partly or totally controlled by *bergsmän* diminished from 81 per cent in 1637 to 30 per cent in 1750.[139]

It is likely that there was a development, similar to that seen at the furnaces, whereby the *bergsmän* had also made the bar, or *osmund*, iron themselves in the forges, but were gradually replaced by a more professional group of forgemen. We still do not know much about this process. It can be said, however, that the 'take-over' of the forges by the ironmasters ought to have speeded the pace at which the forgemen came into the forges to make bar iron for their masters. Professionalisation of bar iron production thus took place a long time before the same process occurred at the furnaces or mines.

The objectives of the state, executed by the Board of Mines, were also an active agent in this process. The decrees concerning forgemen stipulated that, in forges using the German method, bar iron was to be made by a forge crew of three men at each hearth, the master forgeman, the forge hand and the apprentice. An alderman was also elected among the master forgemen in each district. The

137. Hildebrand, *Swedish Iron,* 24–26; A. Florén, M. Isacson, G. Rydén and M. Ågren, 'Swedish Iron Before 1900', in *Ironmaking in Sweden and Russia. A Survey of the Social Organisation of Iron Production before 1900,* eds G. Rydén and M. Ågren, Uppsala 1993, 22–24. The investment of merchant capital followed the typical pattern described by Fernand Braudel, among others, whereby merchants strived for a monopoly of the strategic part of the chain of production, from which he could control the international market. Braudel, *Marknadernas Spel,* 299.

138. This sort of peasant-based iron production also survived in several regions in the German lands until the industrial revolution. H. Kellenbenz, 'Europäisches Eisen. Produktion – Verarbeitung – Handel, Vom Ende des Mittelalters bis ins 18. Jahrhundert', in *Schwerpunkte der Eisengewinnung und Eisenverarbeitung in Europa 1500–1650,* ed. H. Kellenbenz, Köln 1974, 409; M. Mitterauer, 'Produktionsweise, Siedlungsstruktur und Sozialformen im Österreichischen Montanwesen des Mittelalters und der frühen Neuzeit', in *Österreichisches Montanwesen. Produktion, Verteilung, Sozialformen,* ed. M. Mitterauer, Wien 1974, 313.

139. A. Florén, 'Järnproduktionens sociala mångfald. Brukspatroner, bergsmän och hammarsmeder i Nora och Linde bergslager 1620–1750' in *Bergshantering i Örebro län en arkivinventering,* ed. M. Essemyr, Örebro forthcoming, 8.

forge crew was to function in a guild-like manner, with the master bearing full responsibility for the quality and quantity of production, at the same time as he taught the other members of the crew. In bar iron making, however, the forge hand had more freedom than a journeyman in a town guild. He was most often married and took full responsibility for production when the master was away.

Apart from stating who was to work in the forges, the decrees regulated two other major matters. The most important of these was the level of skill among the forgemen. It was very clearly stipulated, as early as in the first decree of 1637, what the different categories of workers were to be able to do, and what they were to have learnt in order to be promoted. Their actual graduation was also to be preceded by a formal test under the superintendence of the alderman. The reason for all this was of course a desire to maintain the quality of Swedish bar iron. If there were plenty of skilled forgemen making bar iron, Sweden's reputation on the international market should remain good.[140]

The other matter of importance dealt with in the decree was wages. These were not negotiated between the ironmaster and his forgemen, they were stipulated by the Board of Mines. One reason for this might have been a wish to prevent competition between ironmasters for the most skilled forgemen.[141] This means that the organisation of Swedish bar iron production was based on a wage labour relationship, although no true labour market actually existed.

A separate jurisdiction for the *bergsmän* was mentioned as early as in medieval times. In the seventeenth century, this was reorganised under the supervision of the Board of Mines, and at the same time regional courts were created for the forgemen. In many respects these courts were reminiscent of traditional artisan guilds. There were also important differences, however. The mining courts were ruled not by the industry's corporate producers, but by the regional authorities answering to the Board of Mines. Nor were they strictly professional, as they also included both ironmasters and forge-hands and apprentices.[142]

The clear intention of the state was obviously that bar iron should be *made in* forges owned by ironmasters, and *made by* forgemen who had passed certain tests proving their ability as

140. Hammarsmedsordning 1637 and subsequent ones, KSFPR, vol. I–II; Montelius, Utterström and Söderlund, *Fagerstabrukens historia*, 29–33.

141. Montelius, Utterström and Söderlund, *Fagerstabrukens historia*, 56–61.

142. Florén et al. 'Swedish Iron Before', 19.

skilled artisans. In that way it would be possible to ensure that Sweden produced high-quality iron. The idea seems, in addition, to have been that the basic working unit in the forges should be a crew of three at each hearth. This meant that six men worked in the commonest type of forge in Sweden, which had two hearths. Together they shared all the work that was done in the forge. The master and the hand were responsible for actually refining the pig-iron into wrought iron and also for shaping the bars, with the master holding the overall responsibility. The apprentice fetched charcoal from the shed and pig-iron from the iron store, although he was also supposed to help the other workers with most of their tasks, while at the same time learning the trade.

It took a very long time for these aims to be achieved, if they ever were fully. One major complication was that in many places the *bergsmän* refused to abandon their forges and went on producing what was considered inferior iron, using workers who did not fulfil the requirements of the Board of Mines. For most of the seventeenth century there was a steady stream of complaints about iron made by the *bergsmän*, and the blame was often attached to their forgemen. In many places, iron was also made by a smaller workforce than was intended and it is likely that members of the *bergsman's* household helped the forgemen with the work in the forge.[143]

The working situation in the forges owned by ironmasters was more like the ideal pattern outlined by the state. It was in the ironmasters' interests to check on the formal status of their forgemen and thus to secure a workforce that was able to produce iron of a high quality. It is, however, far from certain that the forge crews in ironworks always lived up to the objectives of the state. We must also try to avoid the anachronistic trap of analysing the relationship between ironmaster and forgeman as a modern wage-labour relationship. Well into the eighteenth century, many of the master forgemen worked rather on a subcontracting basis and were themselves responsible for maintaining the workshop, hiring forge hands and apprentices, and purchasing charcoal and sometimes even pig-iron.[144]

After the middle of the eighteenth century, this pattern gradually fell into disuse, at least at the ironworks, and the forgemen

143. Florén, 'Järnproduktionens sociala mångfald'.

144. A. Florén and G. Rydén, *Arbete, hushåll och region. Tankar om industrialiseringsprocesser och den svenska järnhanteringen*, Uppsala 1992, 46.

became more like wage labourers, although on a severely regulated labour market. They were as a rule provided with the raw materials by the ironmaster and had to have his permission before hiring new workers. But even after this a master forgeman nevertheless often retained a fairly independent position. This was mainly due to the importance of his skills in the making of quality bar iron. The forge thus remained largely the territory of the forgemen and their internal hierarchies, with first and foremost the master forgeman maintaining his leading position at the shop-floor level.[145]

Leaving the small Swedish forges for the bigger Russian ones means that we are also moving from a small-scale organisation to a very large one. But, to begin once again with the similarities, the basic unit of bar iron production in the Urals was, as in Sweden, a forge crew comprising three members: a master forgeman, an apprentice and a worker. There, though, the resemblance ends. Every hearth in the Ural forges was worked by two forge crews, and a shift system prevailed. In his account of his stay in Siberia Petter Schönström wrote in 1728 that the forges 'are always worked with double personnel, namely two master forgemen at each forge hammer, and each master has two journeymen and two hands'.[146]

This means that workforces in the Urals were twice the size of the Swedish. In addition, in Russia bar iron producing organisation did not, as in Sweden, end with the forge crews. It also included a supervisory element and a number of attached assistant workers. There was an *ustavshchik*, a technician, and a supervisor, an administrator, in charge of production. In large forges there were even two *ustavshchiks*, one for each shift, and the supervisor also had an assistant. The assistant workers included joiners/carpenters, responsible for the water-wheels and the bellows, or blowing-machines, blacksmiths, who repaired the forgemen's tools and the hearths, and lastly charcoal carriers.[147]

The workforce of a Ural forge was thus very large indeed. At ironworks of the size mentioned above, sixteen hearths as at Verkh-Isetsk, more than one hundred people were involved in bar iron production, ninety-six forgemen plus supervisors and assistant workers. Thomas Esper gives a figure of 141 workers for the Nizhnii Tagil' forge, with its fourteen hearths, in the 1840s.[148]

145. G. Rydén, 'Skill and technical change in the Swedish iron industry, 1750–1850', *Technology and Culture* 1998.

146. Schönström, 'Berättelse om'.

147. Ustiantsev and Logunov, *British Technological Experience*.

148. Esper, 'Industrial Serfdom', 600.

This organisation remained fairly stable throughout the eighteenth century and also well into the nineteenth century. This can be concluded from a comparison between a description from the 1730s by Wilhelm De Hennin, Commander-in-Chief of the Ural mining works, and a work by Fëodor Tret'iakov from the 1860s, 'On the Duties of Office and Manual Workers'.[149]

The *ustavshchik* was the technical head of production and a repository of technological experience. He was in charge of the *ustav*, that is, the preparation and adjustment of the hearth depending on the charge, the fuel quality, the intended quality of the iron produced as well as the skills of the workers. The *ustavshchik* instructed the master forgemen, apprentices and workers, watched over and controlled their work and received, together with the supervisor, the finished iron from the master forgemen. The iron produced in the forge was marked with the *ustavshchik's* hallmark, no matter which forgeman had produced it.[150]

The *ustavshchik* personally examined any new methods and devices suggested by the forgemen and either approved or disapproved their adoption by other forge crews. The workers had no rights, and no opportunities, to conceal alternative procedures at either the hearth or the hammer from the *ustavshchik*, since the layout of the forge allowed the latter to observe everything and everyone. The *ustavshchik*, together with the supervisor, also put together the forge crews. In the regulations in force in the 1860s it was stated that 'the master forgeman has no right to choose an apprentice or a worker at will, but should work with those appointed by the administration; equally the apprentice and the worker have no right to choose a master forgeman'. De Hennin's regulations stated the same thing, if not as explicitly. The *ustavshchik* also decided on the activities of the forge crew, replaced absent workers, and took care of drunk and sick men. His duties also included safety measures in the forge.

When a new *ustavshchik* was to be appointed, he was taken from the ranks of the master forgemen, and it was often the most experienced one that was picked. However, he was not only chosen for his skills in making bar iron, but was also viewed in the light of his personal loyalty to the ironmaster and the works administration.

149. Hennin, *Opisanie*, 187–215; L.A. Dashkevich, S.V. Ustiantsev and Z.F. Tret'iakova 'Ob obiazannostiakh sluzhashchikh, 25–30.

150. Concerning technical education, mainly during the nineteenth century, see chapter 8, below.

The *supervisor* was not involved in the technological aspects of making bar iron. He was in charge of the documentation of the forge, recording expenditure on pig-iron, charcoal and tools, and income from manufactured iron of different sorts. He was responsible for the accuracy of the data. Of greater importance, however, is the fact that the supervisor was appointed by the works administration, and thus had an even greater influence on the selection of master forgemen, apprentices and workers than the *ustavshchik*. The latter was also obliged to have the consent of the supervisor in certain matters outside the *ustavshchik's* competence. The supervisor, on the other hand, had to obey instructions from the administration, but he was given a very powerful role in the forge. His working hours were not fixed as he was to be at the works whenever needed.

At the level below the supervisor and the *ustavshchik* we find the forgemen, who made the bar iron in their hearths and under their hammers. As in Sweden they were organised in a hierarchical forge crew, but unlike Sweden two crews shared the work at each hearth in a shift system. As in Sweden the crews were headed by a master forgeman, who had under him an apprentice and a worker.

According to the regulations of the 1860s a *master forgeman* was to 'ensure good conditions in the hearth and the hammer; shingle blooms, bear responsibility for the quality and quantity of iron, and for refining, knead blooms, hammer them off into lumps, and avoid waste of coal'. The *apprentice* was to 'assist the master forgeman and, apart from this, he is, with the worker, to carry pig iron to the hearth, charge coal, deliver iron to the hammer, operate the hammer and also undertake various subsidiary tasks ordered by the master forgeman. He is to obey the master'.

The *worker* was also to help the master forgeman with everything. He was to carry pig-iron to the forge, break it into suitable pieces for the hearth, prepare the so-called *zhuki* (solid lumps of spongy semi-refined iron), weld and carry iron to the hammer, weigh manufactured iron and take it to the warehouse.

The skilled workers in the Russian forges, above all the *ustavshchiks* and the master forgemen, possessed a certain amount of technological experience. However, they were not the owners of that skill and experience. The master forgeman had to teach the apprentices and workers appointed by the *ustavshchik*. Crews were very rarely formed on the basis of household ties, and the master therefore had to teach people who might have been strangers to him. The *ustavshchik* simply had to put together forge

FIGURE 3.6 A Forgeman Washing Himself on Saturday Evening, Närke, Sweden, 1827

crews from a workforce given to him by the administration. Both the *ustavshchik* and the master forgeman were thus trapped in the system of a feudal economy. From the second half of the eighteenth century onwards the majority of *ustavshchiks*, master forgemen, apprentices and workers were serfs. They had no rights to their own labour power or to their skills and dexterity.

During the first half of the eighteenth century, the founding period for the iron industry in the Urals, the ironmasters had to hire free labourers as a way of compensating for labour shortages and also in order to teach the serfs to make bar iron.[151] The ratio of hired workers gradually diminished, however, over the century.

Ideally, forge operations, in both Sweden and the Urals, were supposed to go on throughout the week, from Sunday evening to the following Saturday evening, day and night (Fig. 3.6). In Sweden, the master and the hand had an informal shift system, working for periods of between ten and fourteen hours. The apprentice helped

151. A.S. Cherkasova, *Masterovye i rabotnye liudi Urala v XVIII veke*, Moscow 1985, 95–96.

them both for five to six hours at the beginning of each shift. We do, however, have indications that the forges lay idle for a few hours at night in the eighteenth century, and it seems as if this system of working twenty-four hours a day might have been established on a broad scale towards the end of that century or even in the early nineteenth century (Figs. 3.7 and 3.8).

In the Urals, the whole forge crew with its three members worked its shift as a single unit, after which it was replaced by another similar crew for the next shift. Thus there was, not an informal shift system as in Sweden, but a formal one, and the work at each hearth was performed by six workers working three at a time. The working day for a Russian forge crew lasted twelve hours on average. This is an estimate, however, since working hours were, as in Sweden, dictated by the process in the hearth. The shift was over when the last bloom was converted into bar iron, and the length of the working day fluctuated from ten to fourteen hours.

The World Market and Changing Patterns of Work

From the mid-eighteenth century and for the following one hundred years, some very dramatic changes occurred on the international market for bar iron. Russian iron production had risen since the beginning of the century, and from mid-century it was a major threat to the position Sweden had acquired, primarily on the British market. Towards the end of the century the Russians had overtaken the Swedes, but at the same time important things had happened in Britain itself. The use of coal in the making of both pig and bar iron had gradually been introduced, and as a result home production was rising at an ever increasing pace. After the turn of the century both Russian and Swedish iron were losing ground on the British market, and the Russians lost the most. After the end of the Napoleonic wars, virtually no iron from the Urals entered British ports. Swedish Walloon-made iron still had a secure position on the Sheffield market for steel iron, but the rest of the trade trailed in the same way as its Russian counterpart.[152]

152. For an introduction to developments in Britain see J. Harris, *The British Iron Industry 1700–1850*, London 1988. For the latest figures on British production see R.S.W. Davies and S. Pollard, 'The Iron Industry, 1750–1850, in *Studies in Capital Formation in the United Kingdom 1750–1920*, eds C. Feinstein and S. Pollard, Oxford

FIGURE 3.7 **Early Nineteenth-Century Forge, Närke, Sweden, 1827**

FIGURE 3.8 **Forgemen at Work, Uppland, Sweden, 1838**

How did this trend affect the organisation of bar iron production and actual work in the forges in the two regions under discussion? Were attempts made to change the organisational set-up, to impose stricter rules in the forges, to increase the pace of work or to prolong working hours? It was indicated above that a slow but steady process of technological change began to alter the forges during the last decades of the eighteenth century, new blowing machines, more hammers and hot blast were introduced into the forges, which were also enlarged. However, the full impact of these changes was not felt until the second quarter of the nineteenth century. Were there any organisational changes in parallel with the technological changes?

These are very tricky questions to analyse, as the source material available does not deal directly with these matters, and consequently only a limited number of studies have been carried out. As regards developments in Sweden, however, a few things can be said. We know that, during the second half of the eighteenth century, bar iron production remained fairly stable. From this we should be able to assume that no serious attempts were made either to increase the pace of work or to prolong working hours.

For the period around 1800 we know that the number of working weeks was less than a full year, often closer to forty weeks or less than to fifty weeks. At Tolvfors bruk in Gästrikland, for instance, they only made bar iron over thirty-five weeks in 1815. This specific source, among others, also hints that the forgemen sometimes only made bar iron for a few days each week. The total output of these and other ironworks indicates that the situation might have been the same in many places throughout the second half of the eighteenth century. One can assume that, if a given production level could be achieved in less than a whole year around the turn of the century, there is no reason to think that this could not have been done earlier in the eighteenth century, especially as no important technical change had occurred.[153]

If this was the case, it is hardly surprising that the Swedish ironmasters did not try to enforce stricter working arrangements in their forges, with a higher pace of work. If the demand for their iron were to rise, they only had to make their forgemen work a few weeks longer in order to meet the increase. They only needed

1988. For a treatment of foreign iron on the British market see Hildebrand, *Fagerstabrukens historia*, 99–131; Hildebrand, 'Foreign Markets'; E.F. Heckscher, *Sveriges ekonomiska historia från Gustav Vasa*, band 2:1, Stockholm 1949, 407–427.

153. Tolvfors company records, Gävle kommunarkiv, Gävle.

to increase the working hours if they wanted to increase production. For the same reason, there was no use in enforcing new technology on the production process.

The production of bar iron started to rise once again after the war, with a particularly rapid increase during the prosperous years in the second half of the 1830s. The production potential of the country was stepped up during this period by the building of new forges but the lion's share of the increase in output was achieved at older production units. A study of more than ten forges in Gästrikland and Dalarna shows that in the early 1830s the number of weeks of production in the forges had been extended since the beginning of the century. In only a few cases did production go on for less than fort-five weeks, and the stoppages were often necessitated by repairs to the forge. From a detailed survey of the work done at one of these forges it also appears likely that production took place throughout the whole week. Fewer 'short weeks' occurred in the 1840s than earlier. The increase in production was thus a result of longer working hours for the forgemen.[154]

There are, however, clear indications that the pace of work in the forges also increased from the late 1830s onwards. It is difficult to discuss this, though, without first recapitulating the technical changes that occurred and also mentioning an important organisational alteration. As was mentioned above, the 1830s saw some very important changes in the Swedish iron trade. The German forging method was altered in a number of ways. New cast-iron blowing machines were installed and hot blast began to be used in the hearths. The latter were also rebuilt to better encapsulate the refining process and make possible preheating of both pig-iron and air for the *tuyères*. The forges were in addition provided with an extra water hammer, so that the forge crews, with a hearth and a hammer each, could set their own pace of work, independently of the other crews. There was also a change in the forge crew. A fourth member, called the hammer hand, was installed in an intermediate position between the forge hand and the apprentice.

Work in the forges using this new, in both a technical and an organisational sense, German forging method was organised differently from the traditional method. Production was seen more as a process without stops or delays. Pig-iron for the next heat was preheated at the same time as the forgeman worked at the hearth,

154. For the treatment in this and the following two paragraphs see Rydén, *Hammarlag och hushåll*, chapter 6.

refining and welding. The workers also took the still-hot blooms for welding, which had not been done before. Earlier these blooms had been allowed to cool before being taken back to the hearth. There was also a more pronounced division of labour built into the new organisation. This was not, however, a division between different members of the crew, but the forgemen did not have to perform all the various tasks simultaneously. Under the old system a forgeman had to melt pig-iron and weld blooms at the same time as he shaped other blooms into bars. He thus had to be at the hearth and the hammer at the same time. With the new larger forge crews, greater flexibility could be incorporated, and there were always at least two forgemen in the forge at the same time, one working at the hearth and the other at the hammer.

Although the organisation of the 'new' German forging method was both more efficient and more flexible, it was also more demanding on the forgemen. The more process-oriented production system of the 1840s removed many opportunities for breaks for the workers. There were in fact complaints from one of the senior officers of the Ironmasters' Association that the pace of work in the new forges was so high that it was adversely affecting the quality of the bar iron.[155]

It may safely be concluded that work in the forges during the second quarter of the nineteenth century was affected by longer hours for the forgemen at a higher pace and intensity. The forgemen of the 1830s and 1840s had a much more demanding situation than their predecessors of half a century earlier. It is thus very likely that most of the increase in production during these years was achieved by the forgemen working their longer more intense hours.

We have even less information about the corresponding developments in the Urals. We know that technological development began a few decades earlier in the Urals than in Sweden, but we also know that forge organisation remained more or less the same until the German forging method was replaced with puddling or the Comtois method in the mid-nineteenth century. One source that can be used as an indication of how work developed during these years consists of the figures for overall production of bar iron. Russian production rose very rapidly during the eighteenth century. This expansion was interrupted at the beginning of the nineteenth century, and a period of stagnation then followed until the middle of the 1830s, when a slow rise began. During the period

155. JKA 1844, 142–171.

of expansion production rose as the number of ironworks increased, but from the mid-eighteenth century the older works also stepped up their production.[156]

This expansion could be ascribed to a number of different causes. One, which has already been mentioned, is that the forges were enlarged, another is that they were used more efficiently, and yet another that an increased pace of work was enforced. This third factor behind a rising production has in fact also been mentioned. The introduction of an extra hammer meant that all forge crews worked independently of each other, freeing the pace of work from the effects of a few involuntary breaks.

When it comes to the working time, hence to the efficiency with which the forges were used, it becomes clear that there was a major difference between Sweden and Russia. The average number of working days in the private works in the Urals was only 200 as late as 1849. However, the number varied from fifty-five to 340 days.[157] The reason for this seems to have been that the works were closed for roughly two months in the middle of the summer. This was in turn due to a lack of water to power the water-wheels, and was also intended to allow the workers time off to take care of their allotments.

If we try to bring together all the 'technological' factors that affected the production of bar iron in both Sweden and Russia, we end up with a strange mixture of similarities and differences. A suitable starting-point is that they were both affected by the same developments on the market; both Russian and Swedish iron held a strong position on the British market, but they were gradually swept away by increased competition from British producers. They both seem to have dealt with this latter development in broadly the same way, that is, by stepping up technological progress. In both countries, forges were enlarged, new equipment and methods such as blowing machines and hot blast were tried, and extra hammers were built in the forges. The major difference, apart from the fact that works in the Urals remained much larger than their Swedish rivals, seems to have been that an important organisational development occurred in Sweden, with the inclusion of an extra man in the forge crews, longer working hours and a more intense pace of

156. Minenko et al., 'Ural Iron ', 54f. See also Strumilin's figures quoted in H.D. Jr., Hudson, *The Rise of the Demidov Family and the Russian Iron Industry in the Eighteenth Century*, Newtonville, Mass 1986. This trend might be connected with the lack of resources for charcoaling, see chapter 5, below.

157. GASO,F.24. Op.32.D.1638.

work, while the Russian ironmasters relied more or less on the same organisational set-up. The most peculiar feature is that the utilisation of the forges remained fairly inefficient.

Why were there no changes in the organisation of bar iron making in the Urals during this critical period, when survival on the international market was at stake? It is beyond doubt that Russian bar iron production was stagnating during the first decades of the nineteenth century. Was production already so efficient that no changes were necessary? It is important not to forget that very marked differences still existed between the Russian set-up and that of the Swedish industry. The former was a large-scale hierarchical structure, based on unfree labour. This created scope for a far-reaching division of labour. The sheer size of the workshops meant that it was profitable to have specialist workers performing auxiliary tasks. Carpenters and blacksmiths repaired the forge and tools which Swedish forgemen had to do themselves. In the Urals there were also special workers to carry charcoal to the forge.

This made it easier for the Russian forgemen to devote all their time to the main task of making bar iron, which ought to mean that production per hearth would be very high. Compared with Swedish forges it should have been higher, as forgemen there had to perform all the auxiliary tasks themselves. The advantages could have been further enhanced by the fact that two forge crews shared the work at one hearth, by the slightly earlier technological advances, above all the introduction of the extra hammer, and lastly the presence of spare hearths in the forges for emergencies.

Available figures support this view, showing a higher output per hearth in the Ural forges. It is a very intricate matter comparing the industrial 'performance' of different production units, especially if they are organised in different ways. A second problem is that we do not have compatible sources. However, it is possible to give an outline of developments during the period.

During the first quarter of the eighteenth century, almost ninety tons of bar iron were manufactured at each hearth in the Urals. A century later this figure had risen to 140 tons. Production levels differed very significantly, however, between different forges. In the Nizhnii Tagil' mining district, owned by the Demidov family, some works produced only sixty tons per hearth in the first half of the nineteenth century, while others had figures close to 140 tons.[158]

158. Minenko et. al., 'Ural Iron ', 67; Ustiantsev and Logunov, *British Technological Experience*.

Thomas Esper provides additional information about production at the Nizhnii Tagil' works in the mid-1840s. He gives a production figure of about 1,650 tons for the forge equipped with fourteen hearths, corresponding to an output per hearth of under 120 tons. He also gives a figure for weekly production per hearth. On average, 3.6 tons were produced. These figures taken together support the view of a summer break, as fourteen hearths producing three and a half tons of bar iron every week gives a yearly production of well over 2,000 tons. Esper's figures, based on source material collected by the French metallurgist Frédéric Le Play when he was adviser to Demidov, indicate that the forges lay idle for roughly fifteen weeks that year.[159]

Corresponding figures from Swedish forges show an average production per hearth around 1800 of about sixty-five tons a year.[160] Forty years later it had risen to almost 90 tons on average, and some forges made over 110 tons. Figures for weekly production per hearth can be given for the 1830s and 1840s. At the beginning of the first decade only one and a half tons were made, compared with 2.2 tons in the latter decade.[161]

In spite of all the uncertainties connected with comparisons of this kind, it ought to be very clear that the Russian organisation was more efficient than its Swedish counterpart, at least if efficiency is measured as weekly production of bar iron in a hearth. The forgemen in the Urals made about 50 per cent more iron than their Swedish colleagues. (A more accurate measure of productivity would certainly have reduced this difference, as the workforce involved in the Urals was much larger.) Some figures from the Urals tend to indicate a slightly faster rise towards the end of the eighteenth century than during the first half of the nineteenth century. In Sweden, the reverse seems to be true, with fairly fast growth in production in the new German hearths. Swedish forgemen were thus tending to close the gap.

If this is an accurate description of what happened, and the industry in the Urals was more efficient than its Swedish rival, why was it that the Swedish industry expanded and remained on the international market, while Russian iron dwindled and stagnated? It ought to have been possible for Russian ironmasters to

159. Esper, 'Industrial Serfdom', 599–601.

160. According to the new decree concerning permitted bar iron production, from 1803. See B. Boëthius and Å. Kromnow, *Jernkontorets historia, vol. II*, Stockholm 1947–1968, 446ff.

161. Rydén, *Hammarlag och hushåll*, 313–320.

boost their output, as their Swedish rivals did, simply by increasing the number of working weeks. The answer to this question is partly to be found in a return to the market theme, Swedish and Russian iron on the British market, that was established in the introduction to this section.

As was noted earlier the fiercest competition existing on the British market for most of the eighteenth century was not between Swedish and Russian producers, but rather between the latter and British ironmasters producing for the home market. A divided market prevailed, with Swedish iron holding a fairly secure position as, among other things, a supplier to the British navy. Steel production in Sheffield and other places was also mainly based on Öregrund iron from ironworks in eastern Sweden.[162] Swedish iron was probably on average of a slightly higher quality, as reflected in the higher prices paid for this iron compared with both the Russian and the British commodities.[163]

This relationship is clear in the developments that occurred in Britain with the introduction and spread of puddling. Imports of Russian bar iron to Britain fell from over 25,000 tons in the 1790s to around 5,000 tons in the 1820s. In spite of its quality, Swedish iron was not unaffected by this process. Swedish exports to Britain declined as well, from around 18,000 tons in the 1790s to around 9,000 to 10,000 tons in the 1820s, but this was not as large a decrease. It was primarily the market for steel iron that remained stable, this commodity becoming even more essential during this period than earlier. If quality had guaranteed Swedish iron a place on the British market during the second half of the eighteenth century it was only the top brands that were able to hold on to this position in the face of the growing threats from cheap puddled iron during and after the Napoleonic wars.[164]

Swedish ironmasters, under the guidance of the Ironmasters' Association, did, however, rise to the challenge from the British industry. The technological changes during the second quarter of the nineteenth century that were discussed earlier reflected an endeavour not only to produce iron at lower cost, but also to make iron of better quality. The market for steel iron had been an arena for iron made according to the Walloon method, using ore from

162. For an elaboration on this see, Evans and Rydén, 'Iron in Sweden and Britain'; Florén and Rydén, 'A Journey into'.

163. Hyde, *Technological Change*, 104; Heckscher, *Sveriges ekonomiska*, 407–413.

164. Minenko et al., 'Ural Iron', 58; Hildebrand, 'Foreign Markets', 10; Attman, *Fagerstabrukens historia*, 10; Heckscher, *Sveriges ekonomiska*, 424.

the Dannemora mine. No other Swedish iron was thought to match the requirements of steel production. During the 1830s, however, ironworks in the county of Gästrikland did manage to produce an iron of the required quality, and an increased volume of steel iron was shipped to Britain. During the 1830s, exports to Britain increased to nearly 15,000 tons.[165]

The response in Russia was different. Instead of trying to compete on the British market with either lower prices or higher quality, the Russian ironmakers switched their resources into an attempt, not to supply an international market for bar iron, but to meet own growing home demand for a very wide variety of iron and metal wares. The Russian industry stayed within the limits of producing an ordinary iron at low cost, but not as low as puddled iron from Britain, instead of attempting to achieve higher quality as its Swedish counterparts did.[166]

It might thus have been this change on the international market for iron that also led to a more far-reaching development of the methods of iron production in Sweden than in Russia. Swedish ironmasters fought a battle to remain on the international market, and the means were more efficient production of an iron of higher quality. Developments during the second quarter of the nineteenth century were a starting point in this process, which would soon lead to even more rapid change, involving the replacement of the German method with Lancashire forging and rolling and subsequently also the introduction of modern steel processes. This kind of market pressure cannot have been at hand in Russia. There, the internal market had grown and was of considerable size at the beginning of the nineteenth century. The easiest option for the Russian ironmasters should therefore have been to withdraw from the international scene.

An interesting question arising from this discussion, however, is whether the internal structure of the Russian industry prevented it from taking the same steps as the Swedish industry. Was there something in the socio-economic foundations for iron-making in the two countries, the large-scale feudal mode of production versus the small-scale market-oriented mode, that made the

165. Rydén, *Hammarlag och hushåll*; R. Adamson, 'Borrowing and adaptation of the British technology by the Swedish iron industry in the early nineteenth century', in *Technology transfer and Scandinavian industrialisation*, ed. K. Bruland, Oxford 1991. For the export figures see Attman, *Fagerstabrukens historia*, 10.

166. Minenko et al., 'Ural Iron', 54–57 and 66–71; Ustiantsev and Logunov, *British Technological Experience.*

production of high-quality iron more likely in Sweden? A central place in such a discussion must be given to the issues of discipline and conflict, and it is to these matters we will now turn.

Work Discipline and Conflicts

Turning to conflicts and the organisation of the workers, it is obvious that such a cluster of questions is inseparable from the contradictions and tensions that were intrinsic in the social network of production. Mainly owing to the lack of research in this field, however, it is impossible to give a conclusive picture of these matters in the two regions.

As an important tool for such an analysis, the distinction between everyday resistance and organised conflicts will be used, as formulated in recent research conducted by, among others, James C. Scott.[167] With everyday resistance we mean a form of discord which can primarily be seen as an individual or collective non-acceptance of the relations of production. Escapes, frauds and embezzlements were important aspects of this type of conflict, or to use Denis Woronoff's expression, the 'guerrilla permanent'.[168] The intentions of the actor are not crucial for labelling such actions as resistance. The fact that they did create problems, as pebbles in the social machinery of production, is enough for them to be so classified.[169]

Everyday discord could be transformed into explicit and organised protests, a process that required both means of collective organisation and ideological legitimisation. Such a transformation took various shapes. The picture established in previous research, portraying the pre-industrial relationship between employer and employee as one of harmonious familiarity and household-like intimacy, has been revisited by a number of historians.[170] There is,

167. J.C. Scott, *Weapons of the Weak. Everyday Forms of Peasant Resistance*, New Haven 1985, 28–29.

168. D. Woronoff, *L'industrie sidérurgique en France pendant la révolution et l'empire*, Paris 1984, 192.

169. For analogous problems of interpretation see M. Berg, *The Age of Manufactures 1700–1820*, London 1985, 309; J. Styles 'Embezzlement, industry and the law in England 1500–1800, in *Manufacture in Town and Country before the Factory*, eds M. Berg and M. Sonenscher, Cambridge 1983, 207.

170. Florén, 'Klasskamp utan', 3; M. Sonenscher, 'Journeymen's migrations and workshop organization in eighteenth century England', in *Work in France. Representations, Meaning and Practice*, eds S.L. Kaplan and C.J. Koepp, Cornell 1986, 76.

however, still a tendency to discern a development from largely disorganised protests in pre-industrial times to more organised and institutionalised protests in modern society – a conclusion which seems to be based partly on the fact that historians have often chosen a rather short historical perspective for their studies, restricted to the industrial revolution or at least to the nineteenth century.[171] Consequently, a period has been studied in which many of the old collective forms of societal life had already disappeared as a result of the expansion of an individualist bourgeois society. What is then easily overlooked is the way traditional values and organisations mattered to rural workers and artisans.

As has been indicated above, the hegemony of Swedish ironmasters over the production process in the forges, as well as in the iron trades, was based during the seventeenth and eighteenth centuries on control of the input of raw materials and output of refined products. This was the power technology at hand to stop fraud or embezzlement. An important prerequisite for illicit trade by the workers was their position on the shop floor. William Reddy has demonstrated that at the beginning of the nineteenth century the textile workers in France still maintained the view that they were selling not their labour force but their product.[172] The decentralised organisation of production in the forges accentuated this artisan principle. The master and his crew worked alone at the hearth and could, for example, judge when it was possible to go beyond the normal relationship between raw materials and product, laid down by the ironmaster, and produce a surplus. As this surplus product was seen as a result of their skill, it was regarded as their property which they could freely sell. The political economy of fraud and embezzlement, however, was not only a question of the social and technical organisation of production, but also of access to a market. That such a market existed is obvious. During the second half of the eighteenth century a group of scythe smiths in the Dalarna region, despite the risk of being punished, openly declared that they used bar iron that they bought directly from forgemen at nearby ironworks.[173]

171. D. Geary, 'Protest and Strike. Recent Research on "Collective action" in England, Germany and France', *Historische Zeitschrift, Sonderheft 15, Arbeiter und Arbeiterbewegung im Vergleich,* ed. K. Tenfelde, München 1986, 364–369.

172. W.M. Reddy, *The Rise of Market Culture. The Textile Trade and French Society, 1750–1900,* Cambridge 1984, 213, 251; M. Sonenscher, *Work and Wages. Natural Law, Politics and the Eighteenth Century French Trades,* Cambridge 1989, 72.

173. Rinman, *Allmogesmidet,* 26; Isacson and Magnusson, *Protoindustrialisation.*

Naturally this sort of commerce was strictly forbidden. In accordance with the royal charters for the industry, both the seller and the buyer were to be punished, the former by running the gauntlet. However, it was not just the stick but also the carrot that were used to discipline workers and make them accept that the raw materials, even during the production process, and also the end-product were always the property of their patron. In the middle of the seventeenth century, a special payment for any surplus iron was introduced, a payment which often made a significant difference to the total income of the forgeman.[174] Nevertheless, embezzlement continued.

Fraud and embezzlement of this kind are quite naturally to be seen as an important aspect of everyday resistance against the ironmaster. Viewed from a slightly different perspective, however, it is not certain that its consequences were totally detrimental to the industry, as it encouraged the workers to develop their skills in order to make fraud possible or, if applied within the legal framework, to raise their income for surplus iron.

It can also be stressed that this control of input and output from the forge was not only in the interests of the ironmaster. Since wages were paid on a piece-rate basis, the workers were keen to ensure that the quantity and quality of the pig-iron and the weight of the bars were correctly recorded by the bookkeeper.

Preliminary results of a study of the forge court in the Nora and Linde region confirm that conflicts between workers and ironmasters predominantly revolved around the quality and quantity of raw materials, the hiring and dismissal of workers and, finally the end-product (Table 3.1).[175]

Very few cases thus dealt with the labour process as such, i.e. the techniques of production or the productivity of labour. This reveals the typical character of the social organisation of the production process as a 'centralised putting-out system', meaning that the owner or overseer had very limited scope to supervise the forge crew at work. This power to control production was mainly

174. Florén and Rydén, *Arbete, hushåll*, 76–77.

175. A. Florén, 'Patterns of Crime, Protest and Conflict in the Nora and Linde Mining Region 1650–1720', paper presented at the third meeting on iron-making in Russia and Sweden before the twentieth century, Ekaterinburg 1993. The picture that John Rule and John Styles have given of the dominance of these types of tensions and conflicts in the eighteenth century English putting-out industries and protofactories thus also holds true for the Swedish iron industry. Styles, 'Embezzlement'; J. Rule, *The Experience of Labour in Eighteenth-Century Industry*, London 1981, 125.

TABLE 3.1 Subjects of Discord Related to the Work Process in Cases before the Nora and Linde Forge Court, 1650–1720

	1650–1664	1665–1679	1680–1699	1700–1720
Raw materials	15 (11%)	49 (17%)	25 (16%)	44 (20%)
Means of prod	5 (4%)	7 (2%)	3 (2%)	12 (6%)
Workforce	69 (49%)	105 (37%)	63 (41%)	55 (25%)
Habits at work	9 (6%)	19 (7%)	6 (4%)	16 (7%)
End product	18 (13%)	39 (14%)	9 (6%)	26 (12%)
Payments/debts	26 (18%)	64 (23%)	46 (30%)	62 (29%)
Total	**142**	**283**	**152**	**215**

Sources: Records of the forge court of Nora and Linde *bergslager* (Hammartingsprotokoll för Nora och Linde bergslager). Uppsala Landsarkiv, Uppsala.

restricted to monitoring the amount and the quality of the raw materials and the final products.

The corporate nature of the court also makes it relevant to suppose that it made it easier for the skilled workers to organise collective protests. However, even if collective claims were put forward in the courts (though rarely on a more aggregate level than that of the workplace), the general picture is one of the individual master forgeman presenting his complaints against the ironmaster, or more often vice versa.[176]

As a rule, moreover, these conflicts were settled in the court or by intervention by the Board of Mines, and did not lead to more widespread social unrest.[177] To prevent any outbursts of uncontrolled action, however, the authorities felt a need to restrict gatherings of workers in market places both before and after sessions of the court.[178] The court thus functioned as an institution for the resolution of conflicts, rather than as a means for the forgemen to pursue an independent guild policy.[179]

176. Florén, 'Patterns'.

177. P-A. Karlsson, *Järnbruken och ståndssamhället. Institutionell och attitydmässig konflikt under Sveriges tidiga industrialisering 1700–1770*, Stockholm 1990, 152; Florén, 'Patterns', 12.

178. Florén, 'Patterns', 4.

179. Even if the courts were thus unable to act as a proto-trade unions, they could obviously serve other functions for the corporation of master forgemen. In the courts, a system of poor relief for masters and their widows was built up. Control of the masters' skill and competence was, besides being in the interests of the state and ironmaster as a form of quality control, also in the interests of the masters as a token of their exclusive position in the hierarchy of workers.

The possibilities to embezzle bar iron were greatly enhanced by the fact that the producer had an artisan-like position on the shop floor and therefore opportunities to sell products on the commodity market. This should be even more relevant when it comes to manufactured iron wares: nails, axes, spades, wire etc. – products which found buyers among consumers in both the countryside and the towns. In the middle of the eighteenth century a civil servant responsible for the iron trades concluded that one of the main problems of the industry was that 'the workers do not conscientiously deliver their products to the owners, but sneak some of them away to sell them on their own, without any control of their quality, to the surrounding urban or rural populations, by which the just profits of the ironmasters are withheld'.[180]

Jäders Bruk is an illuminating example in this context. Towards the end of the seventeenth century, the volume of embezzlement by blacksmiths here was as high as a quarter of the total output.[181] This meant that the ironmaster's position on the market, and thereby the basis for his power, was seriously challenged by the actions of the producers.

The independent iron trade guilds, unlike the organisation of the forgemen, were eminently well placed to serve the master smiths' interests.[182] At Jäder the guild organised the masters' massive resistance against the proprietor's efforts to take firmer control of the administration of the workshops and against his endeavour to control the market for their products. The everyday practice of embezzlement was transformed into explicit demands for the right to independently sell the products on the market.[183] Legal means were used by the artisans both at a local and at a more central level. The workers at Jäder also gained support for their cause from other groups: from merchants in the nearby towns and from civil servants in the central administration.

To stress their demands, boycotts were organised and the workers also threatened the ironmaster with their collective departure from his works for other employment. At Jäder the threat of a mass exodus from the ironworks was taken seriously. We can easily see why. In the 1640s one fifth of the master smiths had left the enterprise. Some of them succeeded in establishing themselves as

180. Quoted from Florén, 'Klasskamp utan', 109.

181. A. Florén, *Disciplinering och konflikt. Den sociala organiseringen av arbetet. Jäders bruk 1640–1750*, Uppsala 1987, 134.

182. Florén, 'Klasskamp utan', 14–16.

183. Florén, *Disciplinering och konflikt*, 205–209.

independent smiths in the countryside and had, by resigning, resisted the threat of being subjected to closer control and had maintained their position as artisans.[184]

Thus, from a socially and organisationally strong position, the workers could fight for what they perceived as their rights as master artisans. This is also underlined by the fact that they used the boycott, not the strike, as a weapon against the ironmaster. The essence of their action was thus a form of sabotage in the sphere of circulation, rather than in the labour process.

The pattern of conflict at Jäder changed during the eighteenth century, however. A royal decree that gave the ironmaster the role of directing the guild put an end to its independent actions. His ownership of the workshops as well as his firmer control of the market also strengthened his social position, while the structural alteration of the network of iron production made it more difficult for the masters to furnish themselves with raw materials. With the deterioration of their social position, the workers' demands to be respected as artisans could not be successfully defended as a question on the agenda. Calls for a profound shift in the social organisation of production were dropped, and instead workers concentrated on securing better payment.

As a rule, the reformed guild could cope with the workers' demands and became, as was the case at the forge court, an arena for negotiations between workers and employer. In 1737 the last outbreak of open resistance by the workforce occurred at Jäders bruk, and the masters, allied with their journeymen, for the first time used the strike weapon against their employer. This occurrence emphasises the relationship between means of struggle and the social position of the producers. As they had lost their ownership of the means of production and much of their influence on the market, they chose the proletarian weapon of the strike.[185]

At the beginning of the eighteenth century, the socially and religiously more tolerant situation in Siberia attracted peasants and workers from western Russia. This migration was an important precondition for the expansion of the iron industry. As feudal bonds were tightened in the Urals, too, largely owing to the needs of the same industry, the result was massive escapes to the east or south, primarily to the regions under Cossack rule.[186] The flight of peasants, including journeymen, from the mining districts is commonly

184. Ibid., 104.
185. Ibid., 180–183.
186. Portal, *L´Oural*, 46, 271.

seen as the major problem faced by ironworks during the eighteenth century.[187] The peasants had thus not lost touch with their home villages, a point which is further illuminated by their complaints about being prevented from going home to take care of their farms during the harvest.[188] In the early years of the following century, absconding from the estate was becoming rarer. Even if it still constituted a problem, it was of a new kind, with peasants leaving the ironworks for their villages within the district. These actions can therefore be seen as a form of individual and unorganised strike.[189]

In the latter half of the eighteenth century these protests assumed a rebellious form. The best-known uprising was the Pugachev rebellion in the 1770s. As the power of the feudal landlords and that of the ironmasters were often inseparable, ironworks were burned by the rebels, and the peasants' freedom from their obligations to the ironmasters was put forward as an important demand.[190] Even if these rebellions were quelled, the protests were successful in so far as they resulted in increased payments for both rural and industrial workers.[191] This put a certain pressure on the ironworks during the latter part of the eighteenth century to cut other costs and to rationalise production. The real structural crises, however, came a century later, as a consequence of the abolition of serfdom.[192]

187. Cherkasova, *Masterovye i rabotnye*, 163–232.

188. N.A. Minenko and I.V. Poberezhnikov, 'The interaction of industry and agricultural environment', unpublished paper at the third meeting on iron-making in Russia and Sweden before the twentieth century, Ekaterinburg 1993, 3.

189. V.A. Shkerin, 'The process of adoption and social conflicts at the mining and ironworks in the Urals during the feudal period', unpublished paper at the second meeting on iron-making in Russia and Sweden before the twentieth century, Uppsala 1992, 4; V.A. Shkerin, 'Rebellious crowds in social conflicts at the Ural private works in the first half of the XIX century', in *Metallurgical Works and Peasantry. Problems of Social Organization of Industry in Russia and Sweden in Early-Industrial Period*, Ekaterinburg 1992, 266–270.

190. The document is published in *Imperial Russia a source book, 1700–1917*, ed. B. Dmytryshyn, Portland 1974, 96. The opposition that emerged during the eighteenth century also found expressions in popular culture. One song, for example, described the prisonlike conditions at the Demidov works. 'In iron-works of Demidov/ The work is hard/ Oh, yes work is really hard/ How bad does not our backs ache/ They put us in a factory, a prison/ And do not let us out'. Quoted and translated from Portal, *L'Oural*, 289.

191. Portal, 'Manufactures et classes', 349.

192. R. Portal, 'The industrialisation of Russia', in *The Cambridge Economic History of Europe*, vol. vi:2, Cambridge 1978, 812–814. However, Thomas Esper also points to the beneficial economic effects which the abolition of serfdom had for the ironworks, in that it eliminated the costs for the patriarchal care. Esper, 'Industrial Serfdom', 156.

The social unrest in the iron-producing districts seems to a large degree to have been a peasant affair in which the skilled workers, apart from some journeymen and hands of peasant origin, were not involved. When the Pugachev rebels arrived at the state-owned ironworks in Voskresensk in 1773, for example, they were instantly joined by 150 ascribed peasants, while the skilled workers refused to take part in the insurrection. As at several other ironworks, the skilled workers chose to stay at the establishment and to help defend it against the rebels.[193]

The firm patriarchal rule at the ironworks and the fact that skilled workers were divided into small work teams (compared with the gang-work of the peasants, see chapter 5 below) are important factors behind the absence of large-scale and overt worker protest.[194] Another, and more basic circumstance, was that the peasants were only able to abscond, and to rebel, thanks to the dual economy of their households, working both on the land and in the industry, which made it possible for them to survive, or even to be better off, without the income from the ironworks. The workers at the furnaces and forges were in a different situation in this respect, since they were dependent on industrial work. Not to be forgotten, moreover, is the role played by the rural village community in organising the protests of the agrarian part of the labour force.[195] Or as the American historian Bruce DeHart puts it in order to explain the absence of skilled industrial workers from the Pugachev rebellion: 'Their ties with an agricultural existence had long since been severed, as a result of which they probably considered the factories their sole means of subsistence and the factory barracks, their homes. Thus, the skilled workers were inclined to defend their places of employment.'[196]

Large-scale rebellion, however, was not the only possible avenue of protest open to the skilled ironworkers of the Urals. There were also other means. When Salomon von Stockenström arrived in St Petersburg in 1787 with a view to acquire information about the iron industry, he noticed the lack of regulations on labour discipline. The legal framework for the trade which he knew from Sweden seemed to be non-existent and he saw the slack discipline

193. B. DeHart, 'The Pugachev Revolt and Its Effect on the Industrial Workforce of the Urals, a Reconsideration', in *Russia and the World of the Eighteenth Century*, eds R.P. Bartlett, A.G. Cross and K. Rasmussen, Columbus, Ohio 1986, 525–527.

194. Portal, *L´Oural*, 294–295.

195. Cherkasova, *Masterovye i rabotnye*, 163–232.

196. DeHart, 'Pugachev revolt', 527.

of the Russian workers as a major problem for the industry.[197] However, his conclusion was not entirely correct. A similar institution to the Swedish mining and forge court seems to have existed at the state-owned ironworks. At these courts workers are known to have put forward complaints over wages as well as over working conditions.[198] Workers at the state-owned establishments could also plead directly to the Tsar.[199] As the state-owned share of the industry diminished during the eighteenth century, the number of workers who had this possibility diminished.

Salomon von Stockenström's remark seems relevant, however, in that it indicates that in Russia control was primarily in the hands of the individual ironmasters and, unlike Sweden discipline on the shop floor, in the forges, was upheld by *ustavshchiks* and supervisors. The master forgemen were thus not themselves responsible for procuring raw materials, which were supplied to them by their overseers. Nor was there any stipulated relationship between the maximum amount of charcoal and pig-iron that were to be used to produce a fixed quantity of bars. The consequence, according to Stockenström, was that a very high percentage of pig-iron was wasted, burnt off, during production, namely around 30 per cent which could be compared with the 22.5 per cent which the Swedish regulations accepted as a maximum waste.[200]

Stockenström considered the Russian system of production to be less efficient than its Swedish counterpart. This might very well have been the case concerning the amount of charcoal and pig-iron consumed during bar iron making, but as was noted above conclusions on this subject are not that easy to draw. What most certainly did exist, however, was a difference in the power technology used in the two regions. If we remain within the scope of protests and resistance, the very tight control at the workshop level in the Urals probably made everyday resistance, in form of embezzlement and illegal trade, among the skilled workers a highly risky business, particularly as the rewards might not have been as lucrative as in Sweden. It is questionable whether there existed a small-scale market for bar iron in the Urals, as there was

197. Stockenström, 'Anmärkningar rörande'; Hildebrand, *Fagerstabrukens historia*, 155.

198. Cherkasova, *Masterovye i rabotnye*, 163–232.

199. As, for example, the workers at a copper works near Perm' did in March 1767. Ibid., 91–92.

200. Stockenström, 'Anmärkningar rörande'. Petter Schönström thus wrote in the 1720s that no accounts were kept of the charcoal that the master used for the production process. Schönström, 'Berättelse om'.

in Sweden. As we have seen, rural blacksmiths and free peasants were the main buyers of iron from the forgemen. As was seen in chapter 2, during the eighteenth century Siberian blacksmiths became integrated into the business of the ironworks, and could therefore not be potential buyers of embezzled iron.

It is, however, possible to interpret the supervisory structure of Russian ironworks in a slightly different way. In a study of the organisation of production in an agrarian region, on the Petrovskoe estate in the Tambov region, Steven Hoch has touched upon these matters. He has shown that work in the fields was done by unfree serfs in a collective manner, submitted to a bailiff, a set-up similar to that existing in the forge, where the forgemen were submitted to the *ustavshchik* and the supervisor. The peasants, however, did not act according to any notion of cooperation, every household being guided by its own best interests. This meant stealing, lying, cheating and fighting when necessary. Hoch presents evidence of a village in which breaking the rules of the estate and other crimes were commonplace. They were an integral part of the peasants' strategy. Work-related offences and theft from the estate comprised more than 60 per cent of all offences. Was this a pattern that was also prevalent among serf ironworkers in the Urals, or was it a peculiarity of agrarian societies? No research has been undertaken into these matters, but the employment of special guards at the blast furnaces at some works in the Urals suggests that theft at least was a serious problem.[201]

Continuing this line of argument, it is also possible that the Russian power technology made the forgemen less inclined to take initiatives of their own to improve their skills and the technical devices used in production. When knowledge was regarded as the property of the ironmaster and production was strictly supervised, workers had very little personal interest in developing their skills. Speaking about the low efficiency, it seems as if Stockenström had the less skilled Russian forgemen in mind, or perhaps rather forgemen who were less inclined to use their skills. Russian bar iron was, as was seen above, criticised for its irregular shape and quality. The Russian ironmasters and authorities were obviously aware of the problem. In the first half of the eighteenth century, however, De Hennin stated that Swedish forgemen, notwithstanding their longer tradition of iron production,

201. S. Hoch, *Serfdom and Social Control in Russia. Petrovskoe, a Village in Tambov*, Chicago 1986, 160–186, especially 166.

also sometimes had difficulty making bars of standardised weight and length.[202]

Despite De Hennin's argument, however, it seems reasonable to assert that the problem was more serious in Russia. It is likewise probable that it also affected the willingness of the masters to pass on their knowledge to apprentices and workers. Petter Schönström, for example, noted that the iron bars made at the works in Kaminskii, near Uktus, were 'ill forged in short and thick bars and they [the forgemen] do not know their craft exceedingly well, notwithstanding the fact that they have been taught by German masters'.[203]

In this case it is possible that the transfer of knowledge was further complicated by the fact that the 'teachers' were of foreign origin, thus also implying a certain ethnic dimension to the problem. The fact that, as we shall see in chapter 6, fathers and sons more rarely worked together in the same forge crew in Russia than in Sweden probably also made Russian masters less inclined than their Swedish counterparts to pass on their knowledge and skills to the next generation. The methods used by the Russian ironmasters to overcome these difficulties seem to have been supervision and compulsion, with the inherent risk of fostering a rather passive and irresponsible workforce.

We still do not have enough information about the 'inner life' of the Siberian ironworks to draw a final conclusion on this. Although there is a risk in paying too much attention to a single example, and an example with a certain tendency and from another region, namely the late eighteenth-century English traveller Edward Daniel Clarke's account of his visit to the iron trades in Tula south of Moscow, this account nevertheless seems to illuminate the devastating consequences of introducing strict control of production during a phase of industrialisation in which the essential productive and innovative potential still lay in the skill of the producer. Clarke thus noticed that

> Tula in its present situation, is not likely to prove any advantage to the empire; because the inhabitants are unable to raise the water which is wanted to put the whole fabric in motion. The machinery is ill constructed, and worse preserved. Everything seemed out of order. *Workmen, with long beards, stood staring at each other, wondering*

202. Portal, *L'Oural,* 193. Hennin had probably also visited Sweden; cf. Crisp 'Labour and Industrialization', 319.
203. Schönström, 'Berättelse om'.

what was to be done next; while their intendants and directors were drunk or asleep (our emphasis).[204]

Concluding Remarks

On the eve of the Russian Revolution in 1917, very little bar iron was made in the Urals. Production had changed and all the older forges equipped to make wrought iron had been swept away, as had most of the puddling works erected in the second half of the nineteenth century. They had been replaced with modern steel-works, primarily using the open-hearth method. Gone too was the reliance on water-power and charcoal, as the new technology depended more on steam and coke. The most far-reaching change within the Russian iron industry, however, had nothing to do with technology, being connected rather with the upheaval of feudalism. Serfdom was abolished in 1861, and the subsequent process of industrialisation was placed in a totally new socio-economic setting – that of emerging capitalist relations. The iron industry in the Urals became completely dependent on wage labour.[205]

In a technological sense, developments in Sweden were similar. Wrought-iron making dwindled at the expense of the new steel processes. The Bessemer process was perfected in Sweden in 1858, but it was not until the last decades of the century that steel-making had any real impact. The output of steel exceeded that of wrought iron in 1895, and after that there was no turning back. In 1917 only around 15 per cent of all production of iron and steel was made by the old processes, of which the Lancashire method dominated. There was, of course, no marked break in social relations, as there was in Russia 1861, but the agrarian connections of Swedish iron industry were more or less severed and greater emphasis was placed on the market system and the capitalist setting.[206]

In the early twentieth century both the Swedish and the Russian iron and steel industries had adapted to a pattern of industrialisation more or less common to Western Europe and North America. Steel making took place in huge steelworks based on coke and modern production methods and employing large numbers of

204. Clarke, *Travels in Various*, 183.

205. Minenko et al., 'Ural Iron', 89ff.

206. J.-E. Pettersson, *Från kris till kris. Den svenska stålindustrins omvandling under 1920- och 1970-talen*, Stockholm 1988, 25–66; Attman, *Svenskt järn*, 31–114.

workers. The routes to this point of similarity differed, however, and the industrialisation of iron and steel making in Sweden and Russia varied considerably, as this chapter has shown.

It is very easy, when writing the concluding remarks to this chapter, to agree with Martin Daunton in his recent treatment of British economic and social development from 1700 to 1850 that the traditional view of the concept of proto-industrialisation 'excludes too much to provide the key to developments in the eighteenth-century industry'. (We might as well extend this to include the next century as well.) According to Daunton, we cannot understand the process whereby an industrial society emerged if we do not start to analyse developments that occurred in areas and trades outside the traditional boundaries of proto-industrialisation. The iron industry is of the utmost importance. 'In the country, there was a wide variety of mines, mills, forges, and furnaces, and historians have made little effort to explain how these fitted into the agrarian economy.'[207]

The aim of this volume is to do exactly this, to consider the development of the Swedish and Russian iron industries in an agrarian perspective. In relation to Daunton, who deals only with Britain, it is worth emphasising that our aim has not been to demarcate the study of the iron industry from the more traditional areas of research concerning the process of industrialisation, that is the textile trades, which are of course extremely important, but rather to stress different routes to industrialisation within the same sector but in regions with dissimilar social, agrarian and political pre-settings.[208]

Older treatments of the 'industrialisation' of the iron industry have, largely on the basis of British and Belgian developments, discussed the process as one governed primarily by technological factors. According to E. A. Wrigley, as mentioned earlier, the transfer of iron-making from a pre-industrial to an industrial context was the logical outcome of the substitution of a coal-based technology for the older dependence upon charcoal. It was the development from an 'organic economy' to a 'mineral-based energy

207. M. Daunton, *Progress and Poverty. An Economic and Social History of Britain 1700–1850*, Oxford 1995. Throughout the book there are comments on the limitations of the older notion of proto-industrialisation. The quotations are from page 169.

208. It is unlikely that Daunton would disagree with such an agenda, as he himself emphasises the importance of distinct regional developments. See for instance the third section of his book which is called 'Integrating the economy' and within that especially chapter 10: 'Integrating and specialization'.

economy' that Wrigley had in mind.[209] Other researchers have not emphasised the change of fuel to the same extent, but have rather stressed the greater efficiency and cost benefits of the new technology as such.[210] The implications of both these interpretations are that the iron industry underwent an 'Industrial Revolution' in connection with a marked technological break, primarily the use of coal instead of charcoal, and a centralisation of production into large-scale units.

This pattern does not tally with the developments that occurred in either Sweden or Russia, or perhaps more precisely in either Bergslagen or the Urals. In both these regions, what happened was that growth as well as centralisation occurred, in spite of a continuing dependence on charcoal and the resilience of the older methods of making bar iron. Puddling was only introduced on a very limited scale in Sweden and very late in the Urals, to take just one example, but the industry in the two regions nevertheless entered on the path of change.

Leaving aside the developments that took place in Britain, Belgium and any other countries to concentrate on the aim of this particular theme, namely a comparison between iron-making in Bergslagen and the Urals, it can be concluded that we have picked two regions that differed to a large extent. Our starting point, however, was that they did share one important feature, and it is important to emphasise this. The overall structure of the industry in the two regions was the same: bar iron was made to be sold on the international market. The technology used was also the same, as the indirect method of making bar iron was employed. The iron ore was melted together with charcoal in blast furnaces to pig-iron, which was later refined into bar iron using the German method.

Turning now to the differences between the regions, these can be divided into two groups: dissimilarities stemming from the socio-economic settings of the two countries, Russia being a feudal state which Sweden was not, and divergences related more to the technological organisation. It must, however, be underlined that no firm boundary can be drawn between these two groups. The most striking difference between iron-making in the Urals and in Sweden, starting within the last 'group', is the matter of scale. It must have been a minor shock to travellers arriving at

209. Wrigley, *Continuity, Change.*

210. See as a good example Hyde, *Technological Change.* See also Harris, *British Iron,* for an overview.

ironworks in the former region after visiting Swedish works. In the Urals they could witness large-scale, centralised units with furnaces and forges lying together, adjacent to huge dams. Often, especially during the nineteenth century, these installations were joined by rolling mills and other workshops for different metal trades. Not only were the ironworks in the Urals combined enterprises, with most of the production chain centralised in one place, as opposed to the decentralised Swedish system, the different workshops were also much larger than in Sweden. It was common to build two blast furnaces together while the forges could incorporate as many as twenty hearths and as many hammers.

In Sweden a different, decentralised system prevailed based on many small production units. A large number of small scattered blast furnaces supplied an equally large number of forges with pig-iron. As these furnaces were only used for part of the year, as opposed to Russia where blast furnaces were blown for longer periods, there had to be a very large number indeed. Forges, in both Sweden and Russia, were used more efficiently.

Linked to the differences in the size of the workshops was the number of workers. In the Urals a much larger workforce manned the furnaces and forges. This also created opportunities for a much more highly developed division of labour. From the very beginning of the eighteenth century there was a strictly arranged and hierarchical working organisation, with workers, skilled and unskilled, apprentices, and also supervisors and overseers. A number of different auxiliary workers, such as carpenters and blacksmiths were also part of the workforce at furnaces and forges.

The Ural ironworks thus had a structure that was highly reminiscent of the kind of work organisation that existed after the 'industrial revolution'. Production was centralised and hierarchical while a division of labour was essential. There was one important difference though, and that was that all the workers, as well as their supervisors and foremen, were unfree serfs. The organisation was based not on wage labour, but on compulsory work performed by unfree 'souls', legally defined as peasants.

The social organisation of the industry in Sweden was different from that found in the Urals. It was not particularly uniform. Instead it was a combination of different forms of organisation. The cooperative mode of the pig-iron producing *bergsmän* was linked to the more capitalist-organised ironworks, with wage-dependent forgemen. The ironmaster was not, however, merely an early version of an industrialist, but also an important landowner in a more feudal sense of the word, as later chapters will explore, and large quantities of charcoal were acquired as feudal rents.

The differences in social organisation led to a different matrix of social conflicts. While in the Urals we find a more uniform pattern, with several examples of mass insurrection, the Pugachev rebellion being the most famous example, the Swedish context reveals a much wider spectrum of forms of conflict between different social groups. This chapter has emphasised the 'proto-industrial' friction between *bergsmän* and ironmasters as well as the more 'capitalist' discord between wage-dependent (skilled) workers and ironmasters. The rebellions in Russia sought to undermine the peasants' links with industrial production, rather than to transform conditions within the industry, as in Sweden.

It is important to underline the distinctions between iron-making in Sweden and the Urals, stemming either from social or technological factors, but it is likewise important to stress that developments in the two regions gradually converged, especially towards the mid-nineteenth century. In connection with the undermining of the position of the *bergsmän* it is clear that the ironmasters erected blast furnaces themselves. They sometimes built them in centralised units which also included forges and rolling mills. Integrated ironworks comprising furnaces, large forges using the Lancashire method and rolling mills were a feature of the third quarter of the nineteenth century. At these units the workforce grew and it also took on a structure involving a division of labour and the introduction of foremen and supervisors.

These developments were not matched by an equally rapid process of change in the Urals. The organisation of production within their furnaces and forges remained more or less the same from the early eighteenth century and well into the second half of the nineteenth, with equally slow change in the physical aspects of production. During the nineteenth century, Swedish ironmasters were thus closing the gap in terms of size and structure.

Part II

Integration of the Agrarian Environment in Iron Production

Chapter Four

The Social Organisation of Peasant Work

Maria Ågren with Nina Minenko and Igor Poberezhnikov

> When an ... industry spreads out from the city into the surrounding countryside, it penetrates an unfamiliar environment and economy which receives industrialisation on its own terms ... the country inhabitants ... introduced preconditions which favoured or impeded, and sometimes prevented, reception of industrialisation.[1]
>
> Rudolf Braun
> *Industrialisation and Everyday Life*

Introduction

In Sweden as well as in Russia, the peasantry played a crucial role in early iron production. Peasants carried out a wide range of tasks relating to transportation and the provision of fuel, raw materials and foodstuffs. They were, to use a metaphor, the blood vessels through which the ironworks were kept alive. The image of a social metabolism whereby the agrarian economy interacted with early metallurgy does, however, threaten to make the relationship seem too smooth, at least if we have large-scale industry in mind. In actual fact, making the ironworks' rural neighbours integrated and reliable partners of the industry was a difficult process, beset with conflict and opposition. Ironmasters had to solve a number of problems in order to make production come

1. R. Braun, *Industrialisation and Everyday Life*, Cambridge 1979 (first edition 1960), 10. See also 34.

about.[2] As we shall see, they had to *combine many different solutions* and be prepared to change the combination, if necessary. Thus, it was a general truth that serf labour would often have to be combined with labour used on a contractual basis in Russia, just as charcoal-selling freeholders would have to be involved alongside charcoal-producing tenants in order to make the system work in Sweden. But the actual proportions of these different groups, the actual weight of different work organisations could and indeed did vary from time to time and from place to place. This observation no doubt calls for some caution whenever general statements are made about Russia or Sweden. It is hard to formulate such conclusions with due account taken of all the variations, regional and chronological.[3] On the other hand, a recognition of the multiform character of early modern iron production in both countries can be regarded as one of the main results of the comparative approach, and one which dovetails neatly with recent historiographic developments which emphasise the importance of regional diversity in early modern economic life.[4]

The need to master a wide range of problems when organising work socially was acute for Swedish and Russian owners of large-scale ironworks but it was not unique to iron production. On the contrary. The early modern period witnessed an immense restructuring of many sectors of economic life. Iron production differed only in its pointed military importance and, possibly, in that it was the subject of greater interest on the part of the state. Swedish and Russian ironmasters only had to solve two specific variants of this general problem of early industry. The problems they encountered had to do with the fact that they needed to rely on peasants who were often recalcitrant or even hostile to performing the tasks they were given, and who could muster strength and resistance from peasant traditions and peasant institutions predominant in their region. The relationship between ironmasters and peasants was often marked by contradicting interests, and this made it dynamic. It was this relationship between early industrialisation

2. K.-G. Hildebrand, *Swedish Iron in the Seventeenth and Eighteenth Centuries. Export Industry before the Industrialization*, Stockholm 1992, 98.

3. B. Odén, 'Komparation över tid. Till de långa linjernas metodologi', in *Den kritiske analyse. Festskrift til Ottar Dahl på 70-årsdagen den 5. januar 1994*, Oslo 1994.

4. E.g. *Gewerbe- und Industrielandschaften vom Spätmittelalter bis ins 20. Jahrhundert*, ed. H. Pohl, Stuttgart 1986; P. Hudson, 'The regional perspective', in *Regions and Industries in the Industrial Revolution in Britain*, Cambridge 1989; S. Pollard, 'Regional markets and national development', in *Markets and Manufacture in Early Industrial Europe*, ed. M. Berg, London, New York 1991. See also chapter 1 this volume.

and rural institutions that Rudolf Braun had in mind when he wrote the words quoted above, and it leads to certain strategies for research. In order to understand and compare the situation of Swedish and Russian ironmasters, peasant rationality and peasant institutions have to be analysed. This means that we have to widen the historiographic perspective a little, to include something of what has been said about peasants in previous research.

It has long been assumed by historians and anthropologists that peasants often behave and think differently from other social groups. The idea of a special type of economic behaviour was formulated explicitly in the 1920s by the Russian economist A. V. Chaianov.[5] The 'peasant mode of production' has occasioned considerable debate since then, being in fact one of the roots of the theories about proto-industrialisation. The idea of a special peasant mentality was put forward for example by the anthropologist George M. Foster in the 1960s and has also had wide repercussions.[6] Both approaches set out to explain a certain backwardness among peasants, which impedes their own economic development and also development in general. Foster wanted to show that peasants always consider 'the good things' to be limited by laws of nature, and Chaianov maintained that peasant households are less prone to work more than to satisfy their immediate physical and social needs. They are even said to be reluctant, in fact, to act in a way which means working hard, albeit for profit. On the other hand, they can willingly muster astonishing amounts of energy if necessary to retain hold of their land. This was the mechanism of 'self-exploitation'.

Since then, other scholars have questioned these interpretations, if not the underlying observations. Greater emphasis has been put on the social structures within which peasants live and which may be the true causes of their behaviour and view of the world, not inherent peculiarities in the peasants themselves. Thus, Eric Wolf discerned a wide range of factors which may affect a peasant population, both natural (weather, soil) and social (taxes, wars, laws).[7] The emphasis on social structures has also led to an

5. A.V. Chaianov, *A.V. Chayanov on the Theory of Peasant Economy*, Homewood, Illinois 1966.

6. G. M. Foster, 'Peasant Society and the Image of Limited Good', *American Anthropologist*, 1965, vol. 67, no. 2; cf. J. R. Gregory, 'Image of Limited Good, or Expectation of Reciprocity?', *Current Anthropology*, 1975, vol. 16, no. 1.

7. E. Wolf, *Peasants*, Englewood Cliffs 1966, 78ff. Similar arguments have been put forward in Swedish historiography by e.g. E. Österberg, *Kolonisation och*

acknowledgement of different peasant behaviours. Some peasants have indeed acted as Chaianov's theory would predict, but not all. Many forms of peasant rationality are conceivable. For instance, Jan De Vries discerned two peasant ways of adjusting to external pressure in the form of a rising population. One peasant alternative would, according to De Vries, be to partition farms among all the children and to intensify the exploitation of the soil. In the long run, this alternative would force the peasantry onto the market, to buy grain and to sell their labour to obtain extra incomes. But another peasant alternative would be to specialise in agriculture and to keep farms as large and as compact as possible. In this case, it would be possible to make a profit from agriculture itself and to enter the market as a buyer of labour and a seller of grain.[8] It is clear that the latter alternative is at odds with Chaianov's original idea, since it does not restrict peasant economic behaviour to strategies of sheer survival and 'backwardness'. Rather, peasants are seen as exponents of a new, profit-oriented rationality.

Thinking in terms of two peasant rationalities – at least – allows for tensions among peasants following different paths. That such tensions did exist has been shown empirically for England by C. G. A. Clay, for example. In the early modern period, English rural society was divided into two sectors, one dominated by market-oriented commercial farmers and a more traditional one, upon which the rationality of peasant families put its imprint. In the words of Clay, the activities of the former 'threatened to reduce the resources available to the peasant family farmers, whilst the system of communal rights and communal controls … interfered with the efforts of the farmer to take full advantage of market opportunities'.[9]

The first of De Vries's alternatives – partitioning the land and supplementing the rural economy with various extra sources of income – has been linked to the appearance of early industrial activities. Weaving and spinning were common ways of gaining the extra income necessary to make life possible in regions where partible inheritance was practised and where people had to survive on

kriser. Bebyggelse, skattetryck, odling och agrarstruktur i västra Värmland ca 1300–1600, Lund 1977, and J. Lindegren, *Utskrivning och utsugning. Produktion och reproduktion i Bygdeå 1620–1640*, Uppsala 1980.

8. J. De Vries, *The Dutch Rural Economy in the Golden Age*, London, New Haven 1974, chapter 1. See also M. Sjöberg, *Järn och jord. Bergsmän på 1700-talet*, Stockholm 1993, 196ff.

9. C. G. A. Clay, *Economic Expansion and Social Change: England 1500–1700*, vol. 1, Cambridge 1984, 72.

very small pieces of land.[10] Moreover, access to common rights and common land was extremely important in such regions.[11] But how does work performed by peasants for early iron industries fit into this picture? Should burning of charcoal and transporting of semi-refined products be regarded as parallels to weaving or not? Obviously, the question cannot be answered in the affirmative until we have looked more closely into the conditions under which such work was performed. To what extent was charcoal-burning a duty imposed on the peasantry by a feudal lord/the state, and to what extent was it something they chose to take up of their own free will? Did peasants retain a certain independence as charcoal producers or were they gradually made economically dependent on a merchant or buyer-up? What other resources did peasants possess, with which they could make a living, and did they prefer to use these other resources? The three questions once again emphasise Wolf's argument about the importance of analysing social structures. It is crucial to understand extra-economic and economic compulsion directed against the peasantry and, finally, to grasp the material resources which put peasants in a position to withstand such compulsion.

The questions above hint at *three possible strategies* by which an ironmaster could endeavour to integrate peasants in a work organisation. He could rely on feudal compulsion. He could establish commercial ties. Finally, he could try to deprive the peasantry of the material resources which would have allowed them to choose not to produce charcoal. Once again, we are faced with the potential variety of social organisations.

Even though it is an accepted manner of speech to talk about how peasants 'have chosen to act', as we have done here, it can be misleading since it implies that peasants are to be regarded as individuals. In many cases, however, the analysis must focus on the peasant household or even on a larger peasant community (such as the village). Often, these units or institutions were the real parties interacting with the ironmasters, not individual peasants. It was the internal rules and functions of household and community that put an individual in a position to choose (or not

10. See e.g. J. Thirsk, 'Industries in the Countryside', in *Essays in the Economic and Social History of Tudor and Stuart England in Honour of R.H. Tawney*, Cambridge 1961; P. Hudson, 'Landholding and the organization of textile manufacture in Yorkshire rural townships c. 1660–1810', in *Markets and Manufacture*, 280ff.

11. Cf. J. M. Neeson, *Commoners: Common Rights, Enclosure and Social Change in England, 1700–1820*, Cambridge 1993.

to choose) to work for the ironworks, and it was also those very same rules and functions which could serve as brakes when iron-masters attempted to integrate the rural neighbourhood in his production process. This was shown admirably clearly by Rudolf Braun when he demonstrated how early industry in Switzerland was located in mountain areas where the village communities did not put up barriers (in the form of entry fees) to immigrants. Only there was it possible to recruit and maintain a labour force which could form the basis for a putting-out industry. In the lowlands, material resources were effectively guarded and monopolised by traditional communities.[12] Recently, other historians have also paid closer attention to the way in which agrarian institutions shaped rural industry.[13]

To be able to compare the way peasants and ironworks interacted in two different regions (as we shall do in this chapter), we must be aware that institutions such as the household and the community may have played different roles in different social settings and at different times. It is also necessary to clarify that the contemporary meanings of the crucial concept 'peasant' were not exactly the same in Sweden and Russia.

Comparing Peasants and Peasant Communities in Sweden and Russia

The people whom eighteenth-century Swedes and Russians called 'peasants' were very different from each other. Their socio-economic and legal positions varied to an extent which threatens to mislead the historian who tries to compare them. Indeed, it could even be argued that the only thing that a Swedish and a Russian peasant had in common was the *name* 'peasant', whereas the name referred to totally different realities. In order to make the comparative approach viable, we have to address two problems linked to the concept of 'peasant'. First, there is the trap of a broad versus a narrow definition. Then there is the difference in legal status.

Krest'ianin (peasant) was a broadly defined concept in eighteenth-century Russian. It referred to the overwhelming majority of the population, regardless of occupation, property, residence

12. Braun, *Industrialisation*. See also W. Rösener, *The Peasantry of Europe*, Oxford UK, Cambridge USA 1994, 164.

13. S. Ogilvie, 'Proto-industrialization in Europe', *Continuity and Change*, 1993, vol. 8, no. 2, 166ff. See also chapter 7 this volume.

etc. It could for example be applied to someone living in a garrison town on the periphery of the country, but it could also refer to someone living in an ordinary town. It could be used to talk about someone working at an ironworks. This broad definition meant that the vast majority of the Russian population (approximately 90 per cent in the eighteenth century) belonged to the peasantry, but it also engendered the need for a more precise terminology to designate subgroups within the peasantry. And such a terminology did indeed develop. Russian society was full of different kinds of peasants, just as society at large was full of various kinds of social groups, as Roger Bartlett and Gregory L. Freeze have pointed out.[14] The other side of the coin was that the Russian peasantry were never able to constitute themselves as an estate, in the same sense that, say, Swedish peasants or German burghers were. They remained the somewhat amorphous group consisting of every man who was not ennobled.

For the sake of clarification, a line of demarcation could be drawn between state peasants (including ascribed peasants) and serfs (including *posessionnye* and *votchinnye* peasants).[15] However, it would be unwise to envisage this line as very sharp. There were several intermediate groups. Moreover, the position of the 'free' state peasantry deteriorated during the eighteenth century,[16] especially for those who were used as labour within the industrial sector. At the same time, many serfs were obviously able to act *as if* they had no lord, engaging in trade and holding property of their own.[17] The distinction was thus somewhat blurred, if not totally erased. The fact that many serfs obviously desired to become state peasants, since they were believed to be better off economically and personally,[18] would be incomprehensible if state peasants had been totally equated with serfs. The best way of putting it may be to say that the large Russian peasantry encompassed a vast variety of subgroups, spread over a spectrum of gradually increasing social subordination.

14. R. Bartlett, 'The Russian Peasantry on the Eve of the French Revolution', *History of European Ideas*,1990, vol. 12, no. 3, 397f, and G. L. Freeze, 'The *Soslovie* (Estate) Paradigm and Russian Social History', *The American Historical Review*, 1986, vol. 91, no. 1.

15. Bartlett, 'Russian Peasantry', 398.

16. J. Blum, *Lord and Peasant in Russia from the Ninth to the Nineteenth Century*, Princeton 1961, 277.

17. Blum, *Lord and Peasant*, 292, 299, and Bartlett, 'Russian Peasantry', 398.

18. I. V. Poberezhnikov, *Obshchestvenno-politicheskie vzgliady russkikh krest'ian Sibiri v period pozdnego feodalizma*, Novosibirsk 1989, 64–84, and Bartlett, 'Russian Peasantry', 400.

The Swedish word coming closest to the Russian *krest'iane* is probably *allmoge*, which refers to the vast majority of the people, living in the countryside and making a living at least partly from agriculture.[19] It is clearly this kind of broad definition of peasantry which is taken for granted in theories of a peasant mode of production (in Chaianov's sense of the word) or of peasant life (as discussed by Eric Wolf or G.M. Foster for example). It is also this kind of definition which Russian historians usually apply.

The Swedish word *bonde* (peasant) is normally associated with a narrower definition in, say, eighteenth-century language.[20] *Bonde* was used to designate those rural inhabitants who had a farmstead which was large enough to pay certain taxes.[21] Swedish historians normally conform to this practice and do not include other rural groups when they talk about the peasantry; for these, other terms are used. The term *bonde*, as used by historians, excludes some rural inhabitants, more precisely the semi-proletarian groups which grew considerably in the period 1750 to 1850.[22]

These conceptual differences have important implications for a comparative analysis. According to Russian linguistic habits, peasants who followed the first of De Vries's two alternative paths and became highly dependent on incomes from non-agrarian occupations would still be called peasants, even if they were to lose all access to land. In Sweden, by contrast, such people would no longer be regarded as peasants. Of course there was a 'twilight zone' in which, typically, some peasants/crofters with little land would fall. But someone who had no land at all at his disposal would not qualify as a peasant. The Ural forgemen are a case in point, being still looked upon as peasants because of their social origins. By contrast, Swedish forgemen were regarded as typical non-peasants, despite the fact that they sometimes had access to small plots of land.

No legal differences of the kind that we can find between state peasants and serfs in Russia existed between Swedish rural groups. All inhabitants of Sweden (those named *bönder* and those given other labels) were legally free. No serfdom existed in the early modern period. Of course, this is not to deny that there were

19. *Ordbok över svenska språket utgiven av Svenska Akademien*, Lund 1826– (publication still in progress), A 1021 4c.

20. Ibid., B 3845 1.

21. These farmsteads were called *hemman*, the general fiscal unit since the sixteenth century. Cf. the term *hide*, denoting a standard peasant family endowment which often turned into a fiscal concept; Rösener, *Peasantry*, 21, 116f.

22. See C. Winberg, *Folkökning och proletarisering. Kring den sociala strukturomvandlingen på Sveriges landsbygd under den agrara revolutionen*, Göteborg 1975.

important differences between the Swedish peasantry and the nobility, for example, which sprang from the privileges of the latter and which entitled them to certain prerogatives. Noblemen could (for instance) hold land practically free from ordinary taxation and have their legal cases tried directly in the Royal Courts of Appeal instead of in peasant-dominated hundred courts. It is also important to note that only some Swedish peasants, the freeholders, wielded political influence through their representatives in the Diet. There were thus considerable social, political and economic differences within the early modern Swedish population, but, unlike in Russia, there were no legal differences attached to individuals within the rural population at large.[23]

Having thus emphasised the difference in legal status between Swedish peasants in general and Russian serfs, however, we have to take one step back again. The Russian region with which we are going to deal here, the Urals, was atypical, since state peasants predominated.[24] They accounted for almost 90 per cent of the entire Ural peasantry in the seventeenth century, and in the nineteenth century they still made up approximately 75 per cent of the population in some Ural provinces.[25] This should be compared with the national picture: in 1719, the state peasantry constituted approximately 22 per cent of the entire Russian population.[26] This means that the average Ural peasant differed from the average west Russian one, having a somewhat better social and legal position. State peasants were regarded as subjects of civil and public law and state functionaries tried to intervene when they made complaints about their situation. State peasants were also granted a seat on the 1767 Law Commission.[27] They enjoyed personal freedom and paid taxes themselves – unless they were ascribed to an ironworks.

23. Swedish historians often make a distinction between *skattebönder* (freeholders), *frälsebönder* and *kronobönder* (tenants of the nobility or of the Crown). These terms are actually somewhat misleading, since they suggest differences in personal status. The proper distinction should be made between the three different kinds of land (*skatte, frälse, krono*) which peasants used.

24. Bartlett, 'Russian Peasantry', 401.

25. *Istoriia Urala s drevneishikh vremën do 1861 g.*, Moscow 1989, 198, 299–301; B. G. Pliushchevskii, 'Imushchestvennoe i sotsial'noe rassloenie krest'ianstva gubernii Bol'shogo Urala v 1-i polovine XIX veka', in *Problemy sotsial'no-ekonomicheskogo razvitiia derevni Srednego Povolzh'ia v period feodalizma*, Kazan' 1986, 90.

26. H. Bagger, 'Reformernas tidevarv – och reaktionens', in *Ryssland, ett annat Europa*, Stockholm 1995, 119.

27. M. Raeff, *The Well-Ordered Police State. Social and Institutional Change through Law in the Germanies and Russia, 1600–1800*, Yale 1983, 243f.

Living in community with others was of the utmost importance in olden times. This has been underlined most strongly by medievalists. Medieval Europe was indeed characterised by a vast range of community forms,[28] and one important to peasants – the village community – is believed to have arisen in the High Middle Ages.[29] In Russia, the peasant community (*obshchina, mir*) was of paramount importance and massive scholarly attention has been devoted to its different manifestations.[30] The *obshchina* has become famous most of all for its power to redistribute land within the village. The following quotation sums up some of the most widespread notions about the Russian village community:

> The Russian husbandman was not only bound to his lord but also to the *mir,* a village community characterised by joint possession of land and cultivation by individual families. The terrain turned over to individual members of the community, according to the size of the family, was periodically taken back and redistributed.[31]

Here, the author equates community with village community. Yet it should be kept in mind that *obshchina* was a more general term, which could refer to communities of different sizes and at different levels of society. The village community was but *one* form of communal peasant organisation in Russia,[32] just as Western Europe knew a number of forms of community. Furthermore, west Russian conditions are taken for granted when the village community is linked to practices of land redistribution. An *obshchina* could choose different forms of communalism; redistribution of land was but one. There were parts of Russia where the *obshchina* no doubt existed and played an important role, but where land redistribution was not practised at all or was only a marginal phenomenon.[33] Finally, the function of

28. See e.g. A. J. Gurevitj, *Feodalismens uppkomst i Västeuropa,* Stockholm 1970, 138ff.; G. Althoff, *Verwandte, Freunde und Getreue. Zum politischen Stellenwert der Gruppenbildungen im früheren Mittelalter,* Darmstadt 1990; D. Lindström, *Skrå, stad och stat. Stockholm, Malmö och Bergen ca. 1350–1622,* Uppsala 1991.

29. Rösener, *Peasantry,* 158f.

30. See e.g. *Land Commune and Peasant Community in Russia. Communal Forms in Imperial and Early Soviet Society,* ed. R. Bartlett, London 1990. See also the works of V. A. Aleksandrov, e.g. *Sel'skaia obshchina v Rossii (XVII-nachalo XIX v.),* Moscow 1976, and M. M. Gromyko, e.g. *Traditsionnye normy povedeniia i formy obshcheniia russkikh krest'ian XIX v.,* Moscow 1986.

31. Rösener, *Peasantry,* 26. See also 168ff.

32. J. Channon, 'Regional Variation in the Commune: the Case of Siberia', in *Land Commune.* In the Urals, there was a village community and a *volost'* community.

33. Channon, 'Regional Variation'; M. Lewin, 'The *Obshchina* and the Village', in *Land Commune;* Rösener, *Peasantry,* 169 (about the Ukraine and Belorus); N. A. Minenko,

the *obshchina* changed over time. Originally, it may have been more of a unique peasant institution (though sources are sparse),[34] but in the late eighteenth century it became more firmly subordinated to state and landlord interests.[35] The land redistribution practices proved to be beneficial from a fiscal point of view, since they guaranteed that all households would be in an economic position to fulfil their feudal obligations. After the reforms of Alexander II, the communities regained some of their independence.[36]

In the Urals and western Siberia, the state claimed rights of ownership to all land, but the *obshchiny* exerted actual power over the territory (arable, woods and fishing waters). The state did attempt to lay down rules about how land was to be used and to allot certain areas to newly arrived colonists, but very often these orders were simply disregarded by the local peasantry, who regarded all land which they had traditionally used as their own. Practices of land redistribution within the Ural *obshchina* are unknown. Instead, the *obshchina* fulfilled other tasks, all with the aim of securing each family a living.[37] It was also a unit of jurisdiction and local self-government, even though its independence was somewhat restricted by the local ironmasters.[38] Finally, it was the unit from which state taxes were exacted and military recruits were taken.[39]

Turning to Sweden, the peasant community was also important. Swedish peasants organised themselves communally, sometimes through the village and sometimes through the parish. It seems that the village was more important as the basis for communities in southern and south-eastern Sweden, where settlement was denser and villages larger. But in more northerly parts of Sweden there

---

Russkaia krest'ianskaia obshchina v zapadnoi Sibiri XVIII-pervaia polovina XIX veka., Novosibirsk 1991. See also R. D. Givens, 'To Measure and to Encroach: The Nobility and the Land Survey', in *Russia and the World of the Eighteenth Century*, eds R.P. Bartlett, A.G. Cross and K. Rasmussen, Columbus, Ohio 1986, 537 (and the comment by W. M. Pinter on page 549), about villages that were divided between several owners and the difficulty of bringing about a redistribution in such cases.

34. Cf. R. Pipes, *Russia under the Old Regime*, Harmondsworth 1977, 18, in which the author maintains that there is no evidence for the *obshchina* before the eighteenth century.

35. Minenko, *Russkaia*, 50–88. See also S. Hoch, *Serfdom and Social Control in Russia*, Chicago, London 1986, 188.

36. Bartlett, 'Russian Peasantry', 402; Rösener, *Peasantry*, 25, 169.

37. Minenko, *Russkaia*, 89–102.

38. Ibid., 125–134.

39. Ibid., 163–204.

were often no villages proper, only hamlets of, for example, two, three or four farmsteads. Some farmsteads were even completely isolated. Here, local self-government was concentrated at the level of the parish, which could comprise areas that were vast by European standards. The situation could vary, however, and no simple map can be drawn.

Work and life within the village (regardless of form and size) was regulated by the national legal code. It laid down rules, often in amazing detail, as to what rights each member of the village was to enjoy and, also, what his duties were. In addition to this, village members could choose to have certain by-laws, which became legally binding after having been announced and sanctioned by the local hundred court.[40] Laws and by-laws were both aimed primarily at regulating the use of land in a situation where compact and individually owned farms were very rare.

Beginning in the sixteenth century, the Swedish fiscal system was reorganised so as to lay the main burden of regular taxes on the peasant farmstead (*hemman*) and not on larger communities, such as groups of households or even entire parishes, as had been the case previously. Assessing each farmstead's capacity to pay taxes presupposed a good knowledge of where taxable resources were to be found and necessitated a survey of the whole country. This was carried out thoroughly and successfully in Sweden, whereas in Russia taxes remained tied to individuals and land was not surveyed until the late eighteenth century.[41] To retain control of taxable resources, the state had to lay down rules as to how farmsteads were to be used. The Swedish state strongly opposed any alienation of plots of land from farmsteads, since this was believed to endanger the prosperity of each farmstead and its capacity to continue paying taxes. Similar processes were obviously at hand in other parts of early modern Europe, where princes strove to prohibit the subdivision of farmsteads (e.g. in the German territories).[42] By contrast, Russian rulers enforced the importance of the community by charging it with responsibility for the taxes of its individual members. As will be shown in the following, this is important in understanding

40. W. Ehn, *Mötet mellan centralt och lokalt. Studier i uppländska byordningar,* Uppsala 1991.

41. Cf. C. Peterson, *Peter the Great's Administrative and Judicial Reforms. Swedish Antecedents and the Process of Reception,* Stockholm 1979, 297–302, and Givens, 'To Measure'.

42. Rösener, *Peasantry,* 116f.

the social organisation of peasant work in the Swedish and Russian iron industries.

Peasant Production for the Iron Industry

The Needs of the Industry

In chapter 2 it was shown how, in the early modern period, the Swedish iron industry relied heavily, and for a long period of time, on small-scale pig-iron production by *bergsman* communities. Likewise, the Ural metalworks were initially dependent on ore and blooms supplied by the local peasantry, and in spite of prohibitions this way of organising ore extraction lingered on. V. I. De Hennin, the manager of all the state works, urged his subordinates (in the 1720s) to be 'nice, not harsh, towards people selling ore, in order to encourage them'.[43] P. S. Pallas, who travelled in the Urals (in the early 1770s), noted that near the Kosotursk ironworks even women and little children could easily dig soft, dry iron ore and sell it to the ironmasters.[44]

However, the local peasantry were not only needed to procure the basic raw materials. They also catered for other essential needs of the ironworks. They provided foodstuffs for the iron-making regions, for example.

When large-scale industry was first set up in the Urals, there were attempts to make the peasants supply agricultural products to the manufactures as a way of discharging their fiscal duties.[45] But as early as the eighteenth century, artisans and workers were provided with grain that ironmasters had bought in nearby peasant settlements. The management of the ironworks had special agents who traded in the villages, their main instruction being not to overpay the peasants, but to buy their products at market prices.[46] To prevent potential problems over food supplies, mining authorities and private

43. V. De Hennin, *Opisanie ural'skikh i sibirskikh zavodov 1735 g.*, Moscow 1937 (written in 1735), 89.

44. P. S. Pallas, *Puteshestvie po raznym mestam Rossiiskogo gosudarstva*, St Petersburg 1786, Part 2, Book 1, 108.

45. B. B. Kafengauz, *Istoriia khoziaistva Demidovykh v XVIII–XIX vv.*, Moscow-Leningrad 1949, 63.

46. S. M. Tomsinskii, 'Rassloenie ural'skoi derevni v sviazi s razvitiem manufaktury (pervaia polovina XVIII v.)', in *Ezhegodnik po agrarnoi istorii Vostochnoi Evropy*, Leningrad 1972, 142f.

ironmasters even established grain stores,[47] and Ural ironmasters played an active part in the grain trade in the Zaural'e region.[48]

In due course, the role of peasant food suppliers grew. They brought grain, meat, fat, fish, honey etc., and sold it to the people living at the ironworks. The authorities wished to encourage this trade and exempted these peasants from paying taxes.[49] Neighbouring peasants were not able to meet all the needs of the Ural works, though,[50] which is why residents of remoter agricultural regions appeared in the works' markets. In the second half of the eighteenth century, stable ties were established between the agrarian grain-producing *uezdy* and the mining ones.

Turning to Sweden, it is clear that when large-scale iron production was introduced in Bergslagen in the early seventeenth century, ironmasters relied heavily on the inter-regional peasant trade to obtain sufficient amounts of foodstuffs. Unlike the Urals, there were hardly any good grain-producing districts within Bergslagen in the seventeenth and eighteenth centuries, and peasants from parishes outside Bergslagen proper, which were endowed with better soils, would sell their grain to the population at the mining sites. It was not until the early nineteenth century that some parishes in southern Bergslagen achieved levels of grain production which reduced the need for inter-regional trade.[51]

47. For instance, the Main Board of the Siberian and Kazan' ironworks ordered (in 1734) the managers of the Kamensk, Krutikha, Kamyshlov and Alapaevsk districts to purchase 1,000 *chetverts* of rye, oats and barley each. This was done in view of the small reserves in the Ekaterinburg state store, and the grain was paid for by state money (GASO, F. 572, Op. 1, D. 8, L. 914–920); N. A. Demidov also had his own store in Dalmatov in the early 1760s. A. A. Kondrashenkov, *Krest'iane Zaural'ia v XVII-XVIII vekakh.*, Part 2, Cheliabinsk 1969, 180f.

48. A. S. Cherkasova, 'Sotsial'no-ekonomicheskie sviazi gornozavodskikh tsentrov i dereven' Urala v seredine XVIII v.', in *Derevnia i gorod Urala v epokhu feodalizma: problemy vzaimodeistviia*, Sverdlovsk 1986, 26–35.

49. For instance, in 1734 the ascribed peasants Anika Khorkov and Matvei Kazantsev reported to an office of the Main Board of the State Works that on the way to Ekaterinburg, carting food products, they had been required to pay tax. 'But according to the instructions, such food products should be tax-free', the peasants complained (GASO, F. 572, Op. 1, D. 8, L. 888).

50. A. S. Cherkasova has shown that in the middle of the eighteenth century the Nizhnii Tagil' ironworks needed about 1,638 tons (100,000 *pud*) of grain per year. The state-owned ironworks at Kushva, Turinsk, Polevskoi, Baranchinskii, Verkh-Isetsk, Kamensk and the Empress Anna's together required about 1,321 tons (80,660 *pud*) of grain per year. Cherkasova, 'Sotsial'no-ekonomicheskie', 32f.

51. M. Isacson, *Ekonomisk tillväxt och social differentiering 1680–1860. Bondeklassen i By socken, Kopparbergs län*, Uppsala 1979; M. Ågren, *Jord och gäld. Social skiktning och rättslig konflikt i södra Dalarna ca 1650–1850*, Uppsala 1992.

Peasants from distant regions like Småland, near the Danish border, would take their oxen all the way up to Bergslagen and sell them there. These animals were important not only as food. Their hides were used to make bellows and ropes, with which water and ore could be wound up from the deeper mines.[52]

In Russia, there was also a demand for ropes, pitch, tar, cords, mats, boxes for coal, bricks, candles etc., all of which was met by the local peasantry,[53] as was the demand for cloth, shoes and soap.[54] Though domestic industry flourished all over Russia at this time, it was certainly outstanding in the Urals and partly so because of the needs of the iron industry. No doubt the establishment of a large-scale iron industry offered valuable opportunities to make extra income for some members of the local peasantry. Domestic industry and large-scale industry were interdependent, interacting through networks.

There are indications that Russian ironmasters actively tried to free themselves from this overwhelming dependence on the peasantry. This was done first and foremost by gradually monopolising ore resources. Prohibitions against unauthorised ore digging within the ironworks domains were laid down in the early eighteenth century. These were not immediately successful,[55] but later on new regulations were more strictly enforced.[56] Outside the boundaries of the ironworks ore digging was not prohibited, which accounts for the continuing trade in ore, but since the richest ore sites were generally located within these domains, peasants had a much less favourable position from this point on. They were faced with increasing competition from the large-scale works, which reduced small-scale iron production to the role of catering for local needs. By contrast, Swedish ironmasters did not as a general rule attain exclusive rights of ownership of ore resources until the nineteenth century.[57] Many ironmasters remained

52. J. Myrdal and J. Söderberg, *Kontinuitetens dynamik. Agrar ekonomi i 1500-talets Sverige*, Stockholm 1991, 483ff.

53. Hennin, *Opisanie*, 352; Cherkasova, 'Sotsial'no-ekonomicheskie', 24–25; GASO, F. 105, Op. 2, D. 29, L. 8-8 ob.

54. N. S. Popov, *Khoziaistvennoe opisanie Permskoi gubernii*, Perm' 1804, Part 1, 259–285, Part 2, 351–398; TsGIA, F. 383, Op. 27, D. 50, L. 137; *Svod materialov po kustarnoi promyshlennosti v Rossii*, St Petersburg 1874, 617; V. I. Baidin, 'Formirovanie burzhuazii v srede gosudarstvennogo krest'ianstva Srednego Urala vo vtoroi polovine XVIII v.', in *Krest'ianstvo Urala v epokhu feodalizma*, Sverdlovsk 1988, 81, 88.

55. A. A. Preobrazhenskii, *Ural i Zapadnaia Sibir' v kontse xvi- nachale xviii veka*, Moscow 1972, 272–273.

56. See chapter 7.

57. See chapter 7.

heavily dependent upon pig-iron deliveries by *bergsmän*. It was not until the pressure of international competition forced ironmasters to improve the quality of their products that the extraction of ore and production of pig-iron were integrated with the ironworks.[58]

As for food supplies, the situation was different, many ironmasters being evidently quite content to continue buying grain. At some Swedish ironworks, though, the ironmasters are known to have exhorted their tenants to reclaim new arable land and to ameliorate their farming techniques with the aim of enhancing the self-supporting potential of the plant.

Apart from supplying ore, blooms, food and various goods, the local peasantry would be assigned the crucial tasks of producing charcoal and carrying out transportation. The latter was no less important than the former. Here, however, we will focus our attention on charcoal.[59] More to the point, this chapter will concentrate on the fundamental design of three methods commonly used by Swedish and Russian ironmasters to make the peasantry – this recalcitrant human material – procure fuel. In chapter 5, the focus will be on the actual work process.

Landed Estates and Serfs, Ironworks and Tenants

The establishment of large landed estates, worked by peasants who were increasingly burdened by feudal obligations (such as *corvée* duties) and finally enserfed, was a general trait of social development in Eastern Europe after the late medieval crisis. At the same time, peasants in the West freed themselves from serfdom and gained stronger rights of possession.[60]

Understanding the way a feudal landed estate worked is thus vitally important if we want to analyse the way iron production was organised socially in the Urals. This was pointed out first by V. V. Adamov and, later, by T. K. Gus'kova. They both emphasised that Russian ironmasters had no other experience and no other alternative than that of the landed estate, the *votchina*. Therefore, they could either embark on iron production within the boundaries

58. T. Omberg, *Bergsmän i hyttelag. Bergsmansnäringens utveckling i Linde och Ramsberg under en 100-årsperiod från mitten av 1700-talet*, Stockholm 1992, 122, 124.

59. See Hildebrand, *Swedish Iron*, 109ff., on the transport system in Sweden.

60. Rösener, *Peasantry*, 104ff.; *The Brenner Debate: Agrarian Class Structure and Economic Development in Pre-industrial Europe*, eds T. H. Aston and C. G. E. Philpin, Cambridge 1985, 10–63; W. Kula, *An Economic Theory of the Feudal System. Towards a Model of the Polish Economy 1500–1800*, London 1976.

of an existing landed estate *or* they could attempt to build up an estate around a new iron production plant. The latter alternative was the one most commonly applied in the Urals, where there were few estates proper around 1700,[61] apart from those of the Stroganov family. In both cases, though, the feudal element was enhanced, epitomised in the institution of serfdom.

In the Urals, the serf system was closely linked to the establishment of ironworks. Some serfs were moved by their masters from their patrimonial estates in western Russia. Others had been purchased especially for the needs of the ironworks. In 1745 there were 408 labourers at the Suksun works of Akinfii Demidov, 342 of them serfs who had been moved from Demidov's estates in central Russia.[62] At the same time, there were over 11,000 serfs at all the privately owned Ural ironworks. According to law, the right to have patrimonial serfs (*votchinnye*) only belonged to ironmasters from the nobility. There are some cases of non-noble ironmasters simply disregarding this explicit prohibition (although this still needs to be investigated further), but in general the existence of such a privilege was a severe obstacle to non-noble owners and it explains why, in 1721, the state initiated a specific form of serf labour legally available to these owners as well. These *posessionnye* serfs were looked upon as part of the landed property and could only be used for iron production.[63]

A considerable proportion of the patrimonial Ural serfs lived on the lands of their ironmaster, making a living in a traditional peasant way, but obliged to fulfil various tasks at the ironworks when called upon. In the middle of the nineteenth century, about 24 per cent of the peasant population of the Perm' province belonged to this category.[64] These serfs resembled the ascribed state peasants in many ways: they were compelled to work at the ironworks and they fulfilled similar tasks, like building new production units,

61. V. V. Adamov, 'Ob original'nom stroe i nekotorykh osobennostiakh razvitiia gornozavodskoi promyshlennosti Urala', in *Voprosy istorii kapitalisticheskoi Rossii: problema mnogoukladnosti*, Sverdlovsk 1972, 225–256, and 'Ob original'nom stroe Urala', in *Nauchnaia sessiia, posviashchënnaia problemam mnogoukladnosti*, Moscow 1969, 71–108; T. K. Gus'kova, 'Okruzhnaia sistema kak forma organizatsii ural'skoi gornozavodskoi promyshlennosti', in *Sotsial'naia i proizvodstvennaia organizatsiia metallurgicheskoi promyshlennosti Urala v XVIII-XIX vv.* Forthcoming.

62. *Ironmaking in Sweden and Russia. A Survey of the Social Organisation of Iron Production before 1900*, eds G. Rydén and M. Ågren, Uppsala 1993, 78.

63. Ibid., 77f. This conditional form of tenure was also applied to landed property. See chapter 7.

64. F. S. Gorovoi, *Padenie krepostnogo prava na gornykh zavodakh Urala*, Perm' 1961, 95.

chopping firewood, burning charcoal, digging ore and transporting various goods. But unlike ascribed peasants, the state did not regulate the use of such serf peasants at all (which, incidentally, shows that state peasants were still somewhat better off in terms of their legal position). Private ironmasters could use these peasants as they thought fit and they could also transfer their other serfs to this category, if for instance the situation on the market changed and called for a stepping-up of production. The system was thus capable of considerable flexibility. On the patrimonial estate of the Vsevolozhskis, there were between 8,700 and 8,800 male peasants in 1796, 1,500 of them permanently occupied at the ironworks (approximately 15 per cent). In 1842 the population had risen to 9,094 men with 5,780 permanently employed in iron production (approximately 39 per cent). The relative growth of the latter figure obviously cannot be explained by the general increase in population, but reflects reorganisations undertaken by the ironmasters.[65]

It is important to notice the discrepancy between the occupation of these serfs and their estate status. They often lived at the works themselves, studied an occupation and became skilled specialists. They gained a living mainly through this job, and only supplemented it with horticulture and some cattle breeding, but in spite of this they were still regarded as peasants.[66]

Sweden had no system of serfs and no Swedish ironmasters are known to have tried to enforce such a regime. There was, however, a distinctive group with which the Russian serfs can be compared, namely the tenants or crofters using land that belonged to the ironworks. The comparability lies in the fact that, in both cases, ironmasters were able to force these people to perform various duties because of certain *property rights* with which the ironmaster was endowed. In the Russian case, ironmasters had far-reaching property rights in land and in (some) human beings. In the Swedish case, they had property rights which were considerably more restricted and only applied to land, but by monopolising the use of this landed property the ironmaster was able to attract a workforce. In return for the usufructuary rights given to the peasant household he thus received feudal dues, often rendered in the form of charcoal. Therefore, the methods by which Swedish tenants and Russian serfs were dragged into the sphere

65. V. V. Mukhin, 'Rabochaia sila na predpriiatiiakh ural'skoi gornozavodskoi votchiny Vsevolozhskikh v pervoi polovine XIX veka', in *Iz istorii rabochego klassa i krest'ianstva Permskogo kraia*, Perm' 1965, 24–27.

66. A. S. Cherkasova, *Masterovye i rabotnye liudi Urala v XVIII veke*, Moscow 1985, 152.

of iron production were comparable, although not identical. In Russia, property rights were used both to monopolise the use of land and to compel people to work; in Sweden, by contrast, only the first function was present.

Moreover, tenants and serfs were regarded in both countries as the core of the labour force. Other ways of attracting labour (ascription or commercial ties) were used not because they were more efficient but because there were too few serfs and tenants to meet the demands of the ironmasters. Charcoal obtained from Swedish tenants, for instance, was very cheap, and an ironmaster who owned vast areas and hence employed many tenants had a considerable advantage over one who had little land and few tenants.[67] Likewise, Russian ironmasters who were not noblemen and who were consequently barred from owning *votchinnye* serfs were strongly disadvantaged, compared with their aristocratic colleagues. Here, the Russian state tried to alleviate their situation by introducing *posessionnye* serfs.

For every Swedish ironmaster, the people living on farmsteads that belonged to the ironworks were the core group within the charcoal-producing organisation, even if they were generally too few to be the sole source of this kind of labour. To take just one example, the small ironworks of Grängshammar (in the heartland of Bergslagen) received its charcoal from forty-five tenants and sixty-one freeholders in 1690. In 1826 the ratio was 46 to 107.[68] Thus, the core group of tenants had remained relatively stable and small.

The tenant group comprised not only peasant households, in the narrow Swedish sense of the word, but also crofter households, i.e. households that did not – either any longer or yet – have such a large farmstead at their disposal as sufficed to be taxed separately. Indeed, crofts became an increasingly important constituent of the settlement structure of Bergslagen, and other parts of Sweden where metallurgical production took place, in the seventeenth and eighteenth centuries. Crofts were often ideal from the point of view of the ironmaster, since they were too small to allow the inhabitants to make a living from them. By promoting crofts and preventing excessive clearing of land around these crofts, the ironmasters were able to subject the crofter households to a sort of economic compulsion which was very difficult to bring about in relation to peasants who had larger holdings. Crofter households were under a double compulsion to work for the ironworks: the

67. Hildebrand, *Swedish Iron*, 88.
68. Ågren, *Jord och gäld*, 225.

terms of their contract demanded a certain amount of charcoal, and their lack of resources made them 'willing' to take on more tasks, remunerated somewhat better than the ones stipulated in the contract. Once again, at the Grängshammar ironworks 73 per cent of the tenants and crofters produced so much charcoal that it exceeded the value of their feudal obligations in 1690 (not counting all the other tasks that they also performed for the ironworks). In 1750, 85 per cent produced 'too much'.[69]

The 'accommodating' attitude of tenants and crofters was not only the result of the economic and legal compulsion to which they were subjected. There may have been a psychological dimension to it as well. Many tenants and crofters were former freeholders (or had freeholding ancestors) who had lost their land because of indebtedness to the ironmasters. Having once experienced an adverse fate at his hands, they were more likely to comply with his wishes in the future. The conflict of interest between ironmaster and crofter should not be exaggerated, though. Often, the inhabitants of the farmsteads/crofts were unable to secure a living from agriculture alone, without resorting to extra charcoal production and other forms of assistance to the ironworks. Ironmasters could also approve land reclamation and urge their tenants to concentrate on agriculture, in order to make the ironworks self-sufficient in grain.[70] Thus, interests could coincide, but they could also diverge.

From the point of view of the ironmaster, the problem was that crofters obviously had to be allowed to clear *some* land to begin with, otherwise it would have been quite impossible for them to make a living and to stay on. Consequently, many ironmasters are known to have encouraged clearance of land to a certain extent.[71] The legal code, which stipulated certain compulsory rights and duties of landlord and tenant, also required the tenant to do a certain amount of maintenance each year and this rule could be interpreted as an exhortation to reclaim land.[72] But where the limit was to be drawn – between land reclamation that was in line with the interests of the ironworks owner and that which threatened to make crofters self-sufficient and, hence, too independent – varied

69. Ibid.

70. M. Isacson, 'The decline of peasants' dependence on iron industry. 17th to 19th century', in *Metallurgical Works and Peasantry. Problems of Social Organization of Industry in Russia and Sweden in Early-Industrial Period*, Ekaterinburg 1992, 212.

71. Ågren, *Jord och gäld*.

72. Jordabalken 16, Byggningabalken 6. Both in the national legal code of Sweden.

from case to case. The limit could not only be defined in terms of acreage. It was also dependent upon number of cattle, opportunities to fish and hunt, access to market places where home-made craft goods could be sold etc.

Towards the end of the eighteenth century, many Swedish crofters had developed a wide range of sources of income. They were definitely not in a position which would have allowed them to choose the alternative of specialising in profit-oriented agriculture, as discussed by De Vries, but having reclaimed land and improved the soil they could at least envisage new possibilities: supporting themselves with incomes from many different sources, including agriculture. In this situation, being tied to the ironworks and compelled to burn charcoal was beginning to trouble them.

In the light of this, the frequent occurrence of disputes over land ownership around 1800 is hardly surprising. It became vitally important for ironmasters to assert their property rights to the land that had long been cleared and tilled by their tenants and crofters, simply because *if* they failed to do so, they would have virtually no means of enforcing the kind of work organisation they preferred. With tenants/crofters better equipped in terms of material resources, the need for extra-economic compulsion grew. At the same time, it became equally crucial for crofters to try to protect their property rights to the land, which they had come to regard as their own because of the labour which they and their ancestors had invested in it.[73]

Disputes of this kind are the agrarian counterparts of the kind of conflict which has been observed within the crafts sector.[74] The master forgeman regarded the tools and the iron he had produced as his own property, which he could sell to whomever he liked. He did not perceive himself as someone selling his labour to an employer in return for wages. According to his interpretation of the legal relationship between himself, the raw materials, the product and the buyer/employer, he had self-evident property rights in the fruits of his labour. Likewise, the crofter regarded the grain he had sown *and* the land he and his ancestors had tilled as his own,

73. Ågren, *Jord och gäld*, 232ff.; M. Ågren, 'Land and Debt. On the Process of Social Differentiation in Rural Sweden, circa 1750–1850', *Rural History* 1994, 5 (1).

74. W. Reddy, *The Rise of Market Culture. The Textile Trade and French Society, 1750–1900*, Cambridge 1984; A. Florén, *Disciplinering och konflikt. Den sociala organiseringen av arbetet. Jäders bruk 1640–1750*, Uppsala 1987; M. Sonenscher, *Work and Wages. Natural Law, Politics and the Eighteenth-Century French Trades*, Cambridge 1989. See also chapter 3 this volume, section on 'Work Discipline and Conflicts'.

and he would cite the labour invested when arguing his right. There was – still – a strong feeling that property rights could ensue from labour and from ancient usage.[75]

Village Ascription and Charcoal Districts

Since there were generally too few Russian serfs and Swedish tenants in a situation where iron production was expanding, new methods of attracting peasant labour had to be devised. In the Urals, ascription of entire peasant villages was put into practice from the beginning of the eighteenth century, forcing state peasants to work for the ironworks. In Sweden, by contrast, freeholding peasants (*skattebönder*) were made to trade with ironmasters on unfavourable terms through a system of 'charcoal districts'. In spite of the many differences between these two solutions, it is nevertheless correct to regard them as comparable, since they were *functionally* similar. They provided the means of procuring additional labour, besides the core groups attracted (or bound) by property rights.

Ascription proved a major change within Ural peasant society and was to taint the relationship between ironworks and their agrarian surroundings from the early eighteenth century until the ascription system was abolished from 1807 to 1814. In western Russia, ascription of peasants had been practised as early as the seventeenth century, but never to the extent which came to be the case in the eastern parts of the country. The scarcity of labour in general and the original lack of serfs put Ural ironmasters in an impossible situation and forced the state to carry out mass ascription of state peasant villages (and to a limited extent monastic ones too)[76] to the ironworks. In this way, the task of producing the commodities, the fuel and raw materials previously bought from the peasants was now turned into a compulsion and imposed upon the villages in the neighbourhood.

Until 1721 the labour of ascribed peasants was used mainly at state works and at the enterprises of the Demidov family, but later the right to use ascribed state peasants was given to other private ironmasters as well. Consequently, the number of ascribed peasants grew rapidly, reflecting the general expansion of Russian iron production in the eighteenth century. According to the first census

75. See chapter 7, and M. Ågren, *Att hävda sin rätt. Synen på jordägandet i 1600-talets Sverige, speglad i institutet urminnes hävd.* Stockholm 1997.

76. In actual fact, there is only one example of the ascription of a monastic village.

(1719), there were just 25,000 males working at the state-owned plants. By the middle of the 1760s there were 58,000 males at the state-owned works and about 63,000 at the private ones. Prior to the reforms of 1807–1814, about 250,000 ascribed peasants lived in the region. In some parts virtually all the villages were allotted to ironworks.[77] Whenever ironmasters requested more labour, the Russian state allotted new villages to this purpose.

In terms of their legal status, ascribed peasants continued to be state peasants: they were recognised as subjects of civil and public law, had the right to trade and to sign contracts, could purchase property in their own name, and rise socially (i.e. become members of other estates or social strata). In other words, they were not serfs. But they were obliged to perform various duties at the ironworks in payment of the poll-tax (unlike ordinary state peasants) and they were jointly responsible for the taxes of all the villagers, having thus to work not only for themselves but also for those who had left the village for one reason or another (death, escape, illness etc.). The state relinquished its direct tax rights regarding these villages and transferred them to the ironmasters. The only costs these incurred for using this form of labour were the sums to be paid in poll-tax. These sums were fairly stable, but the amount of labour demanded from the peasants was not.

Originally the sum of poll-tax to be worked off at an ironworks was equal to 1 rouble 10 copecks (70 copecks for the poll-tax itself and 40 copecks for the quit-rent). Under an order issued by the Senate in 1760 (12 October), the sum of the quit-rent that was part of the poll-tax was increased by 60 copecks, making the sum of money to be worked off at ironworks 1 rouble 70 copecks per capita.

Work was paid for at a fixed, official rate, laid down by the state, but over the course of time this rate changed. When fixing the official prices, the state apparently paid some attention not only to ironmasters, but also to the peasantry. But even after the increase in wage rates the situation did not improve for the peasants, since taxes rose much faster.[78] This led to considerable distress and the

77. In the Perm' province more than 80 per cent of the population of state villages served their duties at ironworks, and in the Perm', Solikamsk, Kurgan, Okhansk, Osa and Ekaterinburg *uezdy* practically no one was excepted. *Istoriia Urala*, 301, 303; S. S. Smirnov, 'Sel'skoe khoziaistvo pripisnykh krest'ian gornozavodskogo Urala na rubezhe XVIII-XIX vv.', in *Gosudarstvennye krest'iane Urala v epokhu feodalizma*, Ekaterinburg 1992, 145.

78. PSZ-1. Vol. VII 4425; Vol. XVIII 13303; Vol. XX 14878; V. I. Semëvskii, *Krest'iane v tsarstvovanie imperatritsy Ekateriny II*, St Petersburg 1901, Vol. 2, 308f; Blum, *Lord and Peasant*, 308–314.

state tried to alleviate the hardship experienced by ascribed peasants, especially after the riots that took place in the 1750s and 1760s. Thus, during the periods when peasants were busy with spring and autumn agricultural work, they were freed from all duties towards the ironworks.[79] By a Manifesto proclaimed in 1779 (21 May), wage rates were doubled in comparison with those laid down by a previous decree of 1724. The Manifesto also released ascribed peasants from tasks such as charcoal-burning and ore extraction, which were regarded as particularly onerous, but retained many of their other duties.[80]

The actual need for labour, as estimated by the ironmasters, also influenced the duties laid upon the ascribed peasantry. The legislation of the first quarter of the eighteenth century made it possible to force peasants to perform extra work at the ironworks, i.e. more than that demanded by the poll-tax, on the sole condition that this work be rewarded in accordance with official price rates. The managements of the works made considerable use of this possibility, especially at the private ones, causing massive unrest among ascribed peasants. The period 1754 to 1767 witnessed a wave of peasant uprisings at the private works. The causes were manifold (e.g. a higher poll-tax, ascription of new and far-off villages, transfer of previously state-owned ironworks to noblemen), but all were connected in some way with the ascription system. In 1763, an attempt to limit the right of ironmasters to force peasants in this way was made,[81] but being no more than a recommendation it was quite often ignored.[82] Private works owners do not seem to have observed the letter of the law very rigorously; as a consequence, ascribed peasants at private works were exploited more than those at the state-owned ones.

As we have shown, the amount of work demanded from the ascribed peasantry was defined by the state (through the poll-tax

79. According to the Statement concerning work at the ironworks, issued by the Main Board of Ironworks on 23 October 1721. A. S. Orlov, *Volneniia na Urale v seredine XVIII veka. K voprosu o formirovanii proletariata v Rossii*, Moscow 1979, 56f; N. I. Pavlenko, *Razvitie metallurgicheskoi promyshlennosti Rossii v pervoi polovine xviii v.: Promyshlennaia politika i upravlenie*, Moscow 1953, 480f.

80. B. DeHart, 'The Pugachev Revolt and Its Effect on the Industrial Workforce of the Urals: a Reconsideration', in *Russia and the World*, 524, 528.

81. Known as 'Statement for ascribed peasants of the late General Field Marshal and holder of decorations Shuvalov, of Izhevsk and Votkinsk ironworks', of 9 April 1763, developed by A. A.Viazemskii, the head of the committee of inquiry into the causes of peasant unrest. See also Raeff, *Well-Ordered*, 225, concerning the role of Viazemskii.

82. Semëvskii, *Krest'iane*, 405–408.

and the official price lists) and by the needs of ironmasters. But to be able to evaluate how the system actually worked, we have to look at how these duties were actually exacted. It is here that the importance of the peasant community (*obshchina*) becomes so clear.

The managements of ironworks, civil authorities and representatives of the community were all involved in organising the work of ascribed peasants. Apparently, the civil administration alone was not able to check that taxes were being paid.[83] Therefore, community representatives were elected from the peasantry for a definite period of time. From these representatives, the ascribed peasants elected a *raskladchik* (a person in charge of apportioning work), who kept 'apportionment books' in which the volume of work expected of each particular community was distributed among its members. These books were used as guidelines when the community decided which households were to go to the ironworks next. The wealth and size of the households were also taken into consideration. The peasants chosen were divided into groups of ten, who elected a foreman.

Peasants belonging to such a group were bound by mutual responsibility, which meant that they had to work for those group members who did not report for work or who absconded before the work had been completed. On his return from work, the peasant had to show the community elder a receipt from the ironworks office confirming that work had been duly performed. The only way an individual peasant could free himself from the tasks given to him was to hire a stand-in. This strategy was used, because peasants detested not just the forced labour as such, but also being compelled to travel long distances and being away from home.[84] This was a common complaint from East European peasants subjected to *corvée* duties,[85] and it should be kept in mind whenever the possibly beneficial effects of early industrialisation on family life are discussed.[86]

An evaluation of the ascription system as a whole must hinge upon how we view the role played by the community and by its elders and representatives. Community representatives were the

83. Minenko, *Russkaia*, 163–181.

84. Blum, *Lord and Peasant*, 312f.

85. Rösener, *Peasantry*, 114.

86. R. Braun, 'The impact of cottage industry on an agrarian population', in *The Rise of Capitalism*, ed. D. Landes, New York 1966, 53–64. See also the critical discussion in G. Gullickson, 'Love and power in the proto-industrial family', in *Markets and Manufacture*, 214ff.

last links within the state and/or feudal bureaucracy, since they ensured that duties were being duly performed. They were in a position of indirect dependence on the ironworks administration.[87] On the other hand, they were also ordinary peasants, who suffered from the ascription system just like any others. Often it was the community elders or representatives who took the lead in popular protests, who played an active part in forwarding complaints, and who organised defence against attacks by troops. Their position was thus somewhat ambiguous. At times their attitude towards their superiors was one of complete loyalty: they sent fellow villagers to the ironworks and betrayed the plans of rebels. In such cases, however, the community often had them replaced by worthier individuals, which may indicate that the community controlled its representatives and not the other way around. If this impression is correct, the ultimate link in the ascription system must have been weak, from the point of view of the ironmasters.

As the system of village ascription extended to remote areas, far from the central iron-producing region, travelling and being away from home became an even greater burden for the peasantry. To avoid this, many peasants definitely resorted to hiring substitutes instead of going to the works themselves. This option was quite expensive, though, since labour prices were substantial. In 1767, the ascribed peasants from the Kazan' district complained about the duties imposed on them by the ironmaster Count Shuvalov. 'We suffer ruination, especially from hiring others to do the work', they wrote, going on to point out that the treasury paid each man twenty copecks (i.e. the official price), but to hire a substitute they had to pay 1 rouble and 50 copecks.[88] In this way, ascribed peasants could evade the actual work at the ironworks, but instead they had to 'sponsor' the production of iron by paying the difference between the poll-tax and the market price for labour.[89]

In general, the situation of ascribed peasants deteriorated during the eighteenth century. There was a general tendency among many ironmasters to treat them as *de facto* serfs: to increase their obligations and to make them stay at the ironworks permanently.[90] The state only made occasional efforts to alleviate their

87. Minenko, *Russkaia*, 7–50.

88. *From Supplication to Revolution. A Documentary Social History of Imperial Russia*, ed. G. L. Freeze, New York, Oxford 1988, 88ff.

89. See also R. Portal, *L'Oural au XVIIIe siècle. Étude d'Histoire Économique et Sociale*, Paris 1950, 271ff.

90. Ibid., 46, also 258ff. (on ascribed peasants more generally).

hardship. This situation gave rise to recurring instances of resistance, sabotage and rebellion, rooted in the fact that from the very beginning the peasantry had regarded the ascription system as an unfair imposition. No doubt the Russian state showed a wavering and ambivalent attitude towards this group. On the one hand, the state peasantry were meant to be the equivalent of the Swedish *skattebönder*: freeholders with no other lord than the king/tsar, vested with legal rights and responsible for their own taxes.[91] On the other hand, in the Urals they were allowed to be used by ironmasters in a way which did not differ much from the way real serfs were used.

The utterly negative attitude of the state peasants made the system work less well. Peasants could be made to go to ironworks under the threat of capital punishment, but they could not be made to work diligently and well. As early as in 1738, i.e. shortly after the initiation of the system, it was severely criticised by Vice-Chancellor Osterman, who tried to prove that it was of paramount importance to abolish ascription because of its poor efficiency.[92] Although his advice was not heeded then, the problem of poor quality seems to have been one of the main reasons why the system was finally abolished. Roger Portal has pointed to another reason. Often, ironmasters would use the ascribed peasants more than the poll-tax level entitled them to. For this extra work, payments seem to have been much higher than the official prices for ascribed work.[93]

In Sweden, the method of obtaining extra charcoal by purchase was put into practice in various ways. Ironmaster and peasant could meet in a relatively free market situation, but more often the peasant would be compelled to sell to a specific ironmaster whom the state had endowed with monopsonistic rights. The second alternative (the system of charcoal districts) dominated in the eighteenth century and it is particularly interesting in a comparative context since, as we will argue, it is the phenomenon that can most aptly be compared with the Russian ascription system. In both cases, it was possible for the ironmasters to extend the range of their power to peasants whom they were otherwise unable to

91. See Bagger, 'Reformernas tidevarv', 119.

92. T. V. Safronova, 'Zakonodatel'nye istochniki o formirovanii i polozhenii rabochei sily na metallurgicheskikh predpriiatiiakh Rossii i Urala vo vtoroi polovine XVIII veka', in *Polozhenie i bor'ba trudiashchikhsia Urala XVIII-nachala XX v.*, Ekaterinburg 1987, 9–10.

93. Portal, *L'Oural*, 312, 383.

reach, and to procure more charcoal than they would have done by merely relying on their ordinary power resources. The main difference was that the Swedish solution did not emphasise the role of the peasant community as strongly as the Russian one did. Nor was the Swedish approach quite as severe on the peasant.

Before a new ironworks could be set up, permission had to be obtained from the Swedish Board of Mines (Bergskollegium). The Board made careful inquiries into the situation in the region where the plant was projected. Special attention was paid to forests and to the attitudes and interest expressed by the local peasantry. An explicit peasant promise to sell charcoal in the future made the mining administration adopt a more benevolent view of the project. But there were other factors to be taken into consideration, too. A new ironworks could not be set up in the close vicinity of other works, since this was thought to give rise to competition for charcoal between the ironmasters. For instance, it was particularly difficult to get permission to build a new plant in southern Dalarna (Kopparberg), since this was where the great copper mine at Falun and the copper-smelting works at Avesta were situated, which were both regarded as being of special interest to the country.[94]

Even if permission to start iron production was finally granted, it did not mean that the ironmaster could force the peasantry within his allotted area to do anything at all. The permission merely gave the ironmaster an exclusive right to purchase. He was the only one who had the right to buy charcoal within a certain area, but he had no say as to how much charcoal was to be produced within this area. A royal decree from 1735 stated that apart from any duties that might have arisen as a consequence of their self-inflicted indebtedness, 'the peasantry are not obliged to produce a fixed amount of charcoal each year'.[95] This was hardly surprising, of course, considering that Swedish peasants were legally free and could not be told what to do. The ironmaster could thus only claim an exclusive right to buy all the charcoal that the peasants chose to produce, *not* tell them how much charcoal they had to supply. The Swedish state never explicitly gave ironmasters the means of controlling the charcoal production of the freeholders. No detailed regulations pertaining to trade in charcoal were ever laid down at a central state level. Further, the seventeenth-century

94. *Industri og bjergværksdrift. Privilegering i Norden i det 18. århundrede*, Oslo, Bergen, Stavanger, Tromsø 1985, 278ff., 292ff.

95. 17 October 1735; quoted from Hildebrand, *Swedish Iron*, 86f; S. Montelius, *Arbetare och arbetarförhållanden 1600–1815. Fagerstabrukens historia V*, Uppsala 1959, 174.

statutes talk about general prohibitions against 'illegal charcoal trade', but they never define this expression.[96]

Even if the sources do not elaborate the details of the system, however, it no doubt included a considerable measure of force directed against the peasants and made the relationship between peasantry and ironmaster very asymmetric in reality. For instance, the price of charcoal was strictly regulated and kept at a low level. Sigvard Montelius, who investigated these matters in great depth, concluded that the peasantry might not have resented the monopsonistic system so much, had it not been for the miserly prices.[97]

The system of charcoal districts opened the way for a wide spectrum of disobedience and disputes. Some forms of disobedience are discussed in the next chapter, but there were other methods and conflicts as well. A crucial point was often whether or not peasants could be said to have made legally binding promises to produce certain quantities of charcoal. If the ironmaster was able to prove this, he was immediately placed in a much better position, since in this way the vague monopsony rights granted by the state could be given a more precise meaning. The ironmaster thus had a very strong motive for interpreting every single action of the peasants as a sign of a promise. As was pointed out previously, the peasantry in the neighbourhood had often expressed an accommodating attitude to the mining administration when the ironworks asked for permission to start. Such an attitude was actually demanded by the Board of Mines before granting the privilege. But once permission had been granted, ironmasters were prone to interpret this generally accommodating attitude as a legally binding promise.

For instance, when the Siljansfors ironworks received its charter in 1738, it was said in the preamble to the deed that the ironmaster Waltherson had made a contract in 1729 with the peasantry of the three villages of Vika, Vinäs and Utmeland. In this contract the peasants had accepted the establishment of the ironworks (which affected their landed property) and had promised to deliver firewood, timber and charcoal at a 'reasonable' price. In 1738 a new contract was made which specified the price rates. Very soon, however, the ironworks was reported to be having problems with its contractors, who would not deliver the goods because of the allegedly low prices. In 1745, a representative of the

96. Montelius, *Arbetare*, 161.
97. Ibid., 165; Hildebrand, *Swedish Iron*, 86.

Board of Mines warned the peasants not to carry on with their stubborn resistance, as did the civil authorities. Apparently, the disagreement had its roots in differing interpretations of the actual meaning and legal consequences of the contract of 1729.[98] Similar cases can be cited from many other ironworks.

From the point of view of the ironmasters the system of charcoal districts often proved insufficient as a means of securing a steady supply of charcoal, the problem being, of course, that the peasants were not really forced to do what the ironmasters wanted them to do. Swedish ironmasters became dependent on the neighbouring peasantry for charcoal, just as their Ural fellows had been dependent on their local peasantry for ore and pig-iron (in the early eighteenth century). In both cases ironmasters developed strategies to free themselves from this dependence, but the Ural solutions were not feasible alternatives in Sweden. Ore resources and woods were difficult to monopolise in the same way as in Russia, mainly because of the strong legal and political position of Swedish peasants. And enserfment of the peasantry was politically impossible.

However, it was not just a question of not being able to subject the peasantry to various kinds of force. Leading statesmen (like Christer Bonde and Erik Fleming in 1652) consciously *chose* to build the social organisation of work within the iron industry upon the free peasant household, since this was thought to reduce costs in an unsurpassed way. Peasant households were believed to be prepared to work on fairly unprofitable terms (e.g. producing pig-iron or charcoal), as long as they retained their personal freedom and their homesteads.[99] This belief of Bonde and Fleming reflects observations of peasant behaviour similar to those made, for example, by Chaianov. They also dovetail with Pat Hudson's conclusion that the household unit of production could be very viable, because of the agricultural holding it possessed and because of the self-exploitative behaviour of early industrial families.[100]

If enserfment was not socially and politically possible nor, it seems, desired, Swedish ironmasters had to develop other strategies to ensure a steadier supply of large quantities of charcoal

98. Montelius, *Arbetare,* 164f.

99. Sjöberg, *Järn och jord,* 22.

100. Hudson, 'Landholding', 285f. See also P. Kriedte, H. Medick and J. Schlumbohm, *Industrialization before Industrialization. Rural Industry in the Genesis of Capitalism,* Cambridge 1981, 46; A. Florén and G. Rydén, *Arbete, hushåll och region. Tankar om industrialiseringsprocesser och den svenska järnhanteringen,* Uppsala 1992, 114f.

from freeholders. Here, two main lines of action may be discerned, lines that were frequently combined in real life. First, an endeavour to bind the freeholders to the ironworks. Second, the strategy of forming a political lobby group.

To achieve the first goal, various methods were devised. The most radical one was no doubt to eliminate every other way of making a living that peasants had available to them, making charcoal-burning and transportation their chief occupation. If the ironmaster was successful in his endeavour to introduce this economic compulsion, the peasantry had no choice but to comply, if they wanted to survive. Was this strategy actually put into practice? It is well known that Swedish ironmasters often sought to have the authorities prohibit peasant saw mills. The overt argument for this would be that such mills hindered the water supply to the forges, but we know that sawing was also believed to compete with and endanger charcoal production.[101] Swiddening was opposed for the same reason.[102]

In general, however, this radical line of action was seldom carried to fruition. Traditional peasant households often derived their incomes from so many and such various sources that it was extremely difficult to attack them all. Moreover, the ironworks needed not only charcoal, but also grain. It was of vital importance that the population at and around the ironworks were well fed. Likewise, many horses were kept in the area and access to grazing was important. Opposing all other peasant activities, including agriculture, would not be in line with the overall aims of ironmasters. Therefore, the attitude adopted by an individual ironmaster to different peasant activities was closely linked to the 'charcoal profile' of his ironworks. The more charcoal the ironmaster could obtain through feudal dues from his tenants, the less he had to rely on strategies of preventing saw mills, swiddening, land reclamation etc.

The ironmasters also used other methods of binding the peasantry to the works. Contracting peasants was one way, making them indebted to the ironworks was another. In both cases it was a question of introducing a legal compulsion. Debts that were overdue had to be repaid and a defaulting debtor ran the risk of losing his property, in accordance with rules laid down in the legal code. Either the indebted peasant failed to pay, lost property and, in the

101. *Iggesunds Bruk 1685–1869*, Iggesund 1985.
102. Montelius, *Arbetare*, 167f.

end, saw no other means of surviving than to start producing charcoal. He could also be so afraid of losing his real property that he was ready to comply with every wish of the creditor/ironmaster.

Emphasising the potential usefulness of debts as a means of gaining greater control of charcoal production does not mean to say that all ironmasters strove to make their charcoal suppliers indebted, nor does it mean that all debts were used to put pressure on the debtor. Very often, debts were cancelled on the death of the debtor. It has also been shown that systematic cancellations of debts could occur while the debtors were still alive.[103] This meant that the payment for charcoal deliveries was raised in retrospect. It would be dubious to depict debts as 'dangerous' or socially subversive, considering the fact that indebtedness seems to have been a central and essential characteristic of many social relations in early modern times. Therefore, when we discuss the role of debts in making peasants do what ironmasters wanted them to do, the stress must be on their *potential*, on the fact that with the support of the law and the courts a creditor could use the existence of debt to exert power. Whether he actually did so or not varied.[104]

The establishment of the ironworks at Horndal in 1652 was followed by massive indebtedness and loss of landed property among the freeholders in the vicinity. Around 1690 about sixty tenants lived on the estate of the ironworks, many of whom were previous freeholders who had lost their land as a result of indebtedness.[105] At eighteenth-century Forsbacka the ironmaster never had to pursue this strategy quite as far as at Horndal, since the heavily indebted peasantry were 'willing' to agree to supply certain quantities of charcoal each year, being afraid to lose their land.[106]

The second strategy pursued by Swedish ironmasters was to organise as a group with common interests. This line of action is noteworthy, since it contradicts one of the basic assumptions

103. Ågren, *Jord och gäld*, 203, 219.

104. On the topic of indebtedness and social change see e.g. Ågren, *Jord och gäld*, 31ff.; Ågren, 'Land and Debt'; Kriedte, Medick and Schlumbohm, *Industrialization before*, 58; A. Bhaduri, *The Economic Structure of Backward Agriculture*, London, New York 1983, 73; B. Poulsen, '"Alle myne rent", bondekredit i 15-1600-tallet', *Dansk Historisk tidsskrift*, 1990; B. A. Holderness, 'Credit in a Rural Community, 1660–1800. Some Neglected Aspects', *Midland History*, vol. III, no. 2, 1975; M. Spufford, *Contrasting Communities. English Villagers in the Sixteenth and Seventeenth Centuries*, Cambridge 1974; P. Blickle, *Landschaften im alten Reich. Die Staatliche Funktion des gemeinen Mannes in Oberdeutschland*, München 1973; Hildebrand, *Swedish Iron*.

105. Montelius, *Arbetare*, 163.

106. Ibid., 165.

underlying the policy of the Board of Mines. It was regarded as necessary to regulate the charcoal trade, the fear being that otherwise ironmasters would compete in an uncontrollable manner for this resource. By increasing the price level, they would outrival each other and make Swedish iron too expensive to sell successfully on the international market. No doubt such illegal competition did occur now and then.[107] But the assumption was also proved wrong in many cases. Ironmasters could display a sense of a common interest and an ability to cooperate. They could agree to keep prices low and not to buy charcoal from each other's peasants, i.e. from peasants living within each other's charcoal districts. Through these local combines, the original fragility of the state-created system of charcoal districts was made considerably more robust.[108]

Similar attempts to organise among ironmasters are known from other parts of Europe as well, e.g. from Britain.[109] In the Low Countries the ironmasters had made a mutual agreement among themselves (in the 1750s) to restrict annual production.[110] In Russia, there were no such attempts in the eighteenth century, probably because the private ironmasters were so few and often had direct access to the Emperor or the Empress, and it was not until the end of the nineteenth century that Russian ironmasters formed the Congress of Owners of Ironworks.[111] When the Swedish traveller Samuel von Stockenström visited St Petersburg in the 1780s, he suggested that both Sweden and Russia would profit if all ironmasters, in Sweden and Russia, could be persuaded to pursue the same policy of restricting iron production.[112] No such agreement was reached, however.

The method of organising was even more striking at the national level. In 1747 the Ironmasters' Association (Jernkontoret) was founded. It was headed by a board (Brukssocieteten) and parallelled by an institution known as the Ironworks Diets (Bruksriksdagar). Both were early examples of lobby groups set up to influence the

107. Ibid., 175.

108. Ibid., 185.

109. C. Evans, 'The Corporate Culture of the British Iron Industry 1650–1830'. Paper presented at the Eleventh International Economic History Congress in Milan 1994.

110. A. Florén and G. Rydén, 'A Journey into the Market Society. A Swedish Preindustrial Spy in the Middle of the Eighteenth Century', in *Societies Made Up of History,* eds R. Björk and K. Molin, Stockholm 1996.

111. See chapter 8, this volume.

112. S. von Stockenström, 'Anmärkningar rörande ryska järntillverkningen och järnhandeln', Bergskollegii Arkiv, RA (National Archives, Stockholm).

members of the Diet and make them sympathise with the goals of the ironmasters.[113] The need to organise such a lobby group becomes fully comprehensible if we consider the fact that many Swedish ironmasters did not have direct access to the Diet (unless they were members of the noble estate). By contrast, the group whose charcoal trade the ironmasters wanted to control, the freeholders, were members of the peasant estate and sent their representatives to the Diet. Thus, Swedish freeholders possessed the formal political power which many ironmasters lacked, and they used this influence to restrict iron production in the eighteenth century.[114]

Just like the Russian system of village ascription, the system of charcoal districts and regulated charcoal prices was abolished in the first half of the nineteenth century. After some partial reforms, the entire iron industry was exposed to free market conditions in 1846, allowing freeholders to sell charcoal to whomever they liked and at a price agreed by the buyer and seller.[115] However, there are clear indications that the prices paid to freeholders for their charcoal had already risen substantially in the late eighteenth and early nineteenth century.[116]

Hired Labour

In the nineteenth century, the social organisation of peasant work changed in the two regions under discussion. Both in Bergslagen and in the Urals, hired labour was used more extensively in the field of charcoal production. On the surface, the two systems seem to have converged as the ascription system and the charcoal districts were abolished. There was one important difference, though. In Russia hired labourers were often (but not always) unfree, whereas in Sweden they were not. Capitalist relations started to put their imprint on the Swedish situation, whereas the Russian iron industry retained a peculiar mixture of free and unfree wage labour.

When large-scale iron production had begun in the Urals, attempts to organise work according to the principle of free hire proved to be unsuccessful, mainly because the region was still very sparsely populated. Ironmasters consequently switched to

113. See chapter 8, this volume.

114. This point is put forward most of all in P.-A. Karlsson, *Järnbruken och ståndssamhället. Institutionell och attitydmässig konflikt under Sveriges tidiga industrialisering 1700–1770,* Stockholm 1990.

115. B. Boëthius and Å. Kromnow, *Jernkontorets historia,* Stockholm 1947–1968, part II, 445ff, 744ff, 783ff.

116. Ågren, *Jord och gäld,* 222.

the use of ascribed peasants or serfs. But as time wore on, hired labour again became increasingly important, coming from three sources. First of all, ascribed peasants were interested in freeing themselves from their personal obligations by hiring a substitute whom they could send to the ironworks. Secondly, some serfs would sell their labour to an ironmaster, with the permission of their own lord. Finally, the ascription system was abolished and the ascribed peasants replaced by 'permanent workers' (*nepremennye rabotniki*).[117] In 1811, shortly before the abolition of the system, there were 252,000 male ascribed peasants in the Ural mining industry. Shortly afterwards, in 1815, they had been replaced by 14,600 permanent workers. In 1850, they numbered 48,200.[118] The group thus diminished, but at the same time it was used more intensively by the ironmasters.

Letting one's serfs as labour at someone else's ironworks could be a way of making them profitable.[119] Many Russian landlords accepted this kind of arrangement, which gave them steady incomes from their property without their having to organise some kind of production themselves. Just like those landlords who consented to their serfs engaging in domestic industry, they were in a sense serf-owning *rentiers*. In reality, the serfs of such landlords were able to act as if they were free, and they were looked upon as free wage labourers by the ironmasters. This freedom, however, was conditional and dependent on their master's good will and explicit permission. The permanent workers, on the other hand, were not regarded as serfs since they were free to trade, sign contracts, marry whomever they liked etc. The only restriction imposed on them concerned their mobility. They were forbidden to move from one workplace to another, being bound to 'their' ironworks for the rest of their adult life. In principle, they were free – but the bargaining power usually conferred on free labourers (at least at times when their labour was required by employers) was drastically reduced, to the advantage of the ironmasters.

The increased use of hired labour in the Ural metal industry illustrates the blurred borders between different social categories in Russian society, to which we have already drawn attention.[120]

117. Portal, *L'Oural*, 275, 382f, 391.

118. S. G. Strumilin, *Istoriia chërnoi metallurgii v SSSR*, Moscow 1954, t. 1, 285, 335, 351–358; *Istoriia Urala*, 428.

119. DeHart, 'Pugachev Revolt', 522f.

120. See above, section 'Comparing Peasants and Peasant Communities in Sweden and Russia'.

Servile wage labourers were not subject to the property rights of their employer and were looked upon as free, despite their having a master somewhere. Permanent workers were free in all respects but one, which made them much more stable and immobile than the former group.

Hired peasant labour was used to carry out the same kinds of tasks as other types of peasant labour, i.e. burning charcoal, transporting goods etc. From the point of view of the peasant household, the extent to which they chose to offer their labour for sale depended on the general economic situation of the village: its capacity to provide food through agriculture and to pay state taxes or feudal dues. S. S. Smirnov has shown that in what was termed the 'mining area' of the middle Urals, agriculture did not meet the requirements of the peasant economy because of its unfavourable climate and also as a consequence of expansion of the metallurgical sector, which had deprived traditional peasant households of land as early as the end of the eighteenth century.[121] In the ascribed settlements of the Ekaterinburg, Krasnoufimsk, Kungur, Irbit and Verkhotur'e *uezdy*, arable farming declined up to 1803. In its place, stock raising often became the basis of the economy for many ascribed peasants, since it left enough time for forced work or for extra work, based on free hire.[122]

Even before the middle of the nineteenth century, Swedish tenants had sometimes worked extra for the ironworks, beyond their feudal obligations (as has been mentioned). A mixture of feudal and capitalist features existed, with a kind of 'proto-wage labour' concealed and couched in a feudal setting. Wage labour in the proper sense of the word was also applied occasionally, but it was not really introduced systematically until the nineteenth century.[123] Wage labourers who made a living solely by producing charcoal (*kolare*) were then introduced at many ironworks. At the same time, systematic experiments to improve charcoal-burning techniques were carried out. Moreover, charcoal-burners appeared at the time the system of charcoal districts was abolished, just as *nepremennye* workers replaced ascribed peasants. Charcoal-burners were often recruited from the tenant and crofter population – just as some of the first tenants and crofters had been recruited from the ranks of indebted freeholders some two hundred years earlier.

121. See also chapter 7, this volume, on traditional uses of forests in Siberia.
122. Smirnov, 'Sel'skoe', 145–158.
123. Hildebrand, *Swedish Iron*, 90.

The transition from tenant to charcoal-burner was gradual and not always easy to perceive. Charcoal-burners would often receive their payment in kind: they could make use of a cottage at which some cattle could be kept. It might seem as if these people differed from tenants only in that they had smaller farms and less land. A comparison of the contracts which were concluded between the ironmasters and their subordinates makes the differences apparent, though. For instance, at the small ironworks of Grängshammar a tenancy contract was signed in 1692, by which the tenants were required to pay an annual rent of thirty-three vessels of charcoal and four days' work. They were also required to take good care of the land entrusted to them, and to reclaim more land. In 1832, a contract was signed regarding the same farmstead. This time, the tenant was obliged to supply 80 vessels of charcoal annually, a substantial increase compared with the late seventeenth century. Explicit prohibitions against becoming indebted were included. The actual work to be done in the forests was regulated in detail, including the length of the working day. No felling was allowed to start until the works staff had given their permission. Likewise, the tenant was strictly forbidden to reclaim new land without the prior consent of the ironmaster.[124]

As should be obvious from this example, the 'tenant' in 1832 was bound to produce much more charcoal than his predecessor in 1692. He was also subject to a number of detailed regulations which diminished his freedom of action and reduced his scope for decision-making. Finally, the ironmaster consciously strove to prohibit what he deemed an unnecessary preoccupation with agriculture. The intention was to make the household concentrate on charcoal production in a way which had no precedent in the seventeenth century contract. Thus, the tenant had been more or less clandestinely transformed into a charcoal-burner. This gradual change of status is also reflected in the many different words used to describe these people. The old concepts of tenant and crofter were still applied, but at the same time 'charcoal-burner' and 'labourer' can also be found in the sources.[125]

The appearance of charcoal-burning wage labourers should be seen in connection with concomitant changes within the sphere of agriculture. In the early nineteenth century it had become possible for many Swedish peasants to specialise as farmers. Even though

124. Ågren, *Jord och gäld*, 251ff.
125. Ibid., 250.

Bergslagen was not outstanding in this respect, it has been shown that in some parishes reclamation of land was carried out on a stunning scale. In By parish, arable increased by 325 per cent from the middle of the eighteenth to the middle of the nineteenth century.[126] In some parts of Tuna parish, a comparable development took place.[127] Peasants/farmers now produced a grain surplus which could be sold on a market and used to feed wage labourers, e.g. at the ironworks. At the same time, these farmers were no longer as interested in burning charcoal or supplying other forest products. Probably they would often be glad to leave it to those who had little or no land of their own.[128] Thus, wage labour and specialised commercial farming had common roots: land reclamation, opportunities to act on the grain market, and increasing social differentiation within the Swedish agrarian population.

The same seems to be true of the Urals. Roger Portal has emphasised that the ascription system was extended to the grain-growing parts of the southern Urals in the 1750s, exactly at the time when prices for agricultural products started to rise. To be ascribed to the ironworks then did not merely mean to have to leave home and family. It also meant having to neglect one's land at a juncture at which time spent on agriculture was proving more profitable. It is against this background of an economically tempting alternative that the recalcitrant attitude of many ascribed peasants should be seen.[129] Likewise, it should be kept in mind when we evaluate their participation in the Pugachev rebellion, when we see their willingness – and *capacity* – to pay their substitutes high wages, when we hear them complain about it being 'impossible to do our work at home', when they have to ship pig-iron 'at the best time of the year, when we harvest the spring wheat and make hay'.[130] These may not always have been the words of a group of destitute peasants, clinging to their meagre soils. They could also be the words of well-to-do peasant farmers, more annoyed by the form of the tax – being forced to work away for practically nothing at all – than by its size.

126. Isacson, *Ekonomisk tillväxt*, 82.
127. Ågren, *Jord och gäld*, 104, 109, 113, 196f, 215f.
128. Isacson, *Ekonomisk tillväxt*.
129. Portal, *L'Oural*, 388.
130. *From Supplication*, 89, 90.

Chapter Five

CHARCOAL

Production and Transport

Maths Isacson with Igor Poberezhnikov

Charcoal was the principal fuel used in the early industrial production of iron, and was predominant in the Urals and in Bergslagen long after the British ironworks started using coal as fuel from the end of the eighteenth century. This had been made possible by the puddling process. It is possible that the iron producers' great dependence on charcoal prevented the concentration and centralisation of iron production, as has been claimed.[1] The early industrial production of iron was, as has been described, spread over a wide area, with an obvious advantage; the decentralised production structure prevented the over-exploitation of the forests. However, this also increased the demands on transportation and made the iron industry heavily dependent on the people in the surrounding agrarian countryside.

Rising production brought with it an increased demand for charcoal. Fuel supplies became more and more of a problem for the ironworks owners during the eighteenth century. Firstly, there was a tendency for the price of charcoal to rise. Secondly, the forests closest to the works were devastated, and thirdly there were protests from the charcoal producers, that is to say peasants and crofters. The protests could be both spontaneous and organised, and they had to be dealt with in order for the works to be able to maintain stable iron production. The basic problem was on the whole similar in both regions, but the effects were to some extent dissimilar. In this chapter we will study the problem in some detail and also examine the measures taken by the works owners and the state in Russia and Sweden. The

1. See the debate covered in chapter 3, this volume.

emphasis will be on their attempts to change the social and technical organisation of the work.

The Nature of the Problem: The Consumption of Charcoal in the Urals and Bergslagen

The growth of iron production during the eighteenth century resulted in a rapid increase in the consumption of charcoal. Admittedly, there is a shortage of reliable figures on the consumption of charcoal in the various works or in total over a period of time in each region, but the scraps of data available to us clearly show that huge amounts were required. At the same time, consumption varied from ironworks to ironworks.[2] In 1723 charcoal accounted for 45 per cent of the total production costs in five furnaces investigated in the Urals.[3] At the end of the century the total annual production of iron in the Urals required more than one million tons of charcoal.[4] In the middle of the nineteenth century approximately 1.59 million cubic *sazheni* were burnt per year in the Urals.[5] At the end of the century 2.46 million tons of wood were used to meet the fuel requirements of the iron industry in the Urals. Approximately a third was consumed in the production of pig-iron.[6]

In the Urals the production of one ton of bar iron at the end of the eighteenth century and in the middle of the nineteenth century normally took between five to six tons of charcoal. Consumption was often greater.[7] The figures available for Sweden indicate that the consumption of charcoal was at least as great. The total Swedish

2. S. G. Strumilin, *Chërnaia metallurgiia v Rossii i v SSSR. Tekhnicheskii progress za 100 let*, Moscow-Leningrad 1935, 184–185, 189 and 193.

3. Strumilin, *Chërnaia metallurgiia*, 147–149.

4. The production figures can be found in *Ironmaking in Sweden and Russia. A Survey of the Social Organisation of Iron Production before 1900*, eds G. Rydén and M. Ågren, Opuscula Historica Upsaliensia 12, Uppsala 1993, 55 and in R. Portal, *The Slavs*, Studies in World History, London 1969, 158, as well as in B.R. Mitchell, 'Statistical Appendix', in *The Fontana Economic History of Europe*, vol. 4:2, ed. C. M. Cipolla, third edition, Glasgow 1976, 773.

5. Dankoviskii, 'Ob uglezhzhenii na Ural'skikh kazënnykh zavodakh v nastoiashchee vremia', *Gornyi zhurnal* 1858, part 1, 455.

6. A.V. Dmitriev, *Osnovnye napravleniia razvitiia ekonomiki Urala v poreformennyi period 1861 – 1900 gg.*, Sverdlovsk 1991, 16–28.

7. At Kyn ironworks at the beginning of the nineteenth century one ton of pig-iron required two tons of charcoal, while one ton of bar iron required a further three tons. See Strumilin, *Chërnaia metallurgiia*, 184–193.

production of bar iron at the beginning of the nineteenth century, 60,000 tons, required at least 360,000 tons of charcoal.[8]

Producing this considerable amount of fuel from the forests and transporting it to furnaces and forges was the work of the peasants and crofters in the countryside surrounding the works (Figs. 5.1 and 5.2). As the charcoal-burning work was mainly carried out during the colder half of the year the iron industry could take advantage of the seasonal shortage of work in the peasant households. For this reason the peasants were not exactly opposed to producing charcoal, but at the same time they were not prepared to do so under just any condition.

With manual tools, horses and sledges, an appreciable number of working days were necessary per ton of charcoal, at a rough estimate more than ten working days as late as the middle of the nineteenth century.[9] One ton of bar iron thus required almost seventy days of charcoal-burning.[10] The 200 tons of bar iron that an average-size ironworks in Sweden produced annually as late as at the beginning of the nineteenth century required approximately 14,000 days of charcoal-burning. This work took the peasants five to six months per year, including the transportation of the charcoal from the forests to the furnaces and forges.[11] Between 100 and 120 people were involved each year in producing the required amount of charcoal. This can be compared with 15 to 20 workers in the bar iron forge (forgemen and helpers). The total Swedish annual production of bar iron in the middle of the eighteenth century, approximately 50,000 tons, required almost 3.5 million days' work, which was carried out, in round figures, by 30,000 peasants and crofters. It should also be pointed out in this connection that Sweden had 2.3 million inhabitants at the time.[12] Almost 10 per cent of

8. G. Arpi, *Den svenska järnhanteringens träkolsförsörjning 1830–1950*, Jernkontorets Bergshistoriska Skriftserie 14, Stockholm 1951, 99, 209.

9. Three days per 19.8 hectolitres and 66.7 hectolitres per ton of charcoal (off-loaded measure). Arpi, *Järnhanteringens träkolsförsörjning*, 209; Strumilin, *Chërnaia metallurgiia*, 66, 68–69.

10. K.-G. Hildebrand, *Swedish Iron in the Seventeenth and Eighteenth Centuries. Export Industry before the Industrialization*, Jernkontorets Bergshistoriska Skriftserie 29, Stockholm 1992, 86.

11. S. Montelius, 'Relationer mellan bruk och bönder i Dalarna. Den äldsta järnhanteringens tekniska och ekonomiska struktur', in *Från kulturdagarna i Bonäs Bygdegård 23–25 juni 1975*, Uppsala 1976, 33.

12. Montelius, 'Relationer', 33, estimates the number of people involved in the production of charcoal in Dalarna in the middle of the eighteenth century as being between eight and ten thousand. The number of forgemen at that time was around 200.

FIGURE 5.1 Peasants Transporting Charcoal, Sweden, 1868

FIGURE 5.2 Peasants Delivering Charcoal, Sweden, 1897

all Swedish households can be assumed to have taken part in the production of charcoal.[13] In the Urals, where 87,000 tons of bar iron were produced at the beginning of the nineteenth century, it can be estimated that the number of days of charcoal-burning was over six million.[14] If the Russian peasants on average did this work for five months per year, then at least 50,000 peasants each year were involved in charcoal-burning. To this figure should be added the appreciable production of copper, silver and ironware in the Urals, which also required large quantities of charcoal – and labour.[15]

As a rule, increased bar iron production forced up the price of charcoal, both in Bergslagen and the Urals. At the ironworks investigated by Strumilin one ton of charcoal cost 56.1 roubles in 1737. In 1769 the price had risen to 68.4 roubles.[16] In the middle of the nineteenth century one ton of charcoal at Lys'va ironworks cost on average 199 copecks.[17] In Sweden the price of *köpkol*, that is charcoal which the ironworks owners bought from peasants who burnt charcoal from their own forests, rose after the middle of the eighteenth century. A sharp rise can be noted during the first half of the 1760s. The price then fell during the latter half of the decade, but not as far as the price level at the beginning of the 1760s. The next sharp rise in prices occurred during the latter half of the 1790s. At the beginning of the nineteenth century there were considerable fluctuations in price, both in absolute terms and in relation to the price of pig-iron. From the latter half of the 1830s there was a sharp rise again. Between the 1820s and the 1860s the price of charcoal more than doubled. At the same time there were also considerable variations in price.[18]

The export of timber during the first half of the nineteenth century and the growth of the saw-mill industry from the middle of

13. Estimate based on P.-A. Karlsson, *Järnbruken och ståndssamhället. Institutionell och attitydmässig konflikt under Sveriges tidiga industrialisering 1700–1770*, Jernkontorets Bergshistoriska Skriftserie 25, Stockholm 1990, 17.

14. Strumilin, *Chërnaia metallurgiia*, 150–151, 10.5 working days per ton of charcoal and 6.4 tons of charcoal per ton of bar iron.

15. In 1763 there were 63 ironworks in the Urals. Twelve of these produced both iron and copper. The number of production facilities was, however, somewhat larger, as several were groups of works under a single owner. Portal, *Slavs*, 158.

16. Strumilin, *Chërnaia metallurgiia*, 66, 152–53.

17. Ibid., 68–69.

18. L. Jörberg, *A History of Prices in Sweden 1732–1914, 2 vols.*, Lund 1972, vol. 1, 517ff. and vol. 2, 218–220, 299; K.-G. Hildebrand, *Svenskt järn. Sexton- och sjuttonhundratal. Exportindustri före industrialismen*, Stockholm 1987, 78; S. Montelius, *Säfsnäsbrukens arbetskraft och försörjning 1600–1865*, Geographica, Skrifter från Uppsala universitets Geografiska institution, Falun 1962, 309.

the century further pushed up the price of charcoal. Between 1840 and 1914 the price of charcoal in Sweden rose by 300 per cent compared to the price of bar iron. As a rule, *köpkol* fetched a two to three times higher price than the price of charcoal produced on the estate of the blast furnace owner.[19]

In both regions it was the case that if bar iron production was to be able to increase, a stable and sufficient supply of charcoal was necessary. The question can be divided into two separate parts: firstly, how much forest the ironworks had access to and could use each year for the production of charcoal, and secondly who was able to produce and deliver charcoal at a reasonable price. The long-term supply of forest land suitable for the production of charcoal was of great importance, but this was not the only factor which decided whether the ironworks could supply their production facilities with enough fuel each year. Labour, technology and the way in which the work was organised were important in this connection, and here the Urals and Bergslagen differed.

The Work Process

In both regions the charcoal was mainly produced in kilns. Production was roughly divided up into five separate stages:

(1) The felling of trees, trimming and stacking (before drying)
(2) The building of the kiln (heaping, the laying of conifer branches, sodding and earth covering)
(3) Charcoal-burning
(4) Extinguishing the kiln and dismantling it
(5) Loading the charcoal and transportation from the forest to the furnace and forge

The first four stages were carried out using manual power and hand-operated tools. The last stage required horses, amongst other things. The various stages placed different demands on skill and effort, with the first stage being the most labour-intensive. Up until the 1830s the trees in the Urals were felled using axes alone. At the state ironworks saws were used from the middle of the

19. Arpi, *Järnhanteringens träkolsförsörjning*, 173–206; P. Norberg, *Forna tiders järnbruk i Norr- och Västerbotten*, Stockholm 1957, 219ff.; S. Montelius, 'Iggesunds Bruk 1685–1869', in *Iggesunds Bruks Historia 1685–1985*, ed. G. Utterström, Uppsala 1985, 158ff.; A. Attman, *Svenskt järn och stål 1800–1914*, Jernkontorets Bergshistoriska Skriftserie 21, Stockholm 1986, 120; Jörberg, *History of Prices*, vol. 2, 218ff., 332ff.

1840s. The hand-operated saws had great advantages, as they reduced waste and improved productivity from 2 to 2.5 times. Furthermore, the saws made the work easier and meant that youths and women could be employed. In the middle of the nineteenth century, however, the saws were only used in the Urals.[20]

Swedish peasants and crofters also used simple hand-operated tools in the production of charcoal. The felling of trees, trimming and barking were carried out with axes. Different models were used in different regions, made as they were by village smiths. The axe kept its place in the forest up until the 1860s. At this time saws, many of which had been imported from Britain and the United States, started to be used. Other new tools for handling timber in the forest and for barking entered upon the forest scene at this time.[21]

In both the older, small-scale Ural production of iron and in the production of iron by the *bergsmän,* the iron producer's household carried out all the stages of the charcoal-burning in the kiln. The producer had at his disposal the raw materials, the means of production, the labour and the working hours, and he also inspected the quality of the finished product. However, the large-scale iron production that developed in the Urals from the beginning of the 1720s adopted centralised and specialised charcoal production. But in Bergslagen, in spite of the fact that the ironworks took over bar iron production after the middle of the seventeenth century, the small-scale, household-based production of charcoal continued. The organisation of work was decidedly different in the Urals and in Bergslagen from the beginning of the eighteenth century, and to some extent this had repercussions for technical developments and the forms that protests took.

In the Urals it was not the producer who decided when the work should be carried out. At the state ironworks ascribed peasants did the hewing of the timber from the middle of the 1730s onwards. They carried out the work from 20 March to the beginning of May. Then the peasants could go home and do farming work. They returned in June to where the trees had been felled, cut up and stacked the wood, which was then left to dry. When the time came for gathering in the hay and threshing they returned home yet again. This was on condition that they had completed

20. G. I. Oskolkov, 'Vnedrenie ruchnykh pil na kazënnykh ural´skikh zavodakh v 30–50-e gody XIX veka', *Voprosy istorii Urala*, Sverdlovsk 1964, Collection 5, 35–41.

21. G. Rosander, 'Skogsarbetaren och hans verktyg', *Kungliga Skogs- och Lantbruksakademiens tidskrift 127*, 1988.

their quota; if not, they had to stay in the forest at harvesting time. From the middle of September to the middle of October it was time for pile-making, sodding and pile-strewing. Then the charcoal-burning itself was begun in the kilns. When the snow had settled and the ice was negotiable transportation down to the furnaces and forges began.[22]

On the whole the work followed the same schedule at the private ironworks. However, there were variations. According to regulations from 1839 for the Nizhnii Tagil' mining and iron district, the cutting up of the wood should start on 1 April and continue until 15 June, while the kiln should be erected and the charcoal-burning done from 1 September to 1 November. Transportation to the ironworks then began, and this was to be completed by 1 April the following year.[23]

In Sweden, on the other hand, it was the peasant who decided when he and his helpers would carry out the work. The work usually started after the snow had melted. The timber was to be hewed before the spring tillage. During the summer and early autumn the timber was allowed to dry out. Then it was barked and cut into suitable lengths. Then came the pile-making, sodding and strewing. The kilns were lit at the end of October, or at the latest at the beginning of November. After three to seven weeks, depending on its size, the kiln was burnt out and it was time to dismantle it. The transportation of charcoal from the forest to the furnaces and forges started when the ground was covered with snow and the water courses had frozen.[24]

Of course, the work periods varied regionally, as they did between individual peasants, according to how much charcoal they produced. However, the stages of the work were the same in Sweden and the Urals, and the work period was on the whole similar. What distinguished the work in the two regions was, as has already been mentioned, above all how the work was organised and the quality controlled.

In the large-scale iron industry in the Urals charcoal-burning was a separate process, where each individual worker was also

22. V. De Hennin, *Opisanie ural'skikh i sibirskikh zavodov 1735 g.*, Moscow 1937, 357–361.

23. V. Ia. Krivonogov, 'Krest'ianskii podnevol'nyi i naëmnyi trud na vnezavodskikh operatsiiakh na Urale v kontse XVIII-pervoi polovine XIX v.', *Voprosy agrarnoi istorii Urala*, Sverdlovsk 1975, 27–28.

24. P. Norberg, *Engelsbergs Bruk. En historisk framställning*, Stockholm 1934, 205ff.; E. Eriksson, *Greksåsar. En bergslagsby*, Hembygdsföreningen Noraskogs skriftserie n:o 3, Nora 1963, 16ff.; Hildebrand, *Swedish Iron*, 93.

specialised in one of three functions: organisation, inspection/registration or actual production. The workforce that did the work could be permanent or temporary, more or less skilled, free or unfree. A number of factors decided how their actual work relations were: the requirements of the particular stage of the work, the availability of free and unfree labour, where the felling and charcoal-burning took place, the management and how well the work was paid.

The ascribed peasants did not receive proper wages for their work, but were paid by the season. They were not included in the lists of personnel mentioned below, even though they accounted for most of the charcoal-burning work. While the tax-paying peasants, *skattebönder,* in Bergslagen were only bound to sell to a specific ironworks owner if they had voluntarily agreed to do so,[25] the ascribed peasants in the Urals had to produce and deliver charcoal to the ironworks as a kind of feudal duty – as a compulsory share of the rent. In this respect, however, there were various similarities with the situation of the *landbönder* (peasants who tilled land owned by others) and crofters in Bergslagen.[26]

The number of ascribed peasants was, as a rule, considerable. According to De Hennin (1735), the peasants in the villages of Kamensk, Bagariak and Kalinovsk, as well as the peasants in the towns of Kolchedan and Kataisk, were ascribed to the Kamenskii ironworks. The number of this category of peasants from each place is given in a population census as 1,487, 1,353, 966, 1,355 and 1,850, respectively.[27]

Preserved lists of personnel give us a picture of how the charcoal-burning work was organised in the Urals. It appears that the state works had a more uniform organisation than the private works. The list of personnel that V. N. Tatishchev drew up in 1737 for the Ekaterinburg ironworks was used as a model by other state works during the eighteenth century and much later. According to an office register for the Kamenskii state ironworks from 1761 a forest expedition consisted of '1 *unter-schichtmeister*, 1 supervisor, 1 under-office worker, 3 copyists, 1 scribe, 2 tax-collectors, 1 office-watch and 2 grooms'. As can be seen, the management was considerable, which suggests a meticulous supervision of day-to-day work and the finished product. The charcoal-burning itself was carried out by four workers and four apprentices.[28]

25. Hildebrand, *Swedish Iron*, 86–87.
26. Montelius, 'Iggesunds Bruk 1685–1869', 77–78.
27. Hennin, *Opisanie*, 478.
28. GASO, F. 24 op. 1 d. 1658 1, 296–297.

An *unter-schichtmeister* was in charge of the work. He assigned a cutting area, where the peasants could cut trees and burn charcoal, and he supervised the whole work process. He carried out the technical supervision together with 'supervisor, tax-collectors, copyists and foremen of charcoal-burners'.

The list of personnel from N. A. Demidov's ironworks in Nizhnii Tagil' (1763–1764) shows that the charcoal-burning work could be organised in other ways in the Urals. Here a 'steward', an *ustavshchik,* three assistants and two forest inspectors were in charge of the work. The various stages of the work were carried out by different work groups, 708 cutters (probably Demidov's serfs) cut wood for a total of seven months, and 151 workers erected the kilns. Over a period of seven months in all they constructed 1,062 kilns. Eleven workers burned half-burnt logs in small heaps, while 942 ascribed peasants and 195 residents covered the piles with sods and stubble.

The charcoal-burning work was carried out by the same ascribed peasants and residents at the works, who worked together with sixteen charcoal-burners. These workers also extinguished and dismantled the kilns. Teenagers were brought in to gather the charcoal. It was then the work of N. A. Demidov's permanent workers, that is the serfs, to transport the charcoal to the ironworks during the winter. Three stewards supervised and approved the work.[29]

The same type of charcoal-burning organisation was used during the second half of the 1760s at the works in Nev'iansk. Here G. Makhotin, P. A. Demidov's steward, entered into a contract each year with Demidov's 560 serfs. They undertook to supply 21,200 *sazheni* of wood. The cutting lasted ten months – from 1 February to 1 December. The cutters were divided up into eighteen groups, with thirty-one in each group. Each group was responsible for its contract and the cutters borrowed axes and other tools from the works. A supervisor supervised the cutting, and foremen among the ascribed peasants saw to it that orders were followed. They also had the right to flog negligent workers with a rope.[30]

Other ways of organising charcoal-burning also existed in the Urals in the eighteenth and nineteenth centuries. Among the private

29. B. B. Kafengauz, *Istoriia khoziaistva Demidovykh v XVIII-XIX vv.*, Moscow-Leningrad 1949, Tome 1, 307–310.

30. Kafengauz, *Istoriia*, 310–311; D. A. Kashintsev, *Istoriia metallurgii Urala*, Moscow-Leningrad 1939, Tome 1, 243–246; V. Ia. Krivonogov, *Naëmnyi trud v gornozavodskoi promyshlennosti Urala v XVIII veke*, Sverdlovsk 1959, 71–73.

ironworks in particular there was great variation. At the same time extreme centralisation was a common feature. This centralisation had several advantages: primarily it was easier to control the quality of the charcoal than with a decentralised organisation. This was not least important when labour that was underpaid and unfree was used. The wood and the charcoal could be monitored at each stage by specially appointed people. A combined system of punishments and rewards meant that the charcoal workers received extra remuneration when production was high and of good quality, and correspondingly less when production was too low and/or of poor quality. The foremen played an important role in this system. To be taken on as a foreman a man had to pass an exam which showed that he was able to decide whether the charcoal was up to the required standard.[31]

In Sweden the peasant was his own manager. Since as a rule he participated in the work himself, quality control was not a separate element but rather integrated in the process. The peasant and the crofter carried out all stages of the work, including the quality control, together with older sons, daughters and farm-hands. Only very wealthy peasants were able to allow hired workers to carry out every stage of the work. It probably became more common in the nineteenth century, when class differentiation started to develop in the rural areas. However, sometimes the peasants needed to hire people to carry out the more complicated stages, in particular the charcoal-burning.[32]

It would be reasonable to draw the conclusion that charcoal-burning in the Urals required more people than the Swedish small-scale production of charcoal. These extra people in the Urals had preparatory and supervisory tasks. Were, then, quality and productivity higher in the Urals than in Bergslagen? It cannot be ruled out, but there is no evidence that this was the case.[33] Perhaps peasants and crofters worked more efficiently when it was they themselves who were in control of their working day and decided upon the quality. On the other hand it should have been more difficult to cheat with regard to quality in the Urals, although as we will soon see, there was a great deal of cheating there as well.

31. GASO, F. 24 op. 1 d. 1658 1, 38–39; N. Rogov, 'Raznye svedeniia o lesakh Permskoi gubernii i o zagotovke glavnykh lesnykh materialov dlia gornykh zavodov i solianykh promyslov', *Gornyi zhurnal* 1877, Tome 1, 139.

32. Norberg, *Engelsbergs Bruk*, 206.

33. See the information given earlier on regarding the number of days that were needed for charcoal-burning per ton of charcoal.

There were widespread protests by the peasants about the conditions imposed by the ironworks, both in the Urals and in Bergslagen, but the extent and the focus were somewhat different. At the same time, the conflicts surrounding charcoal run right through the early industrial history of iron production and forced the works owners and the state to take a number of measures and steps in order to uphold their interests. This will be examined in more detail later on in this chapter, after an account of the conflicts.

Conflicts Surrounding Charcoal

The ironworks owners' growing need of charcoal led to constant clashes with the producers of charcoal. Because of differences in the two countries' political systems and the social organisation of the work the conflicts and the everyday resistance in part came to take on different forms. The primary difference was that while the Swedish producers blocked the deliveries of charcoal, the Russian producers refused to carry out the work. Strikes and uprisings were a considerable problem in the Urals in the middle of the eighteenth century. In both regions the peasants also used more disguised forms of protest. They cheated the inspectors and traded in charcoal illegally. Differences in property rights led to the Swedish producers having greater political and legal opportunities to protest and present their demands than the Russian producers. The charcoal producers in Bergslagen could go to the mining-court with their complaints and present them in the *Riksdag* (the Diet). In the Urals the peasants were by and large forced to resort to other methods, even though there were legal ways of protesting here too.

In the Urals the producers, especially in the eighteenth century, were as a rule serfs or ascribed peasants and they had little to gain by producing charcoal and delivering it to the ironworks. They resorted to different tricks in order to cheat the works administration. The aim was to have the delivered charcoal approved by the inspectors even though the produced or delivered quantity was less than that stipulated. One way was to use thinner and smaller logs than was allowed when the kiln was erected. They could also reduce the amount of wood through the way in which they stacked the wood. De Hennin noted as early as the 1730s that the peasants could sometimes get the same kiln accepted on two or three occasions by the ironworks inspectors. This was done by taking away the marked log after the inspection. With a view to preventing this De Hennin advocated that the marking should be done on logs on the top, in the middle or at the bottom of the kiln. This made it impossible to take logs away as the kiln would then collapse. But De Hennin also mentions other ways of cheating the works administration. When the peasants delivered charcoal, they tried to use smaller measures than the approved ones. They also loaded charcoal in the charcoal wagons so that they gave a more well-filled impression, for example by mixing in earth and ash. Another way was to get one and the same delivery approved several times. If the inspection station had two entrances, the peasants could go in through one gate and get the delivery approved, leave the area without unloading and then go in through the other gate and have the load approved yet again.[34]

It was not only the peasants who tried to cheat. The administrative staff sometimes abused their positions with a view to lining their own pockets. They not only demanded that the peasants exceeded the stipulated standard but also set production and delivery requirements that were impossible for the peasants to meet. If the ironworks administrators received a bribe from the producers they ignored the deficit. Bribes were also used by the charcoal producers when they wanted a better position for their kilns. But if the works inspectors accepted the bribes, the whole carefully built up control system was at the same time undermined.[35]

The antagonism and disputes were not only covert, however. They sometimes came out into the open, in the form of complaints, absenteeism, a refusal to carry out work, unrest and armed uprisings. Such protests were especially common during the eighteenth century among charcoal-burning ascribed peasants. During the first third of the century and during the 1750s there were large uprisings, and these culminated in the big riot of Pugachev from 1773 to 1774. The peasants were fighting against what they considered to be excessive delivery quotas, against the system of bribery and against the low wage levels.[36]

There is also information about the same type of open disputes from the first half of the nineteenth century in the Urals. According to a study of N. M. Lushinkova there were 275 big uprisings in the

34. Hennin, *Opisanie*, 356–369.

35. Ibid.

36. A. S. Orlov, *Volneniia na Urale v seredine XVIII veka. K voprosu o formirovanii proletariata v Rossii*, Moscow 1979; A. I. Andrushchenko, *Krest'ianskaia voina 1773–75 gg. na Iaike, v Priural' e, na Urale i v Sibiri*, Moscow 1969, 191–208; R. Portal, *L'Oural au XVIIIe siècle. Étude d'Historie Économique et Sociale*, Paris 1950.

mines and ironworks in the Urals between 1801 and 1860, 110 of them in the 1850s. The charcoal producers were the most active, and the most common demand during these uprisings was that the volume of the charcoal-boxes should be reduced. This demand was voiced with particularly great emphasis at the Revda ironworks in 1802, 1824 to 1826, 1833 and 1844. The charcoal producers at several *posessionnye* works (Alapaevsk and Sergino) fought, on the other hand, for the same type of regulations as at the state works, that is a regulated work period, a regulated amount of work and regulated wages.[37]

If we turn to Sweden we also find growing antagonism between the works owners and the charcoal producers during the eighteenth century. The peasants employed a number of different methods to obtain better remuneration for their work and to oppose the ironworks owners' demands. From the beginning of the 1720s the *skattebönder* (freeholders) united and demanded in the *Riksdag*, via their representatives, higher charcoal prices. They were, however, unsuccessful.[38] There are also examples of the charcoal-producing peasants taking their complaints to the mining-court. This was also without much success.[39] Other methods, some of which were illegal, had a greater effect, for example the outright boycotting of deliveries.[40]

There is information on boycotts from around 1720 and in the first half of the 1760s. Both periods saw rapidly rising prices concerning a number of products that were necessary for the peasants. Boycotts occurred above all at the large Walloon ironworks in Uppland, that is in the other large iron district in Sweden.[41] It will be of interest here to give a short description of these boycotts. Even though the Walloon works were considerably bigger and the works owners' power more absolute it cannot be ruled out that similar protests also occurred in Bergslagen's ironworks.[42]

37. N. M. Lushinkova, 'K voprosu o klassovoi bor'be na gornykh zavodakh Permskoi gubernii v 1801–1860 gg. (Uchastie lesorabochikh v massovykh vystupleniiakh)', *Voprosy istorii Urala*, Sverdlovsk 1975, Sb 14, 73–82; M. A. Gorlovskii and A. N. Piatnitskii, *Iz istorii rabochego dvizheniia na Urale*, Sverdlovsk 1954.

38. Karlsson, *Järnbruken och ståndssamhället*, 59ff.; Montelius, '1600–1815', in S. Montelius, G. Utterström and E. Söderlund, *Fagerstabrukens historia V. Arbetare och arbetarförhållanden*, Uppsala 1959, 184–185.

39. Montelius, *Säfsnäsbruken*, 308.

40. K.-G. Hildebrand, *Fagerstabrukens historia. Sexton- och sjuttonhundratalen*, Uppsala 1957, 267.

41. Montelius, 'Relationer', 44.

42. See also Montelius, '1600–1815', 173ff.

By Swedish standards, Leufsta was a very large ironworks. It was owned by the de Geer family and had Walloon descendants working in the forges. Enormous quantities of charcoal were consumed in the works' facilities. Hundreds of peasants supplied the ironworks with the charcoal they produced. From October 1718 to the beginning of 1720 the charcoal producers refused to supply the ironworks with *köpkol*. Neither the authorities nor the ironworks proprietor Charles de Geer could persuade the peasants to lift the boycott. The disagreement was only settled at the beginning of 1720 when the County Governor himself took the side of the peasants. Even though the new price was lower than that demanded by the peasants their boycott was still a success, since Charles de Geer had agreed to a certain increase in the price of charcoal. That the boycott was not more successful and did not inflict a great deal of damage on the ironworks can be explained by the fact that in the summer of 1719 the Leufsta ironworks had been destroyed by Russian troops who at this time were laying waste the Swedish Baltic coast.[43] The forge at Leufsta was kept going by means of charcoal from the works' own forests.

Forty years later the charcoal peasants boycotted the deliveries of charcoal to Söderfors, another major ironworks in Uppland. Anchors were produced at the works, and this required a production method that consumed enormous quantities of charcoal (the 'half Walloon method'). Furthermore, the works' own forests were not large enough to provide the owner of the works with a stable supply of charcoal.[44] At the beginning of the 1760s the works was the target of the Hedesunda peasants' dissatisfaction. They refused to deliver charcoal if the works management did not accept a steep increase in the price of charcoal. While they awaited the works owner's reply, the peasants took the charcoal home to their villages, where it was heaped up. During this boycott the works inspector tried to break the peasants' solidarity by bribing some of the peasants with drink and tobacco. After a month the works management gave in and accepted a price increase. However, it was lower than what the peasants had originally demanded.[45]

43. E. W. Dahlgren, 'Kolböndernas sammansvärjning vid Lövsta 1718', *Med hammare och fackla II*, Stockholm 1930; O. Isaksson, *Vallonbruken i Uppland. Människor och miljöer*, Stockholm 1995.

44. In 1780 the works got only 30 per cent of its charcoal from its own forests. The rest was *köpkol*. See note 40 and Ö. Tigerstedt, *Fagerstabrukens historia. Kavalkad*, Uppsala 1957, 559ff.

45. Montelius, '1600–1815', 190–191; Hildebrand, *Fagerstabrukens historia*, 269.

Boycotts and other direct action for higher prices or better working conditions thus also occurred in Sweden, but it was not at all as common as in the Urals. Indirect methods were used instead in order to get away from the centrally regulated charcoal prices. One way was illegal charcoal trading, that is trading outside the fixed district or with other recipients than those stipulated by the Board of Mines.

Illegal trading occurred as early as the seventeenth century.[46] However, from the 1720s the number of complaints increased, a fact that mirrors the works owners' increasing difficulties from this time on. Not least of all, the owners of the specially favoured copper and silver works complained that they did not receive the promised amount of charcoal. The ironworks were accused of deliberately enticing the peasants to their production facilities in spite of the fact that according to the regulations they were to supply the precious-metal works first. The peasants did not even deliver the quantity they were liable to pay in tax to the copper and silver works.[47] However, even the ironworks were affected by the illegal trading. The various works outbid each other and the *bergsmän* bought charcoal from their peasant neighbours. However, the *bergsmän* probably lost out in the battle with the powerful works proprietors for *köpkol*.[48]

The illegal trading in charcoal was an increasingly great problem both for the owners of iron, copper and silver works and for the Swedish state in the middle of the eighteenth century. The government tried to solve the problem by stricter supervision of the staff collecting tolls in the towns, by having charcoal controllers who checked the yearly charcoal-burning in the districts and by means of harsher sentences against illegal trading.[49]

Charcoal controllers had existed since the end of the 1630s, at least in certain parts of Bergslagen. The system of controllers was officially regulated in 1649 by a royal decree. The decree stipulated,

46. Hildebrand, *Fagerstabrukens historia*, 266; P. Norberg, *Avesta under kopparbrukets tid*, del 1, Avesta 1956, 298; Montelius, '1600–1815', 184.

47. Karlsson, *Järnbruken och ståndssamhället*, 68ff.; Norberg, *Avesta under kopparbrukets tid*, 299, 317.

48. B. Boëthius, *Gruvornas, hyttornas och hamrarnas folk. Bergshanteringens arbetare från medeltiden till gustavianska tiden*, Stockholm 1951, 114; Norberg, *Engelsbergs Bruk*, 152.

49. Karlsson, *Järnbruken och ståndssamhället*, 70–71.

50. Her Majesty's charcoal decree, 6 July 1649. Bergskollegii Arkiv, Letters and resolutions 1637–1663. RA, (National Archives) Stockholm.

amongst other things, the size of the sledge and of the measure used.[50]

In 1722 a comprehensive investigation was carried out of what the peasants in four parishes in southern Dalarna had to pay in tax to Avesta copper works. The quantities were compared with what they in fact delivered. The works had a right to 8,351 *stigar* (sg. *stig*, almost exactly the equivalent of 20 hectolitres), but received just 3,480 *stigar*. The peasants preferred to sell their charcoal at a higher price rather than deliver it in the form of tax to the works.[51]

Time and time again the state had charcoal supplies to Avesta copper works investigated, but without solving the problem. The peasants in southern Dalarna also complained to the authorities about having to deliver tax coal and coal for sale to the two precious-metal works in the area (Avesta and Garpenberg). However, the protests met with no success. Instead the illegal trading continued. Sometimes this was discovered and heavy fines ensued.[52]

The management of the copper works did not play fairly either, however. The works in Avesta bought charcoal in the 1680s from peasants in parishes which, according to the regulations, should transport their charcoal to the silver mine in Sala, in the county of Västmanland.[53]

After the expansion of the seventeenth century it became increasingly difficult to build new ironworks and to extend production in Bergslagen. The peasants' steadily improving financial situation also made them less willing to deliver charcoal below cost price.[54] During the eighteenth century illegal trading became a growing problem for the Swedish works proprietors, a problem they had to solve. One solution was to support the policy limiting production that the *Riksdag* decided on just before the middle of the eighteenth century.[55] Another way was to form secret charcoal associations and purchasing cartels. The works involved came to an agreement on which parishes, villages and farms they had a right to get their charcoal from and what price

51. Norberg, *Avesta under kopparbrukets tid*, 316–319.

52. Norberg, *Avesta under kopparbrukets tid*, 315, 322; M. Isacson, *Ekonomisk tillväxt och social differentiering 1680–1860*, Uppsala 1979, 51–52; Montelius, '1600–1815', 175ff.

53. Norberg, *Avesta under kopparbrukets tid*, 301.

54. Montelius, '1600–1815', 175–176; Norberg, *Engelsbergs Bruk*, 151ff.; Karlsson, *Järnbruken och ståndssamhället*, 68ff, 102ff, 139ff, 254ff; Montelius, 'Iggesunds Bruk 1685–1869' for a thorough study of a works' charcoal supplies from the 1680s to the 1870s.

55. B. Boëthius and Å. Kromnow, *Jernkontorets historia*, del 1, Stockholm 1947, 40–59.

should be paid. Anyone breaking the agreement had to pay a fine. There are instances of such charcoal associations in Uppland and Gästrikland during the eighteenth century, but they occurred elsewhere at other times.[56]

Another controversial issue was the charcoal measures. By cheating with the charcoal measures the peasants could obtain a better price for their charcoal. As was described earlier, the ironworks in the Urals had strict supervision even in the forests. In Sweden, supervision in the forests was almost non-existent, even though there was the odd charcoal supervisor. Controls at the ironworks became even more important, but it was difficult to block all the loopholes. Admittedly the size of the coal wagons was officially fixed and the volume was inspected at the works before the deliveries could be approved. However, it was quite possible, both in Bergslagen and in the Urals, for a supplier to slip through the controls with a smaller volume than was stipulated.

For the people receiving the incoming charcoal it was difficult, in purely practical terms, to inspect every coal wagon. It was especially difficult on days when the charcoal drivers had to queue as they waited to deliver their load. The measure with which the wagons' length, breadth and height were to be measured was not then put to use. The charcoal measurer confined himself to counting the number of wagons, checking that they were well filled and checking the measurements of suspect wagons. He could also be bribed into approving volumes that were too small. Admittedly it was risky to bribe a coal measurer. In Sweden he had sworn an oath before the mining-court that he would measure correctly and not favour either party, but there were always those who were not so scrupulous.[57]

There were several other ways in which the peasants could cheat with the volume. The charcoal wagons could be constructed so that they could be made either bigger or smaller. By making coal wagons without crossbars at the top and bottom they could be bound, thus reducing the volume. The peasants also weighed in the charcoal in barrels that were too small and too few.[58]

56. Karlsson, *Järnbruken och ståndssamhället*, 210ff. On the charcoal associations, see Montelius, '1600–1815', 185–187; Montelius, 'Relationer', 42–44; Montelius, 'Iggesunds Bruk 1685–1869', 84; A. Florén and G. Rydén, *Arbete, hushåll och region. Tankar om industrialiseringsprocesser och den svenska järnhanteringen*, Uppsala Papers in Economic History, Research Report no 29, Uppsala 1992, 80–81.

57. Arpi, *Järnhanteringens träkolsförsörjning*, 21; Norberg, *Avesta under kopparbrukets tid*, 313; Montelius, 'Relationer', 44; Montelius, '1600–1815', 192.

58. Norberg, *Engelbergs Bruk*, 211; Norberg, *Avesta under kopparbrukets tid*, 313–314.

However, even when the charcoal wagons conformed to the stipulated measurements and were loaded in accordance with the current regulations, the charcoal driver could run the risk of his delivery being rejected by the person receiving the incoming charcoal. The charcoal was packed together when transported over uneven terrain. The charcoal suppliers therefore had several tricks to make the wagons look well filled, even when this was not the case. One trick was to sort the charcoal in the wagon according to its size. By placing the good, big pieces of charcoal at the top and the smaller, inferior ones at the bottom, the delivery made a better impression on the person receiving the incoming charcoal. Another method was to place a small spruce tree in the bottom of the wagon. Just before arrival it was pulled up and out of the wagon, thus shaking up the coal and making the volume in the wagon look greater.[59]

The ironworks' inspectors and owners regularly accused the peasants of cheating with the charcoal measures. At the same time there were works owners who accepted the lesser quantities delivered. This was a way for them to unofficially give the peasants a higher price. In this way they could secure their charcoal supply without upsetting the state and other works owners. From the beginning of the nineteenth century, when Swedish iron production began to rise again, the cheating with the charcoal measures increased. It was a 'seller's market' and the ironworks were forced to quietly accept lesser quantities in order to be able to get any charcoal to the works at all.[60]

To summarise, available information indicates that charcoal was a source of recurring controversy between the peasants and the works owners, both in the Urals and in Bergslagen. The peasants in the two regions used similar methods to protect their interests. However, open organised action seems to have been more common in the Urals. At the same time, the Swedish peasants had a stronger grip on the production of charcoal, as it was small-scale and household-based. The work was more rigidly organised, more specialised and more controlled in the Urals, and this resulted in more open protests.

59. Montelius, '1600–1815', 192; Arpi, *Järnhanteringens träkolsförsörjning*, 21.

60. Norberg, *Avesta under kopparbrukets tid*, 312–314; Montelius, '1600–1815', 192–193; Montelius, 'Relationer', 44. On the cheating during the nineteenth century, see Arpi, *Järnhanteringens träkolsförsörjning*, 21.

Attempts to Deal with the Problem

The state and the works owners tried to overcome the problem by means of different strategies, some of which were more successful than others. However, the problem was only finally solved with the advent of new production techniques and new communications after the middle of the nineteenth century. Railways and canals allowed the works to exploit more remote forests and to purchase over a wider area, and this markedly reduced their dependence on the local peasants.

What methods, then, did the state and the ironworks use in the two regions during the eighteenth and early nineteenth centuries to overcome the difficult charcoal question? At the political level the strategy was primarily to limit trading and to keep the prices down. At the company level we can distinguish four separate strategies which all aimed at either increasing supplies or reducing charcoal consumption:

(1) Increasing the amount of forest available for charcoal-burning
(2) Increasing the workforce
(3) Making production and transportation more efficient by means of technical and organisational improvements
(4) Reducing the consumption of charcoal in the smelting house and forges

The last-mentioned strategy was not the least important, but it had no major effect before a couple of decades into the nineteenth century. In particular, the introduction of the Lancashire forge from the end of the 1840s had a decided effect in Sweden. This method of forging when fully developed, together with the Ekman furnace, reduced charcoal consumption by 50 per cent.[61] From 1825 to 1900 charcoal consumption per ton of iron produced in Sweden fell more than three-fold, including all the processes from the smelting of the ore to the forged or rolled iron.[62] As technical developments in the industrial production units have been dealt with in chapter 3 we will leave this question here and refer the reader to this chapter. Instead we will now look at other strategies, starting at the political level.

61. Attman, *Svenskt järn*, 50.
62. Arpi, *Järnhanteringens träkolsförsörjning*, 99.

The State: Trade and Price Regulations

In both Russia and Sweden the state played an active role in attempts to create good conditions for the works' supply of charcoal. In the Urals the government stipulated how the charcoal should be received and inspected at private and Crown ironworks.[63] The wages of the producers of charcoal were set by the government and the works administration. In the resolution of 1724 fixed wages were introduced for the ascribed peasants in each region. The regulations stipulated that the peasants should be paid by the day. In 1737 payment by the piece was introduced for certain stages (cutting, as well as erecting and dismantling of the kiln) at the state ironworks in the Urals. Wage levels were then increased in 1769 and 1779.[64] If the charcoal producers worked more than the given quota they were paid extra. If they did not meet the quota they were penalised with a hefty reduction.[65]

Wages for forced labour were fixed in regulations, but the wages that free labour received were determined by the market. In the middle of the eighteenth century the market price was twice or four times the level of the fixed price. The difference between the market price and the standard price was sometimes as big as one to ten. When iron production rose the situation for the works owners in the Urals was made more difficult by the fact that they were forced to hire free workers and pay them market wages.

As early as the first half of the eighteenth century Tatishchev and De Hennin were concerned about the wasteful and inefficient use of the forests. The Swedish traveller Samuel von Stockenström made the same observation at the end of the century during a trip to Russia.[66] A century earlier the Swedish state had expressed the same concern for the forests.[67] The dwindling forests were probably a question of greater importance to the government than to the indi-

63. Hennin, *Opisanie*, 362–368; O. S. Tal'skaia, 'Iz istorii uglezhzheniia na ural'skikh zavodahk', *Trudy Sverdlovskogo oblastnogo kraevedcheskogo muzeia*, Sverdlovsk 1960, 52–75; O. S. Tal'skaia, 'Pamiatniki uchëta truda zavisimykh krest'ian v metallurgicheskoi promyshlennosti Rossii', *Isochnikovedenie i arkheografiia Sibiri*, Novosibirsk 1977, 134–161.

64. V. I. Semëvskii, *Krest'iane v tsarstvovanie imperatritsy Ekateriny II*, St Petersburg 1901, Tome 2, 307–309, 507–508.

65. GASO, F. 24 op. 1 d. 1658 l, 39.

66. S. von Stockenström, 'Anmärkningar rörande så Ryska Jerntillverkningen som äfwen Jernhandeln', 1787, Bergskollegii Arkiv, RA (National Archives), Stockholm.

67. Hildebrand, *Swedish Iron*, 122.

68. E. F. Heckscher, *Sveriges ekonomiska historia från Gustav Vasa*, del I:2, Stockholm 1936, 478–480.

vidual works owners. The state in Sweden was seriously concerned at an early stage about the devastation of the forests in Bergslagen.[68]

During the first half of the nineteenth century some mining and iron districts suffered from a permanent shortage of charcoal. At the beginning of the century the *kureni* (the areas where charcoal was produced) were normally situated between 16 and 32 kilometres (15–30 *verst*) from the ironworks. If this distance was allowed to increase transport costs rose, and this added to the price of the end products. The widespread felling of pine, which gave the best quality charcoal, forced the works owners to use other kinds of wood (such as spruce and deciduous trees).[69]

As has already been mentioned, the price of charcoal rose in Sweden during the eighteenth century. The government tried to regulate charcoal trading in order to break this trend. The price of charcoal was regulated as early as 1649 in a decree concerning Swedish iron production. Unapproved trading was banned and anyone paying a higher price than the set price was punished. In 1682 a new and more definite price system was introduced. Here, the price reflected how far the charcoal had been transported. At the same time stricter supervision was introduced. The government tried to prevent inflationary competition for charcoal by keeping Bergslagen primarily for mining and the smelting of pig-iron and by inducing the works owners to set up their bar iron forges in other districts.

The Swedish state was not able, however, to guarantee a low and stable price for charcoal. The first half of the eighteenth century saw numerous conflicts over the price of charcoal, and the price rose way above the set level. The production limits that the *Riksdag* decided on just before the middle of the eighteenth century and which were still in force at the beginning of the nineteenth century can in part be seen as a response to the fact that the works owners found it more difficult to secure charcoal supplies. During the 1760s a more locally adapted system was introduced, and this was kept until the middle of the next century, when the restrictions on charcoal trading were gradually lifted.[70]

To summarise, we can see that the Crown in Russia and Sweden played an active role in limiting the works owners' costs for char-

69. N. M. Lushinkova, 'Vliianie toplivnoi bazy na tekhnicheskie usovershenstvovaniia v metallurgicheskoi promyshlennosti Urala', *Voprosy istorii Urala*, Sverdlovsk 1970, Collection 10, 120–121.

70. Boëthius and Kromnow, *Jernkontorets historia*, del 1, 40ff.; Montelius, '1600–1815', 161ff.; Karlsson, *Järnbruken och ståndssamhället*, 46ff.

coal. At the same time the state was a guarantor for the peasants receiving a certain level of remuneration, thus stimulating them to produce good quality charcoal. The methods employed by the Swedish and the Russian Crown displayed similarities but were also different in a number of ways. Even so, the state's measures did not solve the difficult charcoal problem, either in Sweden or in Russia. The works owners were forced to try other solutions themselves.

The Works Owners: Extension of the Forest Supplies

Decentralised production placed special demands on transportation and the infrastructure in the iron districts. As charcoal is a brittle product which can easily be broken by jolts, transportation over stretches longer than twenty kilometres were out of the question before the advent of railways and canals. The volume decreased when it was reloaded and when it was transported over rough terrain by horse and sledge. The works owners preferably had their suppliers within a radius of ten kilometres. In reality they were often forced to take charcoal from a wider radius.[71]

One way of extending the area that supplied charcoal, and this was also done in the Urals, was to transport wood by boat from distant forests and to burn it in the vicinity of the works. Another, more uncertain method, was to transport the charcoal over long distances by barge or boat.[72] Such means of transportation also took place in Sweden. At Iggesund works in Hälsingland charcoal was transported by sea as early as in the 1680s.[73] At the same time this kind of transportation increased the price. Up until the middle of the nineteenth century the works owners were therefore primarily obliged to arrange for their charcoal supplies locally. At this time Swedish ironworks that lay on the coast or on one of central Sweden's larger lakes increasingly transported their charcoal by boat.[74] However, the peasants did not cease to supply charcoal by horse and sledge. Such transportation was still needed over short stretches, not least to and from the railways.[75]

71. Arpi, *Järnhanteringens träkolsförsörjning*, 20f, 61–63; Hildebrand, *Fagerstabrukens historia*, 263. For the Urals see Lushinkova, 'Vliianie toplivnoi', 120–121.

72. Portal, *Slavs*, 153.

73. Montelius, 'Iggesunds Bruk 1685–1869', 78.

74. Arpi, *Järnhanteringens träkolsförsörjning*, 131.

75. Ibid., 131ff.

One further way of dealing with the question of charcoal was to transport the pig-iron to the bar iron forges instead of the charcoal. Through a geographical separation of pig-iron and bar iron production the area supplying charcoal could be extended. In Sweden, as has already been mentioned, the Board of Mines pushed this line of thinking from as early as the middle of the seventeenth century. Bergslagen's forges were, as a rule, at a safe distance both from the works' own blast furnaces and those of the *bergsmän*. This pattern can also be seen in the Urals, where the apparently centralised companies as a rule had their blast furnaces and forges geographically spread out in their 'mining districts'.[76]

In 1818, experiments were started in the Urals where trees were artificially planted in the works' own forests. In the middle of the nineteenth century active attempts were made, as we shall see further on, to improve the charcoal-burning methods, with a view to increasing yield.[77] At the end of the century attempts were made to replace charcoal with coal, oil, gas and electricity, but as this required sweeping changes throughout the iron industry in the Urals, charcoal maintained its importance well into the twentieth century.[78] The situation was similar in Sweden. The new large iron and steel works, led by Domnarvet, went over to coal from the beginning of the twentieth century. Up until the First World War charcoal was still the most important fuel used by the iron industry.[79]

The Works Owners: A Larger Workforce

The works owners needed a large, stable and cheap workforce for the production of charcoal. In Sweden the works proprietors tried a number of methods in order to strengthen their grip on labour. One way was to purchase farms from the Swedish state, which became possible from the beginning of the eighteenth century.[80] The crucial question was, of course, the availability of such farms in the vicinity of the works. Another method often used was to

76. *Ironmaking in Sweden and Russia*, 16ff., 69–70.

77. M. V. Putilova, 'O sostoianii toplivnoi bazy zavodov Ekaterinburgskogo gornogo okruga v 40–70-kh godakh XIX veka', *Voprosy istorii Urala*, Perm' 1966, 34–135.

78. Dmitriev, *Osnovnye napravleniia*, 16–28.

79. Attman, *Svenskt järn*, 122.

80. See chapter 4, this volume, and the literature mentioned there.

lease the taxes that the peasants in a district were obliged to deliver to the state. The taxes were then converted into charcoal. From the 1680s the works owners could also obtain access to commonly owned forests in return for a so-called fee of recognition (*rekognitionsavgift*). Furthermore, during the seventeenth century the government donated state land to the works owners. As a consequence, the peasants became tenants under the works instead of tenants under the state (*kronobönder*) and they were then obliged to deliver charcoal.[81]

The number of charcoal-burning days that the peasants in Sweden had to carry out was decided either by their tax to the state or by the rent the *landbönder* paid for their tenant farm holdings. To this can be added the amount of charcoal that the peasants chose to sell of their own accord, or which they sold in order to pay off a debt to the works.

During the seventeenth century and at the beginning of the eighteenth century the amount of charcoal that *landbönder* and crofters were to deliver to the works was seldom stipulated in a contract, but in the course of the eighteenth century it became increasingly common that the contract contained specified quantities.[82] It was at this time that the works also started to build cottages to which a small piece of land was attached. The crofters who lived there did a number of jobs for the ironworks. The two most important jobs were the burning and transportation of charcoal.[83]

Up until the abolition of serfdom in the 1860s different types of forced labour were used in the production of charcoal in the Urals: serfs, ascribed peasants during the eighteenth century and early nineteenth century, and *nepremennye* (indispensable) workers from about 1813.[84]

As has been mentioned earlier in this volume the amount of work to be done by the ascribed peasants in the Urals was gov-

81. Montelius, *Säfsnäsbruken*, 288ff; Montelius, '1600–1815', 218ff; Montelius, 'Iggesunds Bruk 1685–1869', 84f; Isacson, *Ekonomisk tillväxt*, 34–35; E. T. Ström, *Hällefors och Grytthytte bergslag från år 1640 till omkring 1700*, Lund 1956, 155ff; C. W. U. Kuylenstierna, *Om rekognitionsskogar och under bruk skatteköpta hemman*, Lund 1916, 218, 228–229.

82. Isacson, *Ekonomisk tillväxt*, 32–33; Montelius, *Säfsnäsbruken*, 118; Montelius, 'Iggesunds Bruk 1685–1869', 78–79; M. Ågren, *Jord och gäld. Social skiktning och rättslig konflikt i södra Dalarna ca 1650–1850*, Studia Historica Upsaliensia 166, Uppsala 1992, 252ff.

83. Montelius, *Säfsnäsbruken*, 242; Ågren, *Jord och gäld*, 194ff, 250ff.

84. For a more detailed description of different categories, see chapter 4.

85. *Pamiatniki sibirskoi istorii XVIII v*, St Petersburg 1882, Tome 1, 317.

erned by the size of the poll-tax. It was not obvious at first what work they should do to work off the tax. However, they rapidly became involved in charcoal production.[85] In De Hennin's regulations in the middle of the 1720s the ascribed peasants are mentioned as timber cutters, kiln builders and charcoal transporters.[86] The Siberian Board of Mines laid down in 1741 that ascribed peasants could be used in private and state ironworks.[87] After the Pugachev riot in the middle of the 1770s the number of tasks to be done by the ascribed peasants was reduced. Henceforward, they needed to take part neither in the erection of the kiln nor in the charcoal-burning, that is in tasks that required skilled labour.[88]

Nepremennye rabotniki (indispensable workers), who replaced the ascribed peasants from the beginning of the nineteenth century, also took over their work. It happened that state works and private works made use of these workers for other purposes as well. The number of working days was regulated, however, and this number decreased from the beginning of the nineteenth century. The workers on horse ('mounted workers') had to do 240 days work at first, which in 1820 had fallen to 200 and at the end of the 1840s to 180.[89] In private ironworks the works owners had to make use of serfs or *posessionnye* peasants.[90]

Free peasants were also used for the charcoal-burning. It was not altogether uncommon that ascribed peasants themselves employed workers to carry out the work that they had been ordered to do. The attitude of the works managements to this type of hired worker varied. In 1735 De Hennin argued against this kind of substitute worker, as he considered that they lowered the quality of the work.[91] However, preserved historical documents show that it was common for ascribed peasants, as well as *nepremennye*,

86. N. I. Pavlenko, *Razvitie metallurgicheskoi promyshlennosti Rossii v pervoi polovine XVIII v.: Promyshlennaia politika i upravlenie*, Moscow 1953, 474.

87. Pavlenko, *Razvitie*, 480.

88. PSZ, Tome 20, St Petersburg 1830, 822; Also A. Florén, 'Social Organization of Work and Labour Conflicts in Proto-industrial Iron Production in Sweden, Belgium and Russia', *International Review of Social History* 39 (1994), Supplement, 96.

89. PSZ, Tome 29, St Petersburg 1830, 1074–1076; Krivonogov, 'Krest'ianskii podnevol'nyi', 22.

90. P. A. Vagina, 'Lichnoe khoziaistvo i rabochaia kvalifikatsiia masterovykh chastnykh zavodov Iuzhnogo Urala v poslednei chetverti XVIII v.', *Voprosy istorii Urala*, Sverdlovsk 1969, Collection 8, 24, 32–33, 37, 38.

91. Hennin, *Opisanie*, 357–358.

92. Orlov, *Volneniia*, 192–262.

to hire workers to carry out work that they themselves had been ordered to do.[92]

Hired labour of this kind affected the works' social environment more than the organisation of the work. The capitalist wage system gained a footing in the works, something which was permitted by the authorities. In April 1838 the head of the Perm' ironworks, Falkner, wrote to V. A. Glinka, the government-appointed head of the Ural ironworks: 'I have the pleasure of reporting to your Excellency that the data which have been gathered here and compared with the local situation show that the free hiring of craftsmen is not only appropriate and possible, but also advantageous with regard to the cutting of wood and to a number of other jobs.'[93]

In the middle of the nineteenth century the Ural Mining Board recommended official regulation of the system by which workers hired other workers, who then carried out the work that they themselves had been ordered to do at state ironworks. This regulation of the system meant that the hired labour first had to be approved by the administration of the local works. The workers were only allowed to hire other workers who were their equal in strength and skill, provided that the latter had the permission of their landlords to carry out this kind of work. The transfer of work between the hired workers was explicitly forbidden. The cutting of wood and sawing were tasks for which hiring was permitted, but not the erection of kilns and charcoal-burning.[94]

There were also other forms of hired labour. At Demidov's ironworks serfs worked as hired labour during the 1760s. Sometimes contracts were drawn up where the ironworks' employees, junior office workers, peasants and others played the role of contractors.[95] According to V. K. Rushett, head of the Nizhnii Tagil' ironworks, free hired workers delivered 39 per cent of the charcoal in 1854. At this time the system of hiring labour was in operation at the majority of the private ironworks.[96]

The ironworks' demand for free workers in auxiliary operations rose during the nineteenth century, particularly after the abolition of the ascription system. As social differentiation in the Russian villages grew, the rural population came to need other means of supporting themselves. Charcoal-burning was one such possibility.

93. Krivonogov, 'Krest'ianskii podnevol'nyi', 21.

94. Ibid., 23.

95. Ibid., 24; N.S. Popov, *Khoziaistvennoe opisanie Permskoi gubernii*, Perm' 1804, Part 1, 296, 306, 310.

96. Krivonogov, 'Krest'ianskii podnevol'nyi', 24–33.

The authorities realised that it was necessary to introduce more detailed regulation of the peasants' relations with the ironworks and stricter control of the administration. On 16 June 1810 the Perm' Mining Board thus introduced uniform rules for the hiring of state peasants at the ironworks. The regulations stipulated that the contract should be monitored both by the ironworks management and by the local administration. Two years later these regulations were further simplified. In a decree issued on 10 January 1812 the Senate laid down the conditions of employment of the state peasants at the ironworks in the province of Viatka. The peasants were given the right to work at the ironworks for four months without a pass from the local court. All that was required was permission from the peasant *volost'* administration. In 1815 this right was extended to all peasants.[97]

If the two regions are compared, it seems that the workforce had a more complex structure in the Urals than in Bergslagen. Hired labour, including free hired workers, were used to supplement serfs, ascribed peasants and *nepremennye* workers. In addition, some peasants even hired substitute workers who carried out their tasks. Up until the middle of the nineteenth century in Sweden, on the other hand, there were just three groups involved in charcoal-burning: freeholders, tenants and crofters.

The Works Owners: Production Techniques

In both regions the charcoal was produced in vertical kilns. However, the kilns were not completely identical. During the period of this study works owners and the state also tried to improve production techniques. Great efforts were made, particularly in the Urals, to develop a good charcoal-burning method. The work was made easier by the centralised production of charcoal.

The raw material in the charcoal-burning done by the Swedish peasants was freshly cut timber and trees blown down by the wind. In the Urals the peasants normally used firewood. In both regions they used their own horses and their own tools. The works in the Urals also provided certain tools.[98]

97. G. V. Iarovoi, *Regulirovanie krest'ianskogo otkhoda na gornye zavody Urala v doreformennyi period. Genezis i razvitie kapitalisticheskikh otnoshenii na Urale*, Sverdlovsk 1980, 15–29.

98. Kashintsev, *Istoriia*, 244.

Several methods of production were used in the Urals. The oldest method was the 'hole method', where the charcoal was burnt in so-called wolf holes. The method was used during the seventeenth century and at the beginning of the eighteenth century in the small-scale Ural production of iron, but was used afterwards in certain other cases as well. With the rise of the large-scale iron industry from the 1720s onwards, however, the older method was largely replaced by a new method, where the charcoal was produced in heaps. This method is known as the nomadic method.[99]

Other production methods were also introduced at the beginning of the eighteenth century. The German or Saxon method was introduced from Saxony. The kiln was laid out on the ground and built in three different layers. Using the German method it took, depending on the size of the kiln, five to sixteen days to burn the charcoal.

The commonest method in the eighteenth century was the Ural method, which was also called the ordinary method. This method was a modification of the German method. The volume of the kiln was twenty *kurennye sazheni*, in comparison with a maximum of fifteen with the German method. Other methods were also tried out in the Urals, including the Swedish method with horizontally stacked, tightly packed, large barked logs. Charcoal which was produced in this way was of high quality, but required timber rather than firewood. There are also examples of the American and the Italian methods being applied in the Urals during the nineteenth century. However, neither method became widely used.

From the middle of the nineteenth century the works owners in both regions made greater and greater efforts to improve charcoal-burning techniques. The goal was to save the forests and to increase the quality of the charcoal. Foreigners who visited the Urals brought with them new techniques. Works managements and experts in the Urals travelled to Europe to study the iron industry and charcoal production. In 1848 two French master forgemen, the Grandmontagne brothers, began to try out a French method for producing charcoal. The kilns in this method were considerably smaller than those that were usually used in the Urals (10 *sazheni*). In this way there could be a reduction in the number of workers required to transport the wood to the kiln and to take away the

99. Nomadic as opposed to stationary. Nomadic charcoal-burning was characterised by an annual change of the place where charcoal was produced, in accordance with economic judgements and the state of the ironworks' forests.

charcoal produced. The charcoal-burning process itself also became more even and control of the process was improved.

The Grandmontagne brothers' trials at the Nizhnii Turinsk ironworks gave good results. Production rose by 10 per cent in comparison with the Ural method, and the quality of the charcoal was improved. The French method was then introduced at the Nizhnii-Isetsk works and at a number of other state and private ironworks.

At the same time as the introduction of the French method, the ironworks manager M. Falkner started using the Tyrolese method at the private ironworks in Suksun. In this method the logs were dried a little in advance. The heap was piled on boards; the top of it was covered by coal dust instead of conifer branches and the fire was kept constantly fed by coal. The trials gave good results, with a yield of 70 per cent (in volume), as compared to 40 to 50 per cent with the Ural (ordinary) method. But as the technique also increased the costs it was never widely used in the Urals. Falkner therefore decided to adapt the Tyrolese method to conditions in the Urals and to the older nomadic method. This resulted in the new Suksun method. It required just as large a workforce but gave a 20 per cent higher yield.

The new Suksun method gained ground during the 1850s and 1860s in the Ekaterinburg mining and iron district. From the beginning of the 1860s it made a complete breakthrough, together with the French method, at the Nizhnii-Isetsk ironworks.[100] But in spite of these breakthroughs the older charcoal-burning methods were still widely used at the end of the century. The reason for this was above all the supply of cheap wood in the Urals.

Producing charcoal in furnaces was tried out in the Urals for the first time at the Nev'iansk ironworks in the 1760s. The charcoal furnaces did not, however, become widely used and were, in spite of the fact that they were first used at several works at the beginning of the nineteenth century, rare for many years into the nineteenth century. Investment costs were great and they required a constant supply of wood. In 1891 not more than 25 per cent of the charcoal in the Urals was produced in charcoal furnaces. The technique made a breakthrough at the beginning of the twentieth century, due to a sharp rise in the price of standing timber. In 1917 the majority of the Ural charcoal was produced in

100. Putilova, 'O sostoianii toplivnoi', 137.

101. Kafengauz, *Istoriia,* Tome 1, 311; K.-I. Nogin, 'K istorii uglezhzheniia na Urale', *Ural'skii tekhnik,* 1918, no. 10–12.

charcoal furnaces. At this time kiln charcoal represented a mere 10 per cent.[101]

In Sweden individual works owners, such as the Jernkontoret (the Ironmasters' Association), also carried out a number of experiments with a view to improving quality and reducing costs. At the same time, the small-scale organisation made it more difficult to achieve a more general improvement of the methods and it was not until some way into the nineteenth century that development work began in earnest. Instead, the state and the works owners tried to teach the peasants a more rational way of working. This was particularly important in districts where the peasants lacked previous experience of charcoal-burning.[102]

In this connection it should not be ruled out that the Swedish peasants themselves improved production methods. However, there is virtually no information at all about this. Nevertheless, it should, in contrast to the peasants in the Urals, have been in their interests to improve the techniques used if it reduced the amount of work required and increased the amount of charcoal.

During the seventeenth, eighteenth and early nineteenth centuries there were by and large two types of charcoal kilns in use in Sweden: the horizontal kiln and the vertical kiln. The exact construction of the kilns varied from region to region.[103] The horizontal kiln was the original kiln and dated back to the Middle Ages. The vertical kiln was introduced either by German immigrants during the sixteenth century or by Walloon master forgemen during the following century.[104]

Certain initiatives were taken within the Jernkontoret during the latter half of the eighteenth century. In 1781 a sum of money was reserved to reward peasants who improved their production of charcoal.[105] It was only after the turn of the century that the Jernkontoret began to get seriously involved in the issue. From 1811 to 1813 the famous Swedish metallurgist C. D. af Uhr carried out some important charcoal-burning experiments. They became the starting point for an intensification of the efforts to improve the Swedish ironworks' supply of charcoal.[106]

102. Montelius, 'Iggesunds Bruk 1685–1869', 76–77.
103. Hildebrand, *Swedish Iron*, 95.
104. Norberg, *Engelsbergs Bruk*, 206; Hildebrand, *Swedish Iron*, 95.
105. Boëthius and Kromnow, *Jernkontorets historia*, vol. III:1, 15, 73.
106. Boëthius and Kromnow, *Jernkontorets historia*, vol. III:1, 15, 73, 145, 512–514, 540–541; Arpi, *Järnhanteringens träkolsförsörjning*, 67ff.

The experiment was carried out in Furudal in the north of Dalarna. A few years earlier the Board of Mines had unsuccessfully tried to get the Jernkontoret interested in this type of experiment. In 1811 af Uhr wrote a memorandum, where he discussed the problem and how it could be solved. In particular, he complained about the charcoal-burners' lack of skill, which gave charcoal of varying quality. In the memorandum he suggested, amongst other things, a broad investigation to solve the old controversial question of whether the horizontal or vertical kiln gave the best quality charcoal. The Jernkontoret accepted the proposal and the trials started in Furudal.[107]

The experiments were carried out very carefully and the charcoal was analysed. The investigation came out slightly in favour of the horizontal kiln. Larger quantities of charcoal were produced in them and less work was required.[108]

These successful experiments in Furudal were followed by a number of charcoal-burning experiments from around 1815 onwards. From the 1820s several experiments using charcoal furnaces were also started, with the financial support of the Jernkontoret. At the end of the 1820s a new charcoal furnace was constructed which produced charcoal of good quality and which reduced the amount of labour needed. Furthermore, the by-products, in particular tar, were profitable.[109]

Conclusions

Great efforts were made in both of our two regions to produce sufficiently cheap charcoal of high quality for the furnaces and forges.[110] There are striking similarities between the approaches employed, but there are also clear differences. The problem was almost the same in both regions. Costs were rising at the same time as the peasants opposed over-harsh conditions and protested, either openly and in an organised manner or secretly and in a disorganised way. Collective action, in the form of charcoal strikes and peasant uprisings, were more common in the Urals than in Sweden. Instead, the Swedish charcoal producers protested in the mining-courts and in the *Riksdag*, or chose more secretive and ille-

107. Boëthius and Kromnow, *Jernkontorets historia*, vol. III:1, 521ff.
108. Ibid., 514.
109. Ibid., 515–518.
110. This section is by Maths Isacson and Maria Ågren.

gal methods, such as cheating the works inspectors and in this way getting paid for a larger volume than they had delivered. If even this did not make their work profitable enough, they could choose open collective action, or, as during the nineteenth century when agriculture developed and certain farmers became increasingly wealthy, stop selling charcoal to the works completely.

The peasants' protests, rising charcoal prices, charcoal of poor quality and expensive transportation over long distances were a growing problem for the works owners and the governments in both Sweden and Russia. Counter-measures were necessary. The works owners in the Urals tried above all to develop the techniques used and to strengthen the organisation of the work. In Sweden, where production was small-scale and household-based, the works owners instead tried to obtain a stronger grip on peasant labour and to get deliveries and prices regulated and monitored.

The workforce in the iron industry in the Urals had a more complex structure than in Sweden. Forced labour was gradually supplemented with more or less free workers as iron production developed. The production of charcoal was run on a large scale, and the different stages of the work were divided up and given to different groups of workers. Production was closely supervised by the works' own inspectors. In Sweden it was not until the end of the nineteenth century that larger-scale charcoal production was set up. In the second half of the century the market for charcoal grew. At the same time, the new modern saw-mills made it possible to use the waste from the sawing to produce charcoal and the ironworks were now able to collect their charcoal from distant forests by means of railways and canals.

The works owners' and the state's measures to secure a stable supply of charcoal were not completely in vain in either the Urals or in Bergslagen. Changes in the structure of the workforce, stricter supervision and (above all in the Urals) a more definite organisation of the work were measures that bore results. However, it is difficult to prove this by means of figures or in any other way. Nevertheless, there is information that shows that the volume of charcoal obtained from the wood rose because of better methods, greater knowledge of charcoal-burning and sharper controls. The 'hole method' used in the seventeenth century gave a 33 per cent yield from coniferous trees in the Urals. The method of burning charcoal in heaps did not increase the volume obtained

111. Strumilin, *Chërnaia metallurgiia*, 65 and 72.

immediately (33 per cent 1660), but with improved knowledge the volume of charcoal increased. In 1847, 47 per cent of all charcoal was obtained from the wood (pine). Subsequently the yield rose and at the end of the century it was almost 70 per cent.[111]

There are also figures for Sweden that show that the charcoal yield increased during the nineteenth century. While the peasants in the north of Sweden got 40 per cent of charcoal from the wood using the traditional techniques, the charcoal peasants under close supervision managed to attain the double. C. D. af Uhr and other metallurgists achieved a yield of between 64 and 94 per cent during their experiments.[112]

The forests were better utilised and supplies stabilised; in addition the consumption of charcoal in the furnaces and forges diminished. In the long term this opened up completely new development opportunities for Swedish and Russian iron production. The figures tell the story. At the end of the eighteenth century a normal Swedish ironworks needed on average 400 hectolitres of charcoal to produce one ton of bar iron. At the beginning of the century as much as 500 hectolitres had been consumed.[113] Several decades into the nineteenth century a new production technique rapidly reduced charcoal consumption. Around 1850 just 300 hectolitres were needed for one ton of bar iron. Twenty-five years later the amount had been reduced to 200 hectolitres.[114] In the Urals the amount of charcoal needed to produce one ton of pig-iron fell from 3.4 tons in 1723 to 1.5 tons at the beginning of the twentieth century. During the same period, especially after the middle of the nineteenth century, the amount of charcoal per ton of bar iron produced decreased from 6.4 to 3.0 tons.[115]

Before the ironworks could invest in production methods that consumed less charcoal in the furnaces and forges, the works owners were forced to deal with the charcoal question within a certain socio-economic framework, that is to say they became involved in a power game with the peasants and with the state's kind help. The history of iron production in the Urals and in Bergslagen is filled with open, collective protests and everyday resistance in the form of cheating with charcoal measures, bribes, illegal transportation etc. It was only at the end of the nineteenth century, with the advent of new smelting techniques and the railways, that the diffi-

112. Boëthius and Kromnow, *Jernkontorets historia,* vol. III:1, 539.
113. Hildebrand, *Swedish Iron,* 86.
114. Arpi, *Järnhanteringens träkolsförsörjning,* 99.
115. Strumilin, *Chërnaia metallurgiia,* 72.

FIGURE 5.3 Charcoal-Burning, an Indispensable Prerequisite for the Industrial Revolution, Dalarna, Sweden, 1934

cult charcoal question, even if it was not completely solved, became considerably less of a problem and easier to handle.

Charcoal was thus the key fuel in early iron production, in Sweden as well as in Russia. Huge amounts were consumed in both furnaces and forges. The inherent fragility of charcoal was an obstacle to its long-distance transportation on shaky, horse-drawn carts, though. It was not until the infrastructure had been considerably improved, mainly through the construction of railways, that charcoal's brittle quality ceased to be a severe problem. Likewise, it was not until it was possible to reduce the need for fuel, through modern production techniques, that the iron industry ceased to be strictly limited by the size of adjacent forests. Until these conditions had been met, which happened in the nineteenth century, the industry was forced to rely on nearby timber resources and the labour of neighbouring peasants.

In their role as charcoal producers, peasants made an indispensable contribution to early modern iron production. Each ton of charcoal required approximately ten days' work, in Bergslagen as well as in Russia; on the basis of these figures, it has been shown what enormous amounts of forest work lay behind the bar iron that mostly ended up on the British market, subsequently to take part in the industrial revolution proper. A crucial prerequisite for the industrial revolution was indeed the work carried out by Swedish and Russian peasants at the charcoal kilns (Fig. 5.3), and any argument as to the relevance of studying the social organisation of forest work should therefore be superfluous.

If we focus on the work process as such, it becomes apparent that forest work was organised differently in Bergslagen and in the Urals. In the latter region, charcoal production was large-scale and centrally directed as early as the eighteenth century. Large gangs of peasants assisted a small core of skilled charcoal workers, all supervised by the works administration. In Bergslagen by contrast, the primary work unit was the peasant household, be it the household of a freeholder, tenant, crofter, or *bergsman*. Production was organised on a small scale, with little central supervision. To a large extent, concrete decisions were taken by the work crew, i.e. by the members of the household. It was not until the nineteenth century that central direction and supervision increased among Swedish charcoal producers, as was shown above with regard to charcoal-burners and their work contracts.

If our focus shifts to the strategies used by Swedish and Russian ironmasters to make peasants take part in the work process, as it did in chapter 4, two important similarities come into view. First, as soon as it was possible work was organised on a hire basis. This process started in the early nineteenth century in both regions. Earlier attempts to bring about such a reorganisation had all failed. Secondly, until the change was possible, work was organised according to two different strategies. A relatively small labour force was attracted by way of property rights – stronger in the Urals, encompassing the labour itself, and weaker in Sweden, comprising only the land. In addition to this, a more substantial labour force was attracted by way of looser forms of compulsion. In the Urals, the alternatives that were thus combined were serf labour and ascribed labour. In Bergslagen, the corresponding alternatives were tenant labour and charcoal bought on monopsonistic terms.

Why did Swedish and Russian ironmasters apply these strategies and not others? Why, for instance, was the social organisation of work not built on serf labour/tenant labour only?

This question inevitably forces us to ask whether a system built exclusively on serfs or tenants would have been desirable to the ironmasters. Here, no definite answer can be offered. It should be noted, though, that in Sweden it was regarded as a great advantage to have a greater number of tenants and land at one's disposal. In Russia, the disciplinary and economic problems entailed by the use of ascribed labour, the illegal attempts of non-noble ironmasters to purchase ordinary serfs and the state initiative to institute *posessionnye* serfs all indicate that a mode of social organisation of work built upon property rights was regarded as the most robust alternative. So why was this approach not implemented more fully?

The question being counterfactual, we dare not venture to propose any definite answers. In Bergslagen it seems as if the answer is related to the availability of land and suitable labour, and political considerations. In other words: it was no simple task to extend the premises of an ironworks. Neighbouring areas would often be the property of individual peasant households, of peasant communities (village, parish, hundred) or of the state. Nor would it be straightforward to recruit tenants/crofters. The Bergslagen population was limited and freeholders predominated. Finally, in the seventeenth century Swedish statesmen pursued a conscious political line which regarded small-scale production (of pig-iron or charcoal) in the hands of peasant-owners as economically viable and worthy of state support. All these factors made it difficult, if not totally impossible, to increase the use of tenant labour.

In the Urals, the answer seems to be linked to the labour factor only, land being regarded as a virtually inexhaustible resource, and state policy favouring large-scale solutions all along the line. Labour was not as easily available in the Urals as might be assumed. This was a region to which many people migrated, but many continued even further east, into Siberia proper. There was a local Ural peasantry, but they were state peasants and could not arbitrarily be transformed into serfs, even in Russia. Once again, this emphasises the importance of the social and legal position of Swedish and Russian peasants in the regions we are discussing.

A sequel to the first question is: why was it that a system of charcoal districts was primarily used in the Swedish case, and not an ascription system? This question is also counterfactual and we will merely hint at some conceivable causes. If we try to envisage a system of ascription in a Swedish setting, two conditions immediately catch the eye. The Russian system of ascription was a way of transferring taxation rights from the state to the ironmasters. Similar transfers did take place in Sweden in the seventeenth century, when the state let noblemen receive certain taxes that were due from freeholders (known as *skattefrälse*). During the same period, it was not uncommon for ironmasters to lease taxes due in the form of charcoal. But in Sweden, transferred rights of taxation were not unlimited. The nobleman was not allowed to exact more than a certain, specified quantity. Further, taxes were calculated on the basis of the resources of the farmstead and the responsibility for paying them lay solely with the possessor. The corollary to this was that anyone holding the taxation rights, be they the state or a nobleman, had to accept delays and total failures to pay taxes, e.g. in times of failed harvests. The tax liability could not be transferred to other peasants within the same community. Tax debts are known to have been frequent in early modern Sweden.

The strength of the Russian ascription system was that it allowed ironmasters to demand more (if not totally) unlimited amounts of charcoal. These demands were addressed to the peasant community and not to individual households, the members of the community being mutually responsible for performing the work. Thus, delays and total failure to perform work were, at least in principle, inconceivable. There would always be some village members who could be drafted to the works. But ascription of individual peasant households and having to accept delays and debts – the most likely design of a hypothetical ascription system in an early modern Swedish context – would hardly have been of interest, even if it had been possible. Thus, in a situation where free hire and serf labour were not available and ascription was less interesting, the best way of reinforcing the Swedish supply system was to build on commercial ties and debt dependence, and to exert political pressure through lobby groups.

Even so, the weakness of the Russian ascription system was that it incessantly engendered serious conflicts between ironmasters and ascribed peasants, who detested their forced labour. These conflicts entailed economic setbacks for many ironmasters. This is obvious in the case of the Pugachev rebellion, in which ascribed peasants took part and which has been estimated to have cost more than 5 million roubles (including damage and loss of production). But everyday resistance, in the form of work performed badly and slowly, also had its costs. This is not to say that the Swedish system of charcoal production did not engender social conflict, which would obviously be untrue. However, the Swedish system may have been more robust, in that it retained the household as a unit of production. What Swedish charcoal-burning peasants seem to have resented, and occasionally fought, was the low price paid for the charcoal, whereas Russian peasants also disliked the way their work was organised, which forced them to be away from home and work in large gangs under strict supervision.

The range of strategies and solutions open to Swedish and Russian ironmasters was thus quite restricted and depended on the social composition of the rural population, on the design of taxation systems, and on the kind of support they received from the state. Finally, the attitude of the local peasantry was not unimportant, especially as it did not remain the same over the years. In both Bergslagen and the Urals, the local peasantry (affected by ascription/the charcoal monopsony) became increasingly hostile to the iron industry, as official prices were kept at an artificially low level while, at the same time, agriculture was turning into a highly profitable economic activity.

Referring to the ideas developed by Jan De Vries (see chapter 4, introduction), we may say that at least some of the Swedish and Russian peasants affected by the early iron industry turned into commercial, market-oriented farmers, unwilling to sell their forest products to anyone who did not pay well, indeed unwilling to work in the forests at all if agriculture was more rewarding. On the other hand, other peasant groups were gradually being subordinated to the ironmasters, as serfs, as *nepremennye rabotniki*, or as charcoal-burners. There was a structural conflict between the needs and interests of ironmasters and those of the local peasantry, a conflict which was exacerbated if peasants were badly paid or if they lost other income by being forced to be away from their ordinary work. Some peasants managed to pull through this situation as well-to-do farmers, whereas others emerged as more or less proletarianised workers.

Chapter Six

Households, Families and Iron-Making

Göran Rydén with Svetlana Golikova

Households in Early Industrial Society

Over the last twenty-five years, the study of family history has experienced a phase which, to use Walt Rostow's term, can be characterised as a take-off. Peter Laslett wrote in 1972, in the introduction to the book that was really the starting point to this phase, about the 'unpopularity of the subject'.[1] This could hardly be said today. The discipline has expanded, both in terms of published titles and new journals, and in terms of its penetration into new areas of research. Family history must today be seen as one of the most vital areas of historical study.[2]

The guiding principle seems to be to analyse the family or household as a very important institution in society, or as one recent contributor put it: 'Understanding the family as a mediator between the lives of individuals and larger communities promises to help us understand the social history of the Western world in richer details.'[3] Households thus functioned as a kind of transmitter between individuals and society, socialising the people according to the norms adopted by society, while at the same time

1. P. Laslett, 'Introduction: The History of the Family', in *Household and Family in Past Time*, eds P. Laslett and R. Wall, Cambridge 1972, 2.

2. See T. Hareven, 'The History of the Family and the Complexity of Social Change', *The American Historical Review*, vol. 96, No. 1, 1991, for a survey of recent developments in the field and also an agenda for future research.

3. K. Lynch, 'The Family and the History of Public life', *Journal of Interdisciplinary History*, XXIV:4, 1994, 665.

changing society in accordance with the needs of the people. In line with this, Wally Seccombe has elaborated a model which views economic and social development as an interdependence between 'the mode of production' and the changing patterns of the household, rather than seeing the latter merely as a passive response to changes in the former.[4]

If the household or the family still has an important position in society today, we think it is safe to state that this was even more clearly the case in pre-industrial times. Research has shown that in an agrarian economy, based largely on self-subsistence, there was a close connection between the work of the people, their lives and the land on which they lived. The household was a unit for production, consumption and reproduction, and it clearly constituted the economic foundation of society. Furthermore, it was the institution which represented individual people in relation to the society. It was, for instance, the household and not its individual members who paid tax, and it was through households that the church delivered its messages. The family or the household was also an archetype on which important relations and institutions were modelled. The ruling state often, in fact, laid down its decrees and aims in terms of a family or household. The king or the state was seen as a paterfamilias dealing with his inferiors – the children, and the religious framework was also structured according to the same discourse, based on family and household.[5]

The aim of this chapter is not to do justice to all these new trends in family history. We do nevertheless want to lend our support to a view that gives family and household an important place in the analysis of historical developments. The more limited aim here is to perform a more traditional study based on the household as a production unit. We want to set the different households connected with the iron industry in Sweden and the Urals in the context of an agrarian economy encountering industrial development in the form of expanding export-oriented production. The people involved in iron-making were therefore involved in a mixed economy with a complex structure.

4. W. Seccombe, *A Millennium of Family Change. Feudalism to Capitalism in Northwestern Europe*, London 1992, and W. Seccombe, *Weathering the Storm. Working-Class Families from the Industrial Revolution to the Fertility Decline*, London 1993.

5. See S. Amussen, *An Ordered Society. Gender and Class in Early Modern England*, Oxford 1988, for such a view.

Household Composition

In an important article published in 1983, Peter Laslett tried to summarise the information available about family forms and sizes in Europe. His starting-point is a division of the continent according to different marriage patterns. John Hajnal had already, back in the 1960s, proposed that a 'European marriage pattern', with late marriage for both sexes and neo-location after marriage, prevailed west of a line from St Petersburg to Trieste. This was later re-phrased as the 'north-west European system'. East of that line, marriage occurred earlier for both men and women, while the married couple often stayed in the household of the young husband's parents.[6]

Laslett elaborated further on this, however. A lot of research, primarily during the 1970s, had indicated that there were many important differences within these two large regions, which in some cases were even more pronounced than those between the regions. Laslett therefore proposes a fourfold division of European household composition. This scheme is based on marriage patterns and family forms, but also on kin and work patterns. His regions are 'west and north-west', 'middle', 'Mediterranean', and 'east'.[7]

According to Laslett, to dwell only on the distinctions between the first and the last of his regions, the striking dissimilarity was that households in the west were fairly small, while their equivalents in the east were very large. The main reason for this is to be found in the two features discussed by Hajnal: in western Europe the couple married fairly late, with the female marrying in her mid-twenties, and formed a new independent household after getting married. In eastern Europe the couple married at an earlier age and seldom formed a new household of their own. Instead they remained in a household that consisted of other married couples with their children. This large co-resident group thus contained individuals who were related to each other by either blood or marriage. In the vocabulary already established by Laslett in 1972, western households were 'simple family households', while the 'multiple family household' prevailed in the east.[8]

6. P. Laslett, 'Family and Household as Work Group and Kin Group', 517 and 525, and J. Hajnal, 'Two Kinds of Pre-industrial Household Formation System', 69. Both in R. Wall, J. Robin and P. Laslett, *Family Forms in Historic Europe,* Cambridge 1983.

7. Laslett, 'Family and Household', 513–525.

8. Laslett, 'Family and Household', see foremost table 17.5, 526f. See also Laslett, 'Introduction', 31.

In spite of a fairly limited amount of research the patterns discussed by Laslett, and also by Hajnal, fit very nicely with information about demographic trends in Sweden and Russia. Concerning the latter country it is obvious that households were large indeed. Peter Czap Jr has, for instance, shown that households of serfs in the Riazan area in southern Russia comprised on average around ten members in the late eighteenth century and the first half of the nineteenth century. His study also shows that between 60 and 70 per cent of the population lived in households of more than nine persons and that 15 per cent of these actually lived in households of fifteen individuals or more. Czap has also investigated the marriage pattern and concluded that most couples married before they reached the age of twenty-one. The mean age of marriage for women oscillated between seventeen and nineteen years.[9]

Steven Hoch has shown the same pattern of large households and a low marrying age. The majority of all households in his study of an estate in the Tambov region were 'multiple family households', consisting of more than two conjugal units. The mean household size was around eight, while the mean age of marriage for women was around eighteen.[10] In a study of the area around the (pre-)industrial town of Jaroslav, in the mid-eighteenth century, Michael Mitterauer and Alexander Kagan found serf households of a slightly smaller size. They included on average just over five people. A likely explanation for this was the higher mean age of marriage. They do not give an exact figure, but only 65 per cent of all women in the age group 20 to 24 had married. This did not, though, prevent the forming of complex households, but not with the generational depth described by Czap.[11]

Both Czap and Hoch emphasised in their respective studies that the family or household was not a static feature of society, but

9. P. Czap, 'Marriage and the Peasant Joint Family in the Era of Serfdom', in D. Ransel, *The Family in Imperial Russia*, Urbana 1978; P. Czap, 'The Perennial Multiple Family Household, Mishino, Russia 1782–1858', *Journal of Family History*, vol. 7, 1982; P. Czap, 'A Large Family: The Peasant's Greatest Wealth. Serf Households in Mishino, Russia, 1814–1858', in R. Wall, J. Robin and P. Laslett, *Family Forms in Historic Europe*, Cambridge 1983.

10. S. Hoch, 'Serfs in Imperial Russia: Demographic Insights', *Journal of Interdisciplinary History*, XIII:2, 1982 and S. Hoch, *Serfdom and Social Control in Russia. Petrovskoe, a Village in Tambov*, Chicago 1986, chapter 2.

11. M. Mitterauer and A. Kagan, 'Russian and central European family structures: a comparative view', *Journal of Family History*, vol. 7, 1982. See above all 109–112 and 116–120.

rather an entity that was constantly changing in both size and composition. It must therefore be analysed as a process, either in a life cycle perspective or in terms of the concept of the 'life course'. Czap has shown that the serf households in Riazan passed through different phases over time. They experienced periods as 'single family households', and also as both 'extended' and 'multiple family households'. It is hard to find a 'beginning' to the cycle, but a suitable starting-point could be a single family household, or a nuclear family. As soon as the eldest son reached marrying age, that is his late teens, he found himself a suitable wife, who moved in with him and his parents. We thus have an extended family, which, however, shortly turned into a multiple family household as soon as the other sons also brought wives into the household. The daughters left the households. As soon as the newly wedded couples had children of their own, we thus have households including three generations, and Czap also notes that about 10 per cent of all households actually included four generations. When the head of the household died his position was often taken over by a son, who was then in charge of his brothers and their children. In some cases households became too large and fissions occurred, large households being turned into two or more smaller ones.[12]

Leaving Russian households for their Swedish counterparts it is clear, as was indicated above, that we are crossing a border from one area with a distinct household composition to another area with a different household 'set-up'. Households diminish in size as we travel westwards, and their structure also changes. Multiple family households disappear and are replaced by simple family households, or nuclear families. As with the case of Russia, however, we still lack general information about the development of family and household in Sweden. We therefore have to rely on a number of local studies in order to sketch a picture of Swedish households in general.

Christer Lundh, however, has recently made a review of the literature devoted to these issues. He argues that in spite of wide variations in household size and structure among the Swedish population during the seventeenth, eighteenth and nineteenth centuries, it is nevertheless possible to discern some general features. Lundh establishes four general points. Two of these relate to varying patterns. According to him, there was a regional variation

12. Czap, 'The Perennial Multiple Family Household', 14ff, and Czap, 'A Large Family', 132–141. See also Hoch, 'Serfs in Imperial Russia', 237ff.

in both the composition and the size of Swedish households. Furthermore there were marked differences between different social groups. A third point is the gradual decrease in the size of households from the seventeenth century to the nineteenth. The last point mentioned by Lundh is that the structure and size of households varied over time, according to a life cycle.[13]

A typical life cycle of Swedish peasant households includes the features of late marriage, 'life cycle servants' and some kind of retirement system. If, as in the Russian case, we start with the marriage of a couple, they often lived by themselves with their children as a nuclear family. If they were freeholders it was very likely that their household was extended to include servants, maids as well as farm-hands. As soon as the children grew up they took over the tasks previously performed by the servants, but youths in their late teens also moved away from their parents to take up positions as servants in other households. The phase as a servant ended in marriage and the creation of a family of one's own. This occurred in the mid-twenties, for both men and women. For freeholders, this could take place in connection with the parents of the newly wedded groom handing over power and control of their farmstead to the new couple.[14]

To make possible a comparison with the above description from Russia, it can be said that the mean age of marriage in Sweden during the pre-industrial period was around 25 years for women and 27 years for men, while the average size of a household was between four and six.[15]

It is interesting to take as our starting-point the variable pattern of Swedish household composition and size when turning from the general picture to a discussion dealing with households connected with the iron industry. In Sweden, a system involving regional as well as social variation between different households prevailed. There was in addition very marked specialisation among these households, which performed different tasks in relation to the iron industry. There were *bergsmän*, responsible for mining and pig-iron production, *tenants*, making charcoal, *freeholders*, undertaking carting, *forgemen* etc., each with a different household

13. C. Lundh, 'Households and Families in Pre-industrial Sweden', *Continuity and Change*, vol. 10, part 1, 1995.

14. Lundh, 'Households and Families'.

15. Lundh, 'Households and Families', 40, and T. Bengtsson and R. Ohlsson, 'Sveriges befolkning – myter och verklighet' in B. Furuhagen, *Äventyret Sverige. En ekonomisk och social historia*, Stockholm 1993, 115.

composition. This is in contrast to the Russian situation, where there seems to have been one form of household supplying labour of different kinds to the ironworks. One type of ironworker household, very similar to the agricultural serf households dealt with above, could for instance include one forgeman, one carter and one charcoal-burner. How, for example, can we characterise the household of Evdokim Chistiakov, from the Nizhnii Tagil works? He was a woodcutter, his oldest son was a worker in the forge, his second son was an apprentice in strip-iron production, while his youngest son was a miner. It is more appropriate to talk of 'households, in which ——— resided', specifying the members occupations, than to delineate specific types of households.[16]

There was, however, a trend towards a situation more similar to the Swedish one, and away from the multiple family household. Skilled workers left their parents after marriage and formed nuclear households. We thus find a varying picture of the households at ironworks in the Urals; on the one hand large multiple family households mainly containing auxiliary workers and, on the other hand small nuclear families including skilled workers.[17]

Starting with Sweden and the *bergsmän* it can be said to begin with that, by Swedish standards, they lived in fairly large households. David Gaunt, who has studied a number of different groups or strata in the countryside in the region of Västmanland, showed that *bergsmän* had the largest households. On average, households in Skinnskatteberg, with a large proportion of *bergsmän*, included 6.5 persons. The women in this parish gave birth to more children, its households more often included kin from outside the nuclear family, and they were also more often extended to include servants than households in the other parishes studied. Closest in terms of household size, number of children, kin in the household and servants came the parish of Västerfärnebo. The socio-economic structure of that parish was dominated by freeholders who in winter were busy transporting iron from Bergslagen to the port in Västerås and grain in the opposite direction.[18]

16. GASO, F.643.Op1.1.D.102.L.338 ob.

17. Kh. Mozel', *Materialy dlia geografii i statistiki Rossii, sobrannye sluzhashchimi general'nogo shtaba: Permskaia guberniia* (Geographical and statistical facts about Russia, collected by employees of the General staff: Province of Perm), St Petersburg 1864, vol. 2, 533.

18. D. Gaunt, 'Familj, hushåll och arbetsintensitet. En tolkning av demografiska variationer i 1600- och 1700-talens Sverige', *Scandia* 1976, and D. Gaunt, 'Pre-industrial Economy and Population Structure', *Scandinavian Journal of History* 2, 1977.

Gaunt connects this pattern of differing household composition with 'differences in the local human ecological situation: the economic, social and ecological framework'. The *bergsmän* had a more intensive and more evenly distributed workload over the whole year than the other groups investigated. It was therefore more efficient for them to have larger households. The same could be said, to a slightly lesser extent, of the freeholders engaged in carting and hauling in the winter.[19]

All this could be complemented with information from the more limited demographic surveys of Maria Sjöberg and Ture Omberg. The latter found a correlation between the size of *bergsman* households and the tasks performed by them in relation to pig-iron production. Households with larger shares in blast furnaces also had larger households.[20] According to Sjöberg, the households of the *bergsmän* in Söderbärke had an average size of between five and seven people during the eighteenth century, which agrees with the figures from Gaunt's survey. It is clear from her study, however, that these households went through a life cycle in which some phases entailed larger households. Sjöberg, like Gaunt, shows that 'extended family units' were common, but she also gives a few examples of 'multiple family households'. The commonest form was when a *bergsman* household was extended to include a married son and his children.[21]

Maths Isacson, in his study of the economic development and social differentiation of the parish of By, also dealt with the matter of household composition and size in relation to the iron industry. He analysed *bergsmän*, charcoal-burning tenants and freeholding peasants. His calculations show, as did studies referred to above, that the *bergsman* households were larger than those of the other groups, but they also show that if these groups are sub-divided according to the size of their holdings it can be concluded that wealthier peasants had larger households. Well-to-do *bergsmän* in By had households of on average 7.35 people in 1810, while their less prosperous neighbours had households comprising only 6.2 members. Charcoal-burning tenants had households of the latter size.[22]

19. Gaunt, 'Familj, hushåll och arbetsintensitet', and Gaunt, 'Pre-Industrial Economy and Population Structure'. The quotation from the latter 184.

20. T. Omberg, *Bergsmän i hyttelag. Bergsmansnäringens utveckling i Linde och Ramsberg under en 100-årsperiod från mitten av 1700-talet*, Stockholm 1992, 90ff.

21. M. Sjöberg, *Järn och jord. Bergsmän på 1700-talet*, Stockholm 1993, 47–55, 203–218.

22. M. Isacson, *Ekonomisk tillväxt och social differentiering 1680–1860. Bondeklassen i By socken, Kopparbergs län*, Uppsala 1979, 96–100.

What sort of conclusions about households leaning towards the 'agricultural side' of the Swedish iron industry can, then, be drawn from the material presented above? It ought to be fairly safe to assume that both the size and the composition of the households of the groups discussed so far were well within the overall pattern for Sweden as a whole. The average size remained roughly within the range of four to seven people, while the composition seems to have been that of a life cycle, with alternating phases of simple and extended family households, but with the former predominating. The mean marrying age for women tended to be around the mid-twenties,[23] while the research cited also suggests a system of life cycle servants.

Another obvious feature is the larger size of the *bergsmän's* households. According to Gaunt, they were on average at least one person larger than those of carting freeholders. Isacson also showed that *bergsmän* had larger households than charcoal-burning tenants. These differences were explained by the more intense and more evenly distributed workload of the *bergsmän*. This last conclusion, however, is rather shaky, and more research must be undertaken in order to confirm the relationships between, on the one hand, the household compositions and sizes of different socio-economic groups and, on the other, household composition and size and the working patterns of these groups. No attempts have been made to analyse the life cycle or life course of the charcoaling and carting groups, for instance, while this aspect has only been studied on a limited scale for the *bergsmän*.

Turning now to the households connected with the Russian iron industry, we can conclude that they also fit into the general pattern of European household composition discussed above, this time representing the eastern form in Laslett's categorisation. However, they seem to have been smaller than the households discussed for the southern parts of Russia, even without taking into consideration the smaller households of skilled workers that emerged towards the end of the eighteenth century. We have no average figures on household size at the Ural ironworks, but an 'average' household went through a family life cycle in which either two or three generation lived together. They thus only passed through the phases of simple family household and extended family household. The average size of households in the first of these phases was less than five people while the extended

23. Gaunt, 'Pre-Industrial Economy', 205.

households comprised more than six.[24] There was thus little difference compared with the households of Swedish *bergsmän*.

A simple family household entered the extended phase on the marriage of the eldest son. As in the general pattern, marriage occurred in the late teens for both spouses. The newly-weds lived together with the husband's parents and younger siblings. This phase lasted until the parents died and the household once again became a simple one with the eldest son as its head. Few multiple households, of three or more married couples, are mentioned in the sources implying that younger sons left their parents when they got married. Maybe these sons were the embryo of the nuclear families of skilled workers mentioned earlier, but it might also have been the case that large, that is multiple, households occupied one house, but each married couple was given a part of its own. Kh. Mozel' writes about this in the 1850s: 'Married sons seldom remain in the same house as their parents. Most of them form a household of their own. If they stay with their parents they nonetheless have a household of their own, which means that they eat separately. If there are two married brothers they try to separate.'[25]

This means that it is very difficult to say anything for certain about household composition and size in relation to the iron industry in the Urals. A plausible hypothesis is that multiple family households dominated for most of the eighteenth century. Towards the turn of the century this pattern was gradually broken down by a number of interrelated factors. Steven Hoch's study of the Tambov region, mentioned earlier, implies that the internal tensions of multiple households meant that as soon as there was no economic or coercive force to maintain them they were gradually dissolved.[26] The agrarian economy of that region differed substantially from the more diversified economy of the ironworks, and the need to sustain these large multi-generational households was perhaps not that strong on the part of either ironmasters or ironworkers. The pattern was thus gradually changing.

This trend might also be connected with the process whereby iron production in the Urals, as in Sweden, was gradually loosening its ties to the agrarian world. The abolition of the system of

24. GASO, F. 24. Op. 2. D. 1927.

25. GASO, F. 24. Op. 2. D. 1927. and Mozel', *Materialy*, 533. Quotation from the latter.

26. Hoch, *Serfdom and Social Control*, 128–132, 134, 189. See also Mitterauer and Kagan, 'Russian and central European family structures', 127f.

ascribed peasants, in the year 1800, and its replacement with the *nepremennye* workers, was part of this process.[27]

It is clear, however, that multiple households did not disappear altogether. A divergence seems to have occurred, with skilled workers coming to live in smaller households, while auxiliary workers remained trapped in the larger ones. A process might also have arisen in which multiple households remained, but were divided along internal demarcations. All married couples with young children were allotted some space of their own, in which they lived more or less separately from the other couples.

The gradual development in Russian ironworks towards two different types of household composition, one for auxiliary workers and one for skilled artisans, had its counterpart in Sweden from at least the sixteenth century. As was noted above, a clear division and specialisation of households existed within the Swedish iron industry. Most of these households had a clear connection with the agrarian society in which iron-making was located, and these were dealt with above. One group remains, and that is the specialised iron workers, living in the ironworking communities and working at the blast furnaces, forges and workshops for metal trades.

Little is known about the blast furnaces workers, but a few studies have been devoted to the forgemen, the aristocracy of the Swedish iron industry. Jan Sundin, in a study of ironworks in the north of Sweden, found a lower average age of marriage for forgemen than among other groups, but a somewhat higher age for their wives. According to Sundin the age fluctuated with the business cycles, with a marked fall after the Napoleonic wars. He also tells us that the forgemen's households were often extended to include servants, especially when the children were young, and that their children survived their early years better than other groups. He does not state the average size of these households, but all the information points towards fairly large units.[28]

This picture is largely confirmed by a study of part of the central iron-making area in Sweden. The forgemen of Gästrikland and eastern Dalarna, during the second quarter of the nineteenth century, built families and households when their spouses were in

27. See chapter 4, and *Iron-making in Sweden and Russia. A Survey of the Social Organisation of Iron Production before 1900*, eds G. Rydén and M. Ågren, 71–84 for an elaboration on this theme.

28. J. Sundin, 'Family Building in Paternalistic Proto-Industries: A Cohort Study from Nineteenth-Century Swedish Iron Foundries', *Journal of Family History*, vol. 14, No. 3, 1989.

their mid-twenties. The average marrying age of men and women was around twenty-five. They were thus of the same age; in Sundin's study the women were three years older.[29] The study from central Sweden also shows that households were sometimes very large. On average the women gave birth to more than five children, servants were frequent, and households extended to include relatives were not uncommon.[30]

The most striking features of these forgemen's households, however, was the very marked life cycle connecting the development of the household and work in the forge.[31] The boys started work as apprentices at the age of about seventeen. They remained in that position for seven years, until the age of twenty-four, when they became forgehands. Changes in the forge crews took place in the autumn, towards the end of the year. Simultaneously with his promotion to forgehand, a young man also got married. The first child of the newly wedded couple was born one year later.

After seven or eight years as a forgehand came the final promotion. The forgeman became a master at the age of about thirty-two.[32] With this new position a number of new obligations arose. The most important of them, in this respect, was the responsibility for apprentices. They – often only one at a time – were to be 'fed and clothed', that is, included in the household. At that time in the career, being in his early thirties, the master's eldest son, if he had one, was not older than about eight, so the apprentice had to be found from outside the household. The common pattern was also that a maid was included in the master's household at that time. The household of a newly promoted master forgeman often consisted of nine to ten members, the parents, five or six children, an apprentice and finally a maid. In extreme cases these households could swell to up to fifteen members if they also included elderly relatives.[33]

29. Sundin, 'Family Building', 276.

30. G. Rydén, *Hammarlag och hushåll. Om relationen mellan smidesarbetet och smedshushållen vid Tore Petrés brukskomplex 1830–1850*, Stockholm 1991, chapter 9. See also G. Rydén, 'Iron Production and the Household As a Production Unit in Nineteenth-Century Sweden', *Continuity and Change*, vol. 10, part 1, 1995 for a treatment in English which also deals with the first quarter of the century.

31. The following paragraphs are based on Rydén, *Hammarlag och hushåll*, 196–222.

32. Not all forgehands became masters. They remained in the crews as forgehands for perhaps another decade. The majority of forgehands who did not become masters left the forge in their early forties for a position as a combination of extra forgeman and day labourer.

33. See two examples in G. Utterström, '1815–1870', in *Fagerstabrukens Historia*, del V, 234f. Twelve per cent of the forgemen's households were extended with kin in some way. Rydén, *Hammarlag och hushåll*, 221.

As the children of the master forgeman grew older they took over from the apprentices and maids. The latter disappeared from the household when the oldest daughter was about fifteen years of age, and the apprentices were withdrawn when the boys turned seventeen. They, in turn, were later replaced by their younger sisters and brothers. The household decreased in size during this period. In cases where few children, or few boys or girls, were born it could become necessary to include servants from outside once again. The master forgeman retired at an average age of fifty-six.

No average size has been calculated for the forgemen's households of this region, but it has been established that the mean household size of forgehands was between three and four, while the masters had households as large as between six and seven.[34] A similar pattern, with masters having larger households than journeymen, was also apparent at ironworks involved in the metal trades. Anders Florén, in a study of Jäders bruk, has shown that master households varied between an average of four and five-and-a-half people from the mid-seventeenth to the beginning of the nineteenth century. Journeymen's households comprised around three members.[35]

This section began with a discussion of the contrasts in household composition and size that existed between different areas of Europe. In eastern Europe, people lived in large multi-generational households generated by early marriage, while the north-western part of the continent was dominated by a system of late marriage and neo-localism, with smaller households as a result. Households connected with the iron industry in the two regions discussed in this book conformed to this pattern. It is clear that households in the Urals were larger than those in Bergslagen. The features of marrying age and household composition also seem to fit in with the respective systems of the two regions. Ironworkers in Russia married earlier and remained in their parents' households, thus forming extended or multiple households. Extended households also existed in Sweden, thus supporting the criticism of Laslett's early views,[36] and above all among the *bergsmän*. The late age of marriage, together with life cycle servants and the existence of various

34. Rydén, *Hammarlag och hushåll*, 220.

35. A. Florén, *Genus och producentroll. Kvinnoarbete inom svensk bergshantering, exemplet Jäders bruk 1640–1840*, Opuscula Historica Upsaliensia 7, 1991, 43–49.

36. The first to express this criticism was L. K. Berkner, 'The stem family and the developmental cycle of the peasant household: an eighteenth-century Austrian exemple', *American Historical Review*, vol. 77, 1972.

types of households (*bergsmän*, carting freeholders, forgemen etc.), however, created a household system more open and adaptable than the one existing in the east.

It is important to emphasise that there are signs of a development at the Ural ironworks which involved a gradual crumbling of the system of multiple family households. The skilled workers might have been forerunners in this process, creating a life cycle of alternating simple and extended family households. It is beyond doubt that the number of simple family households, or nuclear families, rose at the expense of the larger multi-generational units.

Houses and Communities

The Germanic word 'household' (in German *Haushalt* and in Swedish *hushåll*) had a dual meaning in pre-industrial times, standing both for the group of people living together, as in the discussion above, and for the dwellings in which this group lived, including the house in which the people ate and slept, but also barns, sheds etc. We are thus dealing with a notion which encapsulates the whole economic, social and ecological sphere of the group of people concerned. The distinction which we make today between households and houses is a novelty of a later date stemming from the development towards a separation of work and the household and the emergence of the smaller, and more private, nuclear family. A similar notion existed in France, the *ménage*, meaning, as Martine Segalen puts it: 'The couple's life together, the couple itself, the house and its interior and also the upkeep of the house.'[37]

In the Russian language we have the word *dvor*, also with this double meaning. On the one hand it meant, as Czap has put it, 'a distinct, usually enclosed, space which encompassed all the structures and appurtenances of a single economy', such as dwellings, outhouses, a garden, etc., 'in short a farmstead'. On the other hand, it meant the people living within the 'boundaries' of this physical structure.[38]

It is important to bear this connection between people and their environment in mind when analysing earlier periods. It has not always been taken into account, and more demographically

37. See M. Segalen, *Love and Power in the Peasant Family*, Oxford 1983, quotation from 8; M. Mitterauer and R. Sieder, *The European Family*, Oxford 1982, 7–10, and Rydén, *Hammarlag och hushåll*, 28–40, for an introduction to this discussion.

38. Czap, 'A Large Family', 112.

oriented family history in particular has tended to see the groups investigated taken out of the context of their everyday life. One way of dealing with this problem has been to make a distinction between family and household, the first notion being used to highlight the aspect of kinship, while the household has been seen more as a loose economic unit. In his categorisation of household composition from 1972 Laslett tried to match the two together by discussing different types of 'family households'. He did not quite succeed, ending up analysing groups of persons residing together.[39]

If we are to analyse households in pre- and early industrial times in an economic, social and ecological setting, to see them as production and consumption units, we thus find that the 'physical structures' occupy an important position. The houses, the farmsteads, were important places where people met to work and engage in other social activities, they were also the places where allocations of resources took place. We might say that houses were the locus for all the activities undertaken by the members of households. David Sabean has put it in a slightly different way, stating that the household was 'a place' for exchange and the house a locus for this exchange.[40]

At the centre of a household's activities at the Ural ironworks was the house. It consisted of a dwelling house for all the members (*izba*), a stable, a cattle shed, various storerooms and a wash-house. The single-roomed *izba* was the heart of the household, with its large earthenware stove used for cooking and heating. Here the members of the household came together to eat and sleep. It is obvious that these living arrangements were very similar to those described for southern Russia. Czap has outlined the structure of a *dvor*, this time meaning the physical structure, in the Riazan region. It consisted of an *izba* with connected sheds and barns. Together these buildings formed a square enclosing an inner court yard. This complex was built on a parcel of land which also included a kitchen garden, fruit trees and summer enclosures for the animals. All this was probably also present in the case of the ironworkers. It is noted that in the cold and snowy area of the Urals it was common to cover the inner yard with a thatched roof.[41]

39. See Rydén, 'Iron Production', 71–77, for a critique of the demographic way of analysing households and families. See also Seccombe, *A Millennium*, 9–36.

40. See D. Sabean, *Property, Production, and Family in Neckarhausen, 1700–1870*, Cambridge 1990, chapter 3 and especially 122f, for a such a view.

41. V. Iu. Krupianskaia and N. S. Polishchuk, *Kul'tura i byt rabochikh gornozavodskogo Urala* (Workers' culture and way of life in the mining Urals at the turn of the

There were, however, more households than houses at these ironworks. A large proportion of the people therefore had to live in the houses of other households. In 1799, for instance, as many as 20 per cent of the households in the village of Miass had to live as lodgers, while an even larger proportion, a fourth of the entire population, at the Zlatoust ironworks lived together with other households. This does of course have an important bearing on the discussion above about the emergence of smaller households. This trend might have been present, but its wider spread was halted as there were not enough houses to accommodate it.[42]

This uniformity of living arrangements at the Russian ironworks was not paralleled in Sweden. The various types of households connected with the Swedish iron industry were coupled with different ways of arranging the 'co-residence' of the household. The different categories of peasants, the *bergsmän*, charcoal-burning tenants, carting freeholders etc., did of course inhabit farmsteads of a more or less standard type. These were furnished with dwelling houses, together with different kinds of barns, sheds and other outhouses. The *bergsmän* also possessed sheds and outhouses adjoining the blast furnaces, for storing ironstone, charcoal, limestone and pig-iron. The biggest difference between these farmsteads was perhaps not whether they were owned, or only used, by different categories of peasants, but really whether the owners/users were wealthy or not.[43]

The living arrangements of the ironworkers proper, those living at the actual ironworks, differed quite significantly from those of the peasants. They kept animals and had some fields, but their looser connection with agriculture was apparent in the way they lived. They had dwelling houses with outhouses, but they only had a share in a common cattle shed. It is therefore not possible to talk about farmsteads in this case. For forgemen and blast furnace workers it was stipulated in decrees from the Board of Mines how

last century) Moscow 1971, 64; E. N. Bubnov, *Russkoe dereviannoe zodchestvo Urala*, Moscow 1988, 55; *Istoriia Urala s drevneishikh vremën do 1861 g.*, Moscow 1989, 433; Czap, 'A Large Family', 112ff.

42. A. S. Cherkasova, 'Demograficheskaia kharakteristika masterovykh i rabotnykh liudei Zlatoustovskogo i Miasskogo zavodov Urala', in *Demograficheskie protsessy na Urale* ('Demographical characteristics of artisans and workers at the Zlatoust and Miass works in the Urals at the end of the eighteenth and the beginning of the nineteenth centuries' in *Demographical Processes in the Urals During Feudalism*), Sverdlovsk 1990, 107.

43. J. Granlund, 'Greksåsars Bergsmän i Hyttelag och Gruvlag', *Meddelanden från Föreningen Örebro Länsmuseum*, 1947.

large their dwellings were to be, including the size of the outhouse and their share in the cattle shed. A master forgeman had the right to a dwelling with one large combined room and kitchen, together with a smaller room. The size of his share in the cattle shed should enable him to house two cows. A forgehand was only entitled to the combined room and kitchen and space for one cow.[44]

The houses providing dwellings for the forgemen were often large, encompassing four or even more separate apartments. Sometimes they were built along a road or street with the cattle sheds and outhouses on the opposite side of the road. We call these ironworks streets, although we are dealing with small communities in the countryside.

Another important difference between the Urals and Bergslagen in this respect was the size of the communities in which different categories of ironworkers lived. It was not only that the households of Russian ironworkers were larger than their Swedish counterparts. They also lived in much larger communities. The households involved in iron-making in Sweden lived in very small communities indeed, and it did not matter whether they were villages of *bergsmän* or ironworks communities. Maria Sjöberg has shown that only about 150 people, in seventeen households, lived in the village of Saxe in the parish of Söderbärke.[45] The ironworks communities were no larger. On the contrary they were often smaller. None of the works belonging to the enterprise of Uddeholm in western Sweden had more than one hundred inhabitants at the beginning of the nineteenth century.[46] In sharp contrast to this was the size of the Russian ironworks. At the far end of the scale was Nizhnii Tagil, the centre of the Demidov empire. In the 1850s it was described by a British traveller, R. I. Murchison, as a town of 25,000 inhabitants. It had brick and stone houses, a hospital, and 'very comfortable dwellings for the workmen and their families'. Nizhnii Tagil was an exception, enlarged by expanding copper, gold, platinum and malachite extraction and production in a period when iron-making was stagnating. Ten thousand people worked in the mines and industries in and around the town.[47]

44. Rydén, *Hammarlag och hushåll*, 233ff.

45. Sjöberg, *Järn och jord*, 48ff.

46. I. Andersson, *Uddeholms Historia till 1914*, Stockholm 1960, 79ff.

47. T. Esper, 'The condition of the serf workers in Russia's metallurgical industry, 1800–1861', *The Journal of Modern History*, vol. 1978:4, 669 (quotation) and 673; T. Esper, 'Industrial Serfdom and Metallurgical Technology in 19th-Century Russia', *Technology and Culture*, vol. 23, no. 4, 1982, 590.

The normal size of the ironworks in the Urals was smaller than that of Nizhnii Tagil, although Ekaterinburg was also of a considerable size. Roger Portal mentions smaller ironworks of around 150 households in the southern Urals in the mid-eighteenth century.[48] In Sweden, only the Walloon works around the Dannemora mine came close to this in size, but even they must be characterised as small by Ural standards. Forsmarks bruk, one of the most important Walloon works, had a population of between six and seven hundred people in the middle of the eighteenth century. This grew to just under one thousand people in the first half of the next century.[49]

Wages, Work and Everyday Life

During the period discussed, the monetary economy was poorly developed in Sweden, and it can hardly be said that the market economy had advanced very far. In the countryside in particular a high level of self-subsistence prevailed. Payments in cash were a rarity and the economy was able to function thanks to widespread networks of credit, payments in kind and barter. The internal economy of the ironworks has to be set in the context of such a local economy. The forgemen and other workers at the ironworks were wage-labourers.[50] The former, together with the blast furnace workers, were paid a piece-rate according to their output, while different categories of day-workers were paid according to the hours they worked. As a result of the poorly developed monetary economy, they did not receive their wages in cash. Their income was recorded in a ledger, and the workers drew on it when purchasing foodstuffs and other commodities from a warehouse belonging to the ironworks. The most important item was grain, but herring and salt were also purchased by workers. Only a relatively small part of their wages was withdrawn as cash and used

48. R. Portal, *L'Oural au XVIIIe siècle. Étude d'Histoire Économique et Sociale*, Paris 1950, 214–218.

49. M. Essemyr, *Bruksarbetarnas livsmedelskonsumtion. Forsmarks Bruk 1730–1880*, Uppsala 1989, 43.

50. In this section we are leaving the different peasant groups in Sweden, and are thus only analysing the ironworkers living in the ironworks communities. Furthermore, we are dealing primarily with the forgemen, as most research has been undertaken on this group.

on the open market. Workers were sometimes allowed to buy commodities on the 'open market', but using bills later cleared by the ironmasters.[51] Coupled to this system was the distribution of a number of emoluments in kind. These included the free housing discussed above, free firewood and the right to some land, the latter comprising both arable land for growing turnips and, from the nineteenth century, potatoes, and pasture and meadows for a few animals, cows, sheep and goats.[52]

The decrees concerning forgemen stipulated not only the size of their houses, but also the size and quality of the allotments of land to which forgemen and blast furnace workers were entitled. According to the 1823 decree, a master forgeman had a right to a meadow providing hay for two cows, while a forgehand was assigned a field half that size. The ironworks also provided arable land for their forgemen, although this was not stipulated by the Board of Mines. The size of these plots probably varied between different ironworks, but the guideline was that the workers should be self-sufficient in vegetables.[53]

We thus find that ironworkers' households had the resources to produce much of their food by themselves: potatoes, milk and dairy products and also some meat. This meant that they could meet their needs from a number of different sources, and that their own economy was complex in structure, involving a mixture of money, markets, payments in kind, subsistence farming and so on.

The system in Russia was similar and, according to Thomas Esper, industrial serfs derived their income from two sources 'the wages and benefits received from the metallurgical firms, and the product of the serf families' labour in their domestic economies'. The benefits concerned were primarily a ration of grain or flour. An industrial worker received two *pud* of rye flour monthly, and his wife as much, while their children were given half that amount. Auxiliary workers received less as they were given larger allotments of land to grow their own grain, but they also received allotments especially for their horses. Esper also noted that workers at the ironworks could earn extra money by

51. For an introduction to the wage system in Swedish ironworks see K.-G. Hildebrand, *Svenskt Järn. Sexton- och Sjuttonhundratal. Exportindustri före Industrialismen*, Stockholm 1987, 94–100, and S. Montelius, '1600–1815', in *Fagerstabrukens Historia*, del V, Stockholm 1959, 257–334.

52. Montelius, '1600–1815', 121–133; Utterström, '1815–1870', 259–267; Rydén, *Hammarlag och hushåll*, 231–243.

53. Rydén, *Hammarlag och hushåll*, 236–242.

being involved in other trades which today we would categorise as proto-industrial.[54]

The tradition of allocating land to Ural ironworkers had its roots in the difficulties in recruiting personnel during the early decades of the eighteenth century. The ironmasters let parcels of land to their workers, on which they could build their houses, create gardens and let their animals graze. In the nineteenth century these rights were codified in state decrees. According to these, a household including skilled workers had a right to two *desiatiny* of meadow per worker. Their land was thus mainly used to provide hay for their animals, the exception being the vegetables that were grown in their gardens. Auxiliary workers received five *desiatiny* of meadow and arable land per worker, and were thus more like peasants in an economic sense, producing their own food. Both groups were also allotted their share of woodland, from which they could take as much firewood as they needed.[55]

Ironworkers in the Urals thus had the means to keep animals, which they in fact did to a larger extent than their Swedish counterparts. They had cows and oxen, horses, pigs and sheep. Seen over the household life cycle, it is clear that the number of animals varied over the different phases. During the phase as a simple family household there were on average one horse, one or two cows and as many oxen. When the household grew, the number of animals did likewise. On average, the extended household kept two horses, three cows and three oxen. This is not all that surprising, as the area of land allocated was tied to the number of workers in each household. The means to keep animals increased with the size of the household.[56]

This economic system, which prevailed at both Swedish and Russian ironworks and involved a mix of market and subsistence, had important repercussions for the way the household allocated its work and resources. It is fair to say that the Russian ironworkers' houses remained very much reminiscent of small farmsteads.

54. T. Esper, 'The Incomes of Russian Serf Ironworkers in the Nineteenth Century', *Past & Present*, No 93, 1981, 138f and 147, and PSZ (Code of law II), vol. 22, no. 21203.

55. *Svod zakonov Rossiiskoi imperii* (Code of the Russian Empire), St Petersburg 1857, vol. VIII, 427–431. See also Esper, 'The Incomes', 152.

56. M. Iu. Nechaeva and S. V. Golikova, 'Household of the Zlatoustovsky okrug residents on the turn of the XVIII-XIX centuries', in *Metallurgical Works and Peasantry. Problems of Social Organization of Industry in Russia and Sweden in Early-Industrial Period*, Ekaterinburg 1992, 303. See also *Sotsial'no-ekonomicheskoe polozhenie kadrov gornozavodskoi promyshlennosti v doreformennyi period*, Sverdlovsk 1989, 52–53.

It was stated earlier that important similarities seem to have existed between the homes of agricultural serfs in the south of Russia and those of the ironworkers in the Urals. They often functioned as small farmsteads as well. The animals had to be taken care of and the garden had to be looked after. There was also a period during the summer when haymaking was the most important activity around the ironworks. What distinguished the ironworkers from peasants was the absence of any grain cultivation. Households thus had to allocate large resources to agricultural work. However, they also had to allocate resources to the industrial activities of the ironworks, and each household therefore became divided into two segments, geared to achieving different purposes.

It was perhaps slightly misleading earlier to state that houses were the locus for allocating work and resources. Russian ironworkers were serfs and their owners, the ironmasters, also had an important say as regards who should do what work when. There was a very clear division of labour within the Ural ironworks, and the dividing line was drawn according to gender. Adult men made pig-iron in the blast furnaces, bar iron in the forges, sheet iron in the rolling mills etc. They also worked in the forests felling trees and burning charcoal, and carting charcoal, ironstone and different types of iron along the roads. The women were very seldom involved in this kind of work. It was in fact stated that they did not work at all. Ia. V. Khanykov, an official who visited the ironworks in Orenburg in 1838, wrote: 'One of the good things about the workers at the ironworks is that the female sex is liberated from work.'[57] Another observer, P. I. Mel'nikov-Pechërskii, noted one year later that 'the female sex is hardly ever used in the work'.[58]

The women did work, but their labour was confined to the subsistence economy deriving from their 'farmstead'. They took care of the animals and worked in their garden growing vegetables and fruit. In addition, they had full responsibility for the upbringing of the children, for all the cooking and laundry, and also for making new clothes. Every lunchtime it was also a common sight to see the women bringing food for the men. There was a break for lunch

57. Ia. V. Khanykov, 'Obozrenie rudnogo proizvodstva chastnykh Orenburgskikh zavodov v 1838 g.', in *Materialy dlia statistiki Rossiiskoi imperii* ('Survey of the mining production at private works in Orenburg 1838', in *Statistics of the Russian Empire*), St Petersburg 1884, part 4, 71.

58. P. I. Mel'nikov-Pechërskii, 'Dorozhnye na puti iz Tambovskoi gubernii v Sibir', in *Otechestvennye zapiski* (Travel notes written on the way from the Tambov province to Siberia), St Petersburg 1841, vol. 19, No. 4, section VIII, 75.

between eleven and twelve, and, at the ironworks, that hour was called 'women's hour'. In all of these tasks the women were helped by children and elderly members of the households.

The men thus worked directly for the ironworks, and their working day meant that they would be away for most of the time. The normal working day in the forges and rolling mills consisted of a twelve-hour shift. Times were not fixed, however, and the next shift would simply be sent for when the work was finished. At the furnaces the men worked twenty-four-hour shifts, while auxiliary workers' hours were decided by the nature of their tasks or the length of the day. In winter they often worked between six and nine hours a day, while during the summer this was often extended to more than twelve hours.[59]

Although they worked away from home for most of the day, these men still had some tasks to carry out at home. They took care of all the repairs on the house and outhouses, and they were also responsible for the supply of firewood and hay for the animals. As regards the latter they were helped by the fact that the works were closed down for a couple of weeks during the summer, to provide a break for haymaking. There were also a fairly large number of holidays over the year, in addition to Sundays, and the average number of working days seldom exceeded 250 days. It was often even less.[60]

Although the Swedish forgemen were not serfs and thus had a larger degree of freedom, the relationship between work in the workshops and the sphere of the subsistence economy within their households was very similar to that found in Russia. It was similarly not possible for forgemen working at Swedish ironworks to allocate the work and resources of their own household as they wanted. The reason for this was that they did not own the premises on which they lived. The houses, outhouses and allotments of land belonged to the ironmasters and were let to the workers as emoluments in kind, as part of the contract between owner and the worker.

The workers in the forges, as well as those at the furnaces and in the metal trades, were tied to iron production for most of the week. Much of the time when they were not at work was probably spent resting. As in Russia, therefore, it became the duty of the rest of the household, and above all the women, to do all the work connected with subsistence farming. The living arrangements for ironworkers in Sweden were similar to those in Russia, but they cannot be

59. *Istoriia Urala*, 149f.

60. *Rabochee dvizhenie v Rossii v XIX v.* (The workers' movement in Russia in the nineteenth century), Moscow 1955, vol. 1, part 2, 521.

characterised as farmsteads in the same sense as their eastern counterparts can. Forgemen and other workers in Sweden had smaller houses, with less space for animals and cultivation. The Swedish households, however, were also smaller. The tasks of the women at Swedish ironworks nonetheless included taking care of animals and a kitchen garden. This involved hard work in the potato fields, milking cows, dealing with the milk, fetching water etc. The women also obtained firewood. Another important task, as for the Russian women, was cooking and supplying the men with their meals. Cooking meant a lot more than just preparing the meals; it involved all the processing of products purchased from the ironworks store and the market.[61]

Apart from the rather obvious connection between the subsistence of the households and the work done in the mines, furnaces, forges, forests etc., in the form of wages and access to a number of non-monetary benefits or emoluments in kind, there is another important link between iron production and the ironworkers' households. In spite of the gradual penetration of the Swedish ironmasters into the sphere of production, through intermediaries such as foremen and supervisors, the forgemen and other skilled workers managed to remain in control of their work.[62] One very clear sign of this was the internal recruitment of apprentices and future forgemen within the closed circle of the household or the somewhat larger circle of kin. Fathers formed forge crews together with sons, while brothers together formed other crews. This gave rise to dynasties of forgemen which came to dominate the organisation of work at the ironworks.[63]

The importance of household and kinship in forming the crews that worked the furnaces and forges probably went back at least to the beginning of the seventeenth century. This was true in any event of the Walloon immigrants, who preferred to work with each other.[64] At the beginning of the nineteenth century this aspect of kinship became important in a new way. The industry was under

61. For an English treatment of this topic see Rydén, 'Iron Production', 91. For a more extensive treatment on women's work in the ironworks see Rydén, *Hammarlag och hushåll*, 246–277. On food consumption at the ironworks, see Essemyr, *Bruksarbetarnas livsmedelskonsumtion*.

62. See A. Florén and G. Rydén, 'Social organisation of the Swedish bar iron production, 1600–1880', Unpublished paper, for an elaboration of this. See also chapter 3, above.

63. See Montelius, '1600–1815', 53–55.

64. B. Douhan, *Arbete, kapital och migration. Valloninvandringen till Sverige under 1600-talet*. Uppsala 1985.

severe pressure due to developments on the international market, and skill became a sought-after asset. The forgemen therefore gained a better bargaining position, including the power to dictate with whom they wanted to work, with the result that kin members were more frequently included in forge crews.[65]

In the Urals no penetration into the sphere of production by the ironmasters was necessary in order to take away the control from the forgemen. From the beginning of the eighteenth century, all power and control were already in the hands of the *ustavshchik* and the supervisors. As regards kinship within the workforce it was clearly stipulated that forgemen had no power at all to decide with whom they wanted to work. It is therefore all the more surprising to find that 20 per cent of all work crews in skilled trades within the iron industry in Nizhnii Tagil in 1763 included members with close kinship ties. Most often, fathers worked with their sons, but brothers also worked together.[66]

This implies two things. First, it points to the importance of the household. This was not only an economic unit striving to make ends meet, it was also a unit of reproduction, in which fathers sought to push their sons into the same occupation as themselves. Households were a pool of labour from which both Swedish and Russian forgemen wanted to recruit new workers. Second, this situation highlights the importance of skills. It was not only in Sweden that forgemen managed to use the importance of their skills as a way of influencing the way production was organised. The Russian ironworkers might have been serfs, but they were also skilled workers, and they made use of this fact to lighten their burdens somewhat.

Conclusions

The aim of this volume is to discuss iron-making in a pre- and early industrial setting, in which industrial activities meet and mix with an agrarian economy. Households have a very important

65. See C. Evans and G. Rydén, 'Recruitment, kinship and the distribution of skill: Bar iron production in Britain and Sweden 1500–1860', in M. Berg and K. Bruland, eds, *Technological Revolutions in Europe 1760–1860*, Edward Elgar, Aldershot 1998, for a lengthy treatment of this topic.

66. Calculations based on the Third Census (1763); GASO, F.643. Op. 1. D.102 and 105.

position in such a society as an institution for allocating resources, and a unit for production, consumption and reproduction. In addition, the household was a kind of social 'transmitter' between individuals and society.

When it comes to the specific households connected with the iron industry in Sweden and Russia, it can be emphasised to begin with that this pattern prevailed in these cases. We have here an economy with a poorly developed market and monetary system. The workers were paid not in cash, but rather with a mix of money and payments in kind. Households in both regions therefore operated within a mixed economy, with subsistence farming coming face to face with the logic of the market.

That is as far as the similarities go between households connected with the iron industries of the Urals and Sweden, regions which both produced iron within the framework of a pre- and early industrial economy. The differences were many and they were often extensive. The most important of them was that Russia was a feudal society and that the households discussed in this chapter consisted of serfs. We are thus comparing households with unfree individuals with households with legally free individuals. The most conspicuous difference was, as we have seen in earlier chapters as well, one of size and structure. Households in the Urals were larger than their counterparts in Bergslagen, and they also had a much more elaborate structure. Multi-generational households prevailed in the Urals, as in other parts of Russia. The common pattern was early marriage for both spouses, and for the newly wedded couple to remain in the groom's household. The result of this was large households with more than one married couple.

In Sweden a different household composition prevailed, based on late marriage, neo-location and life cycle servants. The couple married in their mid-twenties and formed a new independent household. Before marriage, both spouses experienced a period in life in which they lived and worked in households other than those of their parents. They worked as maids, farm-hands, or apprentices of some kind if they belonged to the category of ironworkers.

Another important difference between the two regions is the variable pattern of household composition and size in Sweden, compared with the more uniform pattern of the Urals. In the former country, a situation prevailed in which different groups, with differently composed households, undertook different tasks in connection with iron-making. *Bergsmän* were responsible for mining and pig-iron making, while tenants supplied charcoal and forgemen made bar iron. In the Urals, a more monolithic system existed. One type of ironworker household supplied the whole industry with all the types of workers required. One household could include workers belonging to different areas: forgeman, carter, woodcutter, charcoal-burner etc. The multiple family household was connected to the serf economy in such a way that it was the owner of the works, or rather the administration, that decided the work of all the individual 'souls'.

A shift away from this system can be observed, however, during the period discussed, with two important features emerging. The first is the gradual appearance of smaller households. Towards the end of the eighteenth century, skilled workers more often formed new households after marriage, and the number of simple family households increased. This resulted in a split between auxiliary workers, in large complex households, and skilled workers, in smaller households. The second feature was that skilled workers managed to bring their sons into the same workshops as themselves, thus forming what could be described as dynasties, in spite of the power of the administration to put together work crews without the consent of the workers.

A more diversified system of household composition was thus created at the Russian ironworks, at the same time as Sweden went through the reverse process. As was dealt with in chapter 3 the ironmasters gradually took control of both mining and pig-iron making, reducing the importance of the *bergsmän*. The same process occurred within charcoal making and carting, as these tasks were gradually taken over during the nineteenth century by wage-dependent day workers from the ironworks. The creation of more integrated ironworks from the middle of the century meant a change from a system in which a number of different household types were connected with iron-making in Sweden to one involving only one type: the proletarian household.

PART III

THE INSTITUTIONAL ENVIRONMENT AND HOW IT CHANGED

Chapter Seven

COMMUNITY AND PROPERTY

Maria Ågren with Vladimir Zhelezkin and Vladimir Shkerin

Introduction

When Samuel von Stockenström visited St Petersburg in the 1780s, he praised the quality of Russian iron and the skills of Russian forgemen. In these respects, he maintained, Russia was fully comparable to Sweden. He was less appreciative of the way the Russians had ordered their economy, though. Raw materials were wasted and production control was inadequate; this explained why profits were not higher. Ironmasters ought to live at their works, know more about production and supervise their subordinates, von Stockenström concluded. Moreover, the Russian state should follow the example of the Swedish state and lay down rules as to how much charcoal a forgeman could use to make a certain amount of bar iron. There should also be stricter demands as regards productivity: in Sweden, no more than 22.5 per cent was allowed to be wasted in the course of smelting, but in Russia the proportion could be 30 or even 35 per cent, so von Stockenström was told.[1]

Whether von Stockenström's point – that Russian iron-making was less regulated than its Swedish counterpart – was correct or not is less important. What is interesting is that he clearly considered the rules governing iron-making to be as vital to its success as the quality of the ores or the skill of the workforce. Nor is he the only one to have emphasised the connection between production

1. S. von Stockenström, 'Anmärkningar rörande så Ryska järntillverkningen som äfwen Jernhandeln', 1787, Bergskollegii Arkiv, RA (National Archives, Stockholm).

rules and the outcome of production, between the politico-institutional and the economic level.

Pioneers in the field of early industrialisation were well aware of the impact of institutional arrangements. Rudolf Braun emphasised that traditional village communities could raise obstacles to domestic textile industry by demanding very high entry fees from the rural poor who wanted to move onto the village land. According to Braun's line of argument, the existence of such rules explains why early domestic industry was often located in mountain areas, where community rules were less demanding and an independent labour force could find a living.[2] Joan Thirsk, on the other hand, drew attention to the way early domestic industry often appeared in regions where rural households were free to divide their possessions, including land, equally among all their children. Where such inheritance practices predominated, land holdings were soon fragmented to an extent where the need to take on additional work became urgent.[3]

Both Braun and Thirsk focused on sets of rules that *predated* early industry, the idea being that the presence (or absence) of such rules might explain why early industry was found in some areas and not in others. Likewise, both authors discussed rules that were connected with a specific institution, the rural community. But, as von Stockenström's comments illustrate, there were also other types of social institutions, producing rules and exerting an influence not only over whether domestic industry was established or not, but also over the success or failure of large-scale enterprises.

By social institutions, we mean 'established rules and practices, through which people have organised their economic, social, demographic, political and cultural activities'.[4] In the previous chapters, many such institutions have been elucidated with regard to their impact on early iron production. It has been shown how a special type of peasant farmstead (*bergsmanshemman*) formed the basis for pig-iron production in early modern Sweden (chapter 2).

2. R. Braun, *Industrialisation and Everyday Life*, Cambridge 1979 (first edition 1960), 10, 34.

3. J. Thirsk, 'Industries in the Countryside', in *Essays in the Economic and Social History of Tudor and Stuart England in Honour of R.H. Tawney*, Cambridge 1961. For a critical discussion, see M. Berg, *The Age of Manufactures 1700–1820*, London 1985, 86.

4. S. Ogilvie, 'Soziale Institutionen und Proto-industrialisierung', in *Protoindustrialisierung in Europa*, Vienna 1994, 36. See also chapter 1, this volume.

It has also been shown how Russian ironworks were modelled on the pre-existing landed estate (chapter 4) and how the Russian iron industry integrated the institution of serfdom, many forgemen being legally unfree (chapter 3). In chapter 6, the focus was on the household – possibly the single most important institution in earlier times. In this chapter, we will focus more on institutions, rules and practices directly related to the early modern state and its interest in iron production.

The role of the early modern state with regard to early industry has not been exhausted in historical research.[5] The subject is broad, to say the least.[6] Fernand Braudel has argued that princes or states tended to take charge of early modern mining, since such activities demanded large inputs of capital which small-scale production units were incapable of making.[7] In chapter 3, this line of argument was shown to be an over-simplification. Using Sweden's experience as evidence, it was demonstrated that iron-making cooperatives were indeed able to cope with such investments.[8] State involvement in early industry is therefore better explained by Jürgen Schlumbohm. Schlumbohm emphasises the need for security and clear rules. Neither producers nor capitalists could by themselves create and safeguard the basic institutional framework needed to achieve security, which was necessary to make market production and commodity exchange a reality. Instead, a strong political power would have to guarantee this legal framework.[9]

The role of the state was particularly important in the field of iron production. In textile production, for example, the raw materials were easily moved. In principle, weaving could be carried out anywhere and therefore, according to Rudolf Braun, it would usually be located in areas where the pre-existing institutional framework did not obstruct it. But in the case of iron, raw materials were hard to move and, more importantly, production required huge amounts of energy. It was often deemed better to locate ironworks in close proximity to ore, flowing water and forests, without destroying the latter by over-exploitation. Favourable institutional

5. See S. Ogilvie, 'Proto-industrialization in Europe', *Continuity and Change* 1993, vol. 8:2, 171.

6. P. Kriedte, H. Medick and J. Schlumbohm, *Industrialization before Industrialization. Rural Industry in the Genesis of Capitalism*, Cambridge 1981, 127.

7. F. Braudel, *Civilization and Capitalism: Fifteenth–Eighteenth Century. Vol. 2: The Wheels of Commerce*, London 1982, 321ff.

8. See chapter 3.

9. Kriedte, Medick and Schlumbohm, *Industrialization before*, 126.

conditions thus had to be created in the area in question, including certain restrictions on the use of forest resources. To bring about institutional arrangements of such complexity, state involvement would often be necessary. For iron industries situated on the extreme periphery of the country, as in the Urals, state involvement in the form of troops was also necessary to protect the ironworks and their labour against hostile attacks from outside Russian territory.[10]

The early modern states were indeed willing to be involved. The statesmen of those days were greatly preoccupied with the issue of finding new resources and exploiting them in the most efficient way. They were also keenly interested in transforming the behaviour of their subjects, instilling a spirit of discipline and thrift to make them take good care of resources. Therefore, questions of order (in every sense of the word) were given high priority, including detailed regulation and routinised supervision of economic life. The whole undertaking was deeply influenced by cameralist ideas about how to restructure society in general and economic life in particular. It sought to create a 'well-ordered police state', in the words of Marc Raeff; it can also be described as an early form of social engineering.[11]

However, to implement the project of social order, discipline and more efficient use of resources, of which early industry was often part and parcel, early modern states did not always choose to create completely new institutions. Often, they would take pre-existing ones and reshape them. This kind of symbiosis between the state and institutions such as peasant or town communities would often be the most important form of state influence on early industry. As has been pointed out by Shelagh Ogilvie, the importance of such a symbiosis catches the eye whenever the development of social control on a local level is brought into focus.[12] This method of co-opting society, of giving pre-existing corporate bodies a new function, has been analysed with admirable perspicacity by Marc Raeff. It was used widely and successfully in the German territories, where there were corporate bodies with which states could cooperate. The same method has been shown to have existed in Sweden. Here, the early modern state was 'allied' with the

10. R. Portal, *L'Oural au XVIIIe siècle. Étude d'Histoire Économique et Sociale*, Paris 1950, 25.

11. M. Raeff, *The Well-Ordered Police State. Social and Institutional Change through Law in the Germanies and Russia, 1600–1800*, Yale 1983.

12. Ogilvie, 'Soziale Institutionen', 48.

artisan guilds, which performed state-like functions.[13] The state also actively promoted the establishment of guild-like institutions in the iron industry.[14] As for Russia, however, Raeff makes the general remark that the co-opting method did not work.

In this chapter, we will examine Raeff's theory and discuss it with reference to the social organisation of iron production in Swedish Bergslagen and the Russian Urals. Does the existence/absence of corporate bodies and co-opting strategies add something to our understanding of why the Swedish and Russian experiences differed? The possible relevance of such structures and strategies should be investigated at a concrete empirical level, however. To this end, the focus will be on one of the most crucial questions in economic life: property rights and control of resources. By looking at the kind of property arrangements that were made regarding ore deposits and forests, we will be able to show just how small-scale and large-scale production cooperated – or competed.

An Ordered Society: Communities and Control

The household of the Swedish *bergsman* formed the basic unit of labour in the mines in the seventeenth century.[15] Mines were used in turn by individual households, and the costs of repairs and investments were shared within the *bergsman* community. Thus, rights and duties with regard to mines and woods were spread among several actors: individual households, the community of *bergsmän* using the same blast furnace (*hyttelag*), and the larger community of *bergsmän* within the same mining district (*bergslag*). In order to balance the three one against the other, orderly forms of communication and decision-making were needed. For this purpose, a specific institution was used in the early modern period: the mining court which had existed back in the Middle Ages was now revitalised and given a new function from the 1640s on.

The creation of order was central to the community: order within each unit of labour (household) and, more important, order between the labour units which were to extract ore from the same

13. D. Lindström, *Skrå, stad och stat. Stockholm, Malmö och Bergen ca. 1350–1622*, Uppsala 1991.

14. See chapter 3: decrees from the Board of Mines said that the forge crew was to function in a guild-like manner.

15. See chapter 3.

mine and smelt it in the same blast furnace. At the mining courts, order was a recurring theme. In 1653 and 1655 there were complaints about a lack of order in the mine at Bispberg, and in 1745 an agreement was reached among all the shareholders of the copper mine at Bäsinge, in order to prevent disorder. Likewise, all the participants were interested in regulating grazing rights at the mine, since everyone had to take their horses there. Moreover, dams could not be built just anywhere, without prior consent from all concerned, and the court was the place to ask for and give such permission. The felling of trees and subsequent production of charcoal also required that everyone adhere to certain rules, especially since there were not always clear boundaries in the forests.[16]

Agreements and conflicts about how to enforce order and divide costs were repeatedly brought to the mining court. At the court session of Pershyttan, Åsbo and Lämås in 1695, the part-owners of the Åshytte furnace requested the court to force two *bergsmän* to participate in repairing the furnace. Without their assistance, the others would be quite helpless, they claimed. The two *bergsmän*, who had shares in two other furnaces as well, maintained that they carried a heavy responsibility for these units and asked to be relieved of the present task at the Åshytte furnace. The other owners would not consent to this, however, and in order not to be excluded from Åshyttan one of the two summoned, Nils Israelsson, accepted the duties of which he had been reminded.[17]

The mining court thus remained an institution for local and regional self-government in the seventeenth and eighteenth centuries. But at the same time, it was used as a channel through which state officials passed on demands and inquiries to the population. Mining statutes were read aloud at the inception of each court session, and attendance was compulsory.[18] As the session

16. See e.g. Gruv- och hammarsmedsting vid Bispberg, 10 May 1653, issue 7; ibid., 7 May 1655, issue 21; Bergstinget med Österbergslagen i Hedemora, 10 July 1745, page 651; Gruv- och hammarsmedsting vid Bispberg, 7 May 1655, issue 24; ibid., 26 May 1654, issue 12; ibid., issue 19; Gruvting vid Säter, 27 October 1657, issue 5. Uppsala Landsarkiv. See also *Noraskogs Arkiv: bergshistoriska samlingar och anteckningar,* ed. J. Johansson, Stockholm 1889–1928, vol. 6:3, 234, 236 (permission to build a mill near the furnace).

17. *Noraskogs Arkiv*, vol. 6:3, 247. See also 249: all the part-owners of the Jerle furnace were required to lend a hand in repairing the parts that had been destroyed by the spring flood.

18. G. Haggrén, 'Från brukspatrons disciplineringsredskap till kronans förvaltningsorgan. Bergstingen i västra Nyland under stormaktstiden', *Historisk Tidskrift för Finland* 1994 no. 1, 59.

proceeded, the *bergmästare* would refer to various issues that were important to the state. The quality of the iron was scrutinised repeatedly, and *bergsmän* were forbidden to do their own forging, since they were believed to lack the necessary skill. Prohibitions against paying a forgeman in iron, or selling beer at places like furnaces and forges, were announced, as were the nation-wide rules (laid down by the Board of Mines) as to how much charcoal a forgeman was allowed to use for each unit of finished bar iron.[19] The mining courts were also formally subordinated to the Board of Mines, to which appeals were directed, in the same way as ordinary hundred courts were subordinated to the Royal Courts of Appeal.

Just as in Bergslagen, the population of the Urals was organised communally, through the peasant *obshchina* (manifested at a village or *volost'* level). Like the Swedish mining court, the *obshchina* served the function of securing local order. It could take care of local economic matters, such as organising cooperatives for the production of butter, and was responsible for orphans. It also had some judicial tasks, often relating to border disputes or to a (limited) redistribution of land. It could prescribe crop rotations, and it could act as *de facto* owner of the land possessed by the community members, e.g. by leasing it.

Even so, the Ural *obshchina*'s character of an *agrarian* peasant community, seeking independence, should be emphasised.[20] The *obshchiny* were not peasant communities organised with the explicit aim of producing iron, i.e. they did not have much in common with the Swedish *hyttelag* and *bergslag*. Nor were they used as a kind of meeting place for state representatives and community members, like the Swedish mining courts. The peasant *obshchina* was more or less separate from the state's sphere of influence. Whereas self-government and state control were closely intertwined in Sweden, they were kept apart in Russia.

Sometimes, the state did try to utilise the *obshchina* as a channel through which to spread information, e.g. to warn of epizootics. Local officials would also attempt to make *obshchina* members concentrate on grain cultivation and not to waste time fishing and

19. Gruv- och hammarsmedsting vid Bispberg, 10 May 1653, issue 25 (*bergsmän* and forgemen who had not attended the court session were fined); ibid., 10 May 1653, issue 10; ibid., 24 July 1656, issue 6 (continued the following year); ibid., 10 May 1653, issue 11. Uppsala Landsarkiv.

20. J. Channon, 'Regional Variation in the Commune: The Case of Siberia', in *Land Commune and Peasant Community. Communal Forms in Imperial and Early Soviet Society*, ed. R. Bartlett, London 1990, 78; N. A. Minenko, *Russkaia krest'ianskaia obshchina v zapadnoi Sibira XVIII – pervaia polovina XIX veka*, Novosibirsk 1991, 3–6, 238–240.

hunting.[21] But such attempts never amounted to more than random measures. As regards ensuring that mining and iron production came about, the *obshchina* was only used in one way: it was the unit liable to be ascribed and, therefore, responsible for the labour duties to be performed at the works.[22]

The Russian mining administration thus never really managed to co-opt the *obshchina* and make it a part of its own network. The administrative structure remained a separate network, without solid roots in peasant society. It was a creation from above, the work of Peter the Great like so much else. From 1720, a system of local management was set up, following the creation (in 1718) of the Central Board of Mines.[23] By the 1730s, a three-tier system had been established, with the Siberian *Ober-Bergamt* immediately below the Board of Mines, followed by local *Bergämte*, and, further down, offices at each metalworks.[24] In the middle of the century, the administration dwindled, just as many ironworks were being sold off to private noblemen. It was not until the end of the century that it was re-established; in 1796, the Siberian *Ober-Bergamt* was revived.

Through the Swedish mining courts and the sources they have left, it is possible to outline the character of the interaction between state and community, between control and self-government; we can see the actual operation of what Marc Raeff labelled 'co-optation'. As for Russia, the situation is different, with the *obshchina* rarely displaying the hallmark of a meeting place.

An Ordered Economy: Property and Privileges

At seventeenth-century mining court sessions, internal community control *and* external state control were enforced at the same time. The main functions of the mining courts were not to settle disputes, nor to condemn criminals. In the eighteenth and early nineteenth centuries, though, these features became more accentuated. Most of all, many civil cases were now brought before the courts, cases which often dealt with the indebtedness of *bergsmän*,

21. Minenko, *Russkaia*, 102ff.
22. Cf. chapter 4.
23. C. Peterson, *Peter the Great's Administrative and Judicial Reforms. Swedish Antecedents and the Process of Reception*, Stockholm 1979, 372.
24. *Istoriia Urala s drevneishikh vremën do 1861 g.*, Moscow 1989, 296.

forgemen or ironmasters.[25] Many cases also reflected increasing competition for natural resources. The mining courts were no longer as preoccupied as before with the task of ordering society, of making its members interact smoothly. Instead, they devoted much attention to the definition and interpretation of property rights.

It is no easy matter to decide who to regard as the real owner of Swedish mines in the seventeenth century. Karl-Gustaf Hildebrand quite accurately described the situation as one in which the issue of mine ownership rights was irrelevant. Such questions were seldom raised in the mining courts. When mines were discussed, it was usually when the work there was badly organised and disorder occurred.[26] As for property rights, we can only say that certain social groups had a customary right to use 'their' mine and were thus able to carry out work at the furnace. For instance, in the seventeenth century the Persberg ore deposits in western Bergslagen were at the disposal of the *bergsman* community of Filipstad, which had the right to use the mine or let others use it.[27]

The same applies to forests: they would either be claimed by peasant communities (village commons, parish commons, or hundred commons), by the state, or not all.[28] Often, such claims would be contradictory. The areas concerned could not be effectively supervised, owing to their vast extent and remoteness, and traditional boundaries were seldom officially registered.[29] In the absence of undisputed owners and well-defined parcels of property, woods were used on the basis of customary rights by traditional communities and for traditional purposes. When disputes arose, they often originated in contradictory opinions about how to interpret traditions.

The right to use mines and furnaces ensued from community membership (*gruvlag, hyttelag*). This was, strictly speaking, a close parallel to the situation in the more agrarian peasant village/hamlet. There, being a member of the hamlet entitled the individual household to a share of the hamlet land, proportional to the size of the *hemman*. It also obliged the household to pay

25. M. Ågren, 'Att lösa ekonomiska tvister – domstolarnas främsta sysselsättning på 1700-talet?', *Svensk Historisk Tidskrift* 1988:4.

26. Haggrén, 'Från brukspatrons', 61, 68.

27. K.-G. Hildebrand, *Swedish Iron in the Seventeenth and Eighteenth Centuries: Export Industry before the Industrialization*, Stockholm 1992, 82.

28. C. G. Ihrfors, *Om häradsallmänningar enligt svensk rättsutveckling*, Uppsala 1916.

29. Cf. *Noraskogs Arkiv*, vol. 6:2, 106ff: a dispute about whether a woodland was to be regarded as a royal common or not had been going on for more than a century.

taxes in accordance with the size of the *hemman*.[30] In mining areas, community members had the right to use the mine and furnace for a period of time which was proportional to the size of their *hemman*, and they had to perform certain duties relating to pig-iron production, such as sharing the burden of repairing furnace and dams, and paying taxes.

Apparently, the performance of such duties was accorded great importance by *bergsman* communities. At the mining court of Bispberg in 1653, a prohibition was announced against buying and selling ore to anyone who did not have the right to mine and who did not 'share burdens and expenses'. At the same court session, the *bergmästare* asked the community if more *bergsmän* could be allowed to enter the community. The community answered that anyone could enter, as long as he was willing to share all their expenses.[31]

The emphasis on duties, which can thus be found at the local level, is also present in the royal statutes pertaining to mining. In 1649, a decree was published, saying that many *bergsmän* were known to have tried to make the mines their own property, i.e. to let their sons inherit them. In view of this, the state explicitly reserved all ore deposits for itself, on the grounds of royal prerogative. The decree went on to clarify this:

> … this should be understood to mean that those mining peasants, who use mines and who are in a position to *pay due tithes* to Her Royal Highness and the Crown, and who *perform the duties* that fall on each of them with assiduity and without defiance, they shall *keep and enjoy* the shares of the mines that *they are wont to use*, as long as they pay what is due and perform their work accordingly.[32]

Obviously, what the state wanted to assert was its right to ensure that mines were actually being used, and that it received its share

30. See C. Winberg, *Folkökning och proletarisering. Kring den sociala strukturomvandlingen på Sveriges landsbygd under den agrara revolutionen*, Göteborg 1975, 39. 'A *hemman* did not consist of a specific amount of land. It was supposed to comprise all types of land – arable and meadow – and have access to common waste and other common rights. In a sense, we can regard the *hemman* as a share in the village' (author's translation). In another sense, the *hemman* can be regarded as a unit of taxation.

31. Gruv- och hammarsmedsting vid Bispberg, 10 May 1653, issues 6 and 5. Uppsala Landsarkiv.

32. Kongl. Majt:s Ordning, varefter Bergsmän och andra, som i järnbergslagerna driva deras hantering med järnbruk, skola sig regulera och rätta, 6 July 1649, in *Kongl. Stadgar, Förordningar, Privilegier och Resolutioner angående Justitien och Hushållningen wid Bergwerken och Bruken (KSFPR)*, Stockholm 1736, 155f. Author's translation and italics.

of the returns. It only opposed claims which it was feared could lead to mines being closed down. If the *bergsmän* agreed to carry on working as they had done by tradition and if they paid their taxes, the state had no interest whatsoever in interfering with the arrangement. The decree of 1649 emphasised that the state had a right to supervise mining, but it also enforced the traditional usufructuary rights of *bergsman* communities.

Community members who wanted to retain their rights of usufruct thus had to fulfil duties towards the state and the community: to maintain production, pay taxes, and share community duties. This profoundly affected the way people looked upon the issue of entitlement. Long usage, fulfilment of duties and payment of taxes could be used to support one's right to obtain or keep a *hemman*. In 1653, Mårten Larsson of Långsbyn made complaints at the mining court. Despite the fact that he had been using his mine share for twenty years, his *hemman* had now been taken from him.[33] In 1655, Israel Persson of Bondbyn complained about Johan Hansson, who had tried to encroach upon a mine which Israel and his neighbours had bought some twelve years ago and used continuously. The court found in Israel's favour, as long as he and his fellows continued to pay tithes to the Crown.[34] In a dispute in 1676, Anders Jönsson claimed a right to buy his brother's *bergsmanshemman* before anyone else, partly because of his kinship, but also because for some time he had been paying taxes for the farm.[35] In 1743, a group of inhabitants in Tuna parish complained that three other inhabitants had tried to exclude them from their shares in an iron mine, 'which their ancestors had started and kept going in spite of considerable trouble'.[36] In all these cases, the fact that individuals had used a farm or mine continuously and not been deterred by toil and troubles was seen as a legal ground for possession.

In disputes over resources, the amount of tax paid could serve as a guideline in deciding how much each of the parties was entitled to. In a dispute in 1637 between Anders Pedersson and Olof Kristofersson, it was decided that Olof was to have one fifth of the

33. Gruv- och hammarsmedsting vid Bispberg, 10 May 1653, issue 9. Uppsala Landsarkiv.

34. *Noraskogs Arkiv*, vol. 6:3, 234. Cf. also 240 for a similar case.

35. *Noraskogs Arkiv*, vol. 6:3, 252. A similar argument is used in a case in 1644 (ibid., 266).

36. Bergstinget med Östra Bergslagen i Hedemora, 31 May 1743 (no page number). Uppsala Landsarkiv.

contentious forest, 'since he is paying one fifth of the taxes'.[37] In 1741, two *bergsmän* of Östanfors made complaints about two of their fellow *bergsmän*, who made use of the furnace and forge for a longer period of time than they did, despite the fact that the complainants paid more taxes. The interesting thing is that, instead of asking the state to raise the tax liability of the others, the complainants asked the court to diminish the usufructuary rights of their opponents. They did not look upon the different shares as the absolute property of the individuals concerned, but as something a community member was entitled to hold and to use in strict proportion to the duties he carried out.[38]

The issue of entitlement was at stake once again when Petter Andersson brought a civil case against the Matsson brothers in 1742. The brothers claimed to be 'participants' of half of the iron mine at Hästberg, since they had used it several years previously. Petter claimed a right to the entire mine because of the expenses he had incurred in more recent seasons. Here, new investments, possibly exceeding the ordinary investments demanded of community members, were used as argument for a larger share of the mine and to fight off arguments resting on ancient custom.[39]

The question of entitlement could be difficult to handle, even within the *bergsman* community. Entitlement followed from community membership, and membership could be gained on condition that duties were fulfilled and costs shared. The problem was how to deal with situations in which someone paid or did more than was normally required, and then demanded more. But when mines were no longer used exclusively by *bergsman* communities, the situation became even more complicated. As new types of part-owners appeared, the whole concept of ownership was transformed or, more accurately perhaps, it was these new actors who introduced the notion of ownership as we conceive of it. A dispute in 1745 concerning the iron mine at Nyberg, Skedvi parish, can serve to illustrate this.

The parties involved were Adolph Christiernin, an ironworks owner, and Jacob Hagtorn, a merchant and moneylender. Neither of them was actually present, though, their cases being in the hands of representatives. Hagtorn complained of Christiernin's

37. *Noraskogs Arkiv*, vol. 6:3, 227.

38. See C. Winberg, *Grenverket, Studier rörande jord, släktskapssystem och ståndsprivilegier*, Stockholm 1985, 94.

39. Bergstinget med Östra Järnbergslagen i Hedemora, 1 May 1742, 239. Uppsala Landsarkiv.

illegal use of the mine, producing a deed which proved that Hagtorn had purchased one quarter of the mine from the late Johan Woltersson. Woltersson had received advances from Hagtorn, as had other part-owners of the mine, though their names remain unknown. All this led Hagtorn to claim priority to the mine, which Christiernin had apparently violated by using it for his own purposes. Hagtorn supported his claim by invoking the section of the national legal code dealing with landed estates.[40]

Now, the kind of situation presented here is clearly different from the cases in which members of *bergsman* communities fought over shares in mines. Neither Christiernin nor Hagtorn were personally involved in production proper; they did not even attend the court hearing. Very likely, neither of them was content to have a usufructuary right to the mine, conditional on sharing common costs and duties. Hagtorn claimed a right to the mine, or to a major share of it, on the ground of a legal transaction (a purchase), which was a consequence of his having paid advances to other owners. His whole behaviour with regard to the mine reveals a conception of property in which alienability and exclusivity were central. It was a conception which could find support in the legal code, which contained a short enumeration of (five) legal transactions, but was silent on the value of more customary entitlements.[41] More importantly, it was a view which clashed more with the traditional views of *bergsmän* than with the views of his opponent Christiernin. As was shown previously, many *bergsmän* wanted to retain a symbiotic lifestyle based on the combination of agriculture and iron-making.[42] They were reluctant to become involved in monetary spending and feared indebtedness; they preferred a system where shared use and the shared performance of duties were essential aspects of their 'ownership'.

This does not mean to say, though, that all *bergsmän* had such a conception of property. There are indeed cases where *bergsmän* utilised the precarious situation of their indebted fellows to take over their possessions. Through kinship with merchants from the port towns, such *bergsmän* could advance socially and become members of the capitalist bourgeoisie.[43] How people in the Bergslagen

40. Bergstinget med Österbergslagen i Hedemora, 10 July 1745, 655. In a subsequent case, a similar demand was made on Christiernin by Fredrik Rothoff, ironworks owner and moneylender. Ibid., 661. Uppsala Landsarkiv.

41. Jordabalken 1, in the national legal code of Sweden.

42. See chapter 3.

43. See M. Ågren, 'Nordvikens bergsmän och deras syn på rättten', *Husby-Rocken* 1993, 1994.

region looked upon entitlement was not directly related to their social estate, but rather to their socio-economic background. Some tended to emphasise invested labour, shared duties, tradition and time; others tended to seek support from the fact that they had formal, legal title, ensuing from a transaction involving money.

The appearance of new types of mine owners may have been facilitated by a decree promulgated in 1723, as a clarification of the 1649 decree. The 1723 document has been regarded as a consolidation of the right of private individuals to exploit mines.[44] It explicitly stated that no longer did anyone have to fear that the Crown might revoke mines with reference to the royal prerogative. Possession of mines was thus granted better protection, and the intention was no doubt to attract owners of capital, who had previously been unwilling to risk their money in such enterprises. However, it should be underlined that the right of private individuals was still conditional in a sense, since mineral deposits had to be used 'for the good of the land and its inhabitants'. If a mine fell into disuse, the state could transfer the right to work it to anyone willing to do so.[45]

As regards the forests, these were crucial to iron production and the Swedish Board of Mines was alert to the necessity of safeguarding them from overexploitation. At the beginning of the seventeenth century, it launched the idea of separating mines and furnaces from forges, so as to alleviate the pressure on the Bergslagen woods. It was intended that Bergslagen was to become a purely pig-iron producing region. Henceforth, it became very difficult to obtain a permit to start an ironworks if neighbouring forests were believed to be threatened. The Board constantly kept an attentive eye on forests that were intended for mining purposes, as did the local mining courts and the *bergmästare*.[46]

In 1683, a royal decree was issued which claimed a right to all forests that did not already belong to someone else.[47] Similar

44. B. Ericsson, 'Privilegiegivningen till järnbruk och järnmanufaktur i Sverige under frihetstiden', in *Industri og bjergværksdrift. Privilegering i Norden i det 18. århundrede*, Oslo, Bergen, Stavanger, Tromsø 1985, 275.

45. Kongl. Majt:s Placat och Förordning angående de förmåner, vilka alle de i gemen hava att njuta, som här i Riket och därunder lydande Provinser några Metall- samt mineral streck och nyttiga Bergarter uppfinna, angiva och i gång bringa, 27 August 1723, in *KSFPR*, Stockholm 1736, 628ff.

46. Cf. Gruv- och hammarsmedsting vid Bispberg, 10 May 1653, issue 4: complaints were made about illegal felling in the woods belonging to the mine. The forester is told to fine the guilty parties. Uppsala Landsarkiv.

47. Förordning och Påbud angående skogarne och vad därvid i akt tagas bör, 19 December 1683, *Kongl. Stadgar, Förordningar, Brev och Resolutioner från år 1528 intill*

claims had been put forward before,[48] but the new decree differed in one important respect. The older decrees merely reflected a conviction that the Crown was entitled to lay down meticulous rules for the proper use of crucial resources; by contrast, the decree of 1683 explicitly claimed state ownership of all land that had no owner.[49] All forests were to be divided and each farmstead allotted its due share. The remaining woods were to be regarded as state domain. Moreover, the claim was put forward to protect the iron industry and to prevent rural communities from expanding in an uncontrollable manner. If ironworks needed more woodlands, they could henceforth apply for the right to use state land. In return, they were to pay a small sum of money (*rekognition*). The renting of such land was conditional, though; they could only be used for the purpose of making iron.

All these decrees sprang from the idea that order and stricter rules had to be enforced in order to safeguard important resources and activities. Rural communities and ironworks were both of great value to the country, but they could endanger each other's existence by competing for the same forests. Establishing unquestionable boundaries and giving each piece of land an undisputed owner were seen as appropriate means of achieving the goal of a well-ordered economy. Dividing all the woodlands, with due heed paid to all the interested parties, however, would turn out to be a very protracted process.

In 1734, a new ordinance regarding the use of woods was issued.[50] In order to safeguard the interests of the iron industry, swidden (slash-and-burn) was severely restricted. Peasants could still apply to the hundred court for special permission, but it would only be granted on certain conditions. For instance, swiddening had to be part of a serious project to improve grazing or to

---

1701 angående Justitiae och Executionsährender, ed. J. Schmedemann, Stockholm 1706, 857ff.

48. Gustaf I:s registratur, 20 April 1542; Kongl. Majt:s Ordning och Stadga om skogarne i riket, 22 May 1647; Kongl. Majt:s Ordning och Stadga om skogarne i riket, 29 August 1664. *Kongl. Stadgar*, 246ff and 341ff.

49. The decree only expressed the general claim of the state. Of course, it took a long time to decide which lands were part of traditional peasant communities and which were not. Consequently, it was not until the nineteenth century that the state had secured actual ownership of all areas without an owner. See C. W. U. Kuylenstierna, *Om rekognitionsskogar och under bruk skatteköpta hemman,* Lund 1916, 39; B. Boëthius, *Ur de stora skogarnas historia,* Stockholm 1917.

50. K. Bäck, *Bondeopposition och bondeinflytande under frihetstiden,* Stockholm 1984, 43ff.

reclaim new land.[51] All in all, the ordinance contained a multitude of conditions that had to be fulfilled for a given use of a wood to be lawful. In principle, crofts erected in the forests were deemed unlawful, but they could be tolerated if the crofters were able to produce a royal permit, if they agreed to pay taxes, to contribute to the upkeep of the military forces, and to assist nearby ironworks.[52] The ordinance also emphasised that all use of forests was restricted to meeting the essential needs of the household (*husbehov*), making the selling of trees and logs a crime. Any use of woods had to contribute to the common good.[53]

Conflicts continued to arise in Bergslagen. In 1742, thirty-two Finns living on a royal common which had recently been allotted to the ironworks at Svartnäs (on payment of a fee), were accused of illegal swiddening. In answer to these reproaches, the Finns appealed to the court to respect a letter from 1613, by which the King (Gustav Adolf) had allowed three of their ancestors to build crofts on the royal common. They also claimed to have no knowledge of the fact that Svartnäs had been given the right to use the woodland, and that they would have to leave their crofts unless they could continue to slash-and-burn (a veiled threat, since this would mean that the Crown lost income). But the mining court paid no attention to this. It stated that the royal letter should not be taken to mean that the wood could be used arbitrarily and to the detriment of others; like the rest of society, the Finns were bound by law and ordinances to 'use their property in a way, which does not harm *the common good* and generations to come.'[54]

The subject was repeatedly raised before the mining court by representatives of the local ironworks. In 1745, the foresters of Stora Kopparberg at Falun accused ten peasants of having practised illegal swiddening. To defend themselves, some of the peasants maintained that they had done this to improve grazing, not to cultivate grain. Others argued that they had in fact obtained permission from the hundred court. Further, they argued that they had a *natural right* to protect themselves from wild animals (by clearing the woods) and to improve grazing; otherwise, they, *and*

51. Förordning om Skogarne i Riket, 12 December 1734, § 24. In *Utdrag utur alle ifrån den 7 decemb. 1718 utkomne publique handlingar, placater, förordningar, resolutioner och publicationer*, ed. R. G. Modée, Stockholm 1746, vol. 2.

52. Ibid., § 12.

53. Ibid., §§ 2, 6, 16.

54. Bergsting med Östra Järnbergslagen i Hedemora, 26 April 1742, 193ff. (author's translation and italics). Similar cases 233 and (in 1741) 24, (in 1744) 572, (in 1745) 673, 677. Uppsala Landsarkiv.

the clergy, would be unable to survive. Finally, they denied ever having been allotted to the copperworks at Falun.[55]

It is worth emphasising once again that, just like mines, forests still belonged to peasant communities in the eighteenth century. In this part of Bergslagen, it was not until the 1820s that woods were finally divided and each individual farmstead allotted its share. Nevertheless, this rearrangement initially caused problems, too. It was quite common for a person who had previously been using a certain wooded area to retain the right to do so for some time. Naturally, this led to disputes with the person who had been appointed owner.

In 1825, Anders Matsson and two other *bergsmän* of Långshyttan brought a case against their community fellow Jan Enok Westberg. Westberg was accused of having removed coal dust from a place on his land; on the other hand, the three *bergsmän* were accused of illegal felling. At the court session, it was revealed that the place in contention had 'of old' been used by ancestors of the three *bergsmän*. Lately, it had been used in particular by Anders Matsson, who had the right to continue felling there for two more years; after that, Westberg would have full control of his private property.[56]

The kind of reasoning which can be detected both behind various royal ordinances and behind statements in court was no doubt akin to the ideas that have been unveiled by Michael Sonenscher. Discussing eighteenth century French artisans, Sonenscher has pointed to the existence of the idea of natural rights. Among these natural rights, the right to self-support and freedom were crucial. So were the notions that work was a natural obligation and that workers have a right of property in the things that they produce.[57] Likewise, Margaret Somers has shown how the conception of property to be found in British textile communities was closely linked to community membership, communal obligations and relations within the community.[58]

In Sweden, until the early nineteenth century, natural resources such as ores and forests were not conceived of as private property

55. Bergsting med Österbergslagen i Hedemora, 10 July 1745, 624, 667. Uppsala Landsarkiv.

56. Bergsting i Hedemora, 11 June 1825, issue 24. Cf. also similar cases in 1826, §§ 19, 20, 27, 28. Uppsala Landsarkiv.

57. M. Sonenscher, *Work and Wages. Natural Law, Politics and the Eighteenth-Century French Trades*, Cambridge 1989, 42ff.

58. M. Somers, 'The 'misteries' of property. Relationality, rural-industrialization, and community in Chartist narratives of political rights', in *Early Modern Conceptions of Property*, London, New York 1995.

in the same way as they would eventually be. Instead, the use of such resources was informed by ideas about royal prerogative, free mining, and the common good.[59] Their use was also governed by a mentality saturated with traditional views on entitlement. In this context, invested labour, shared duties and ancient usage were important pillars. So were notions about the right to self-protection and to self-support. No-one could be denied the right to see to his own or his household's indispensable needs. This was a key concept in natural rights theories. It was also an idea which can be seen as the ideological framework of the household-based production so typical of *bergsmän* and ordinary peasants.[60]

However, the obligation to pay taxes and share common duties – central topics in early modern conceptions of property and entitlement – became the subject of debate in the nineteenth century. To construct property rights in a way which made them conditional, dependent on the fulfilment of various duties, was increasingly seen as a serious infringement of security of tenure, since taxes and duties were seldom clearly defined. Clearly defined duties put owners in a position to take them into the reckoning. Conversely, the absence of clear definitions allowed duties to be altered and made ownership rights illusory.

In 1851, the *bergmästare* of Noraskog asked the Board of Mines 'how far are [holders of] *bergsmanshemman* obliged to build workshops for pig-iron production?' Such decisions had previously been made by the mining administrators, 'guided by extremely vague requirements in some royal ordinance'. What usually happened, the *bergmästare* said, was that each time a new building had to be erected, he would summon the *bergsmän* and together they would reach some kind of fair compromise. But this would only rest on agreement, and would not do in the long run, since 'without stated laws and a fixed legal foundation, there will always be someone who feels he is suffering – and is indeed suffering.'[61] The emphasis on shared duties, which had provided small-scale producers with security of tenure in the old days, now proved to be detrimental as 'new interests'[62] manifested themselves among

59. This kind of reasoning, in which *salus populi* was also a key concept, is discussed by C. Peterson in ''En god ämbetsman är bättre än en god lag …'. Frågan om justitiekanslern som en allmogens besvärsinstans i klagomål över kronobetjänternas ämbetsutövning', in Yrjö Blomstedt et.al., *Administrasjon i Norden på 1700-talet,* Oslo 1985, 312ff.

60. See chapters 3 and 6.

61. *Noraskogs Arkiv,* vol. 6:2, 147f.

62. Ibid.

owners of mines and furnaces, i.e. owners willing to spend more money in order to reorganise production.

A conception of property according to which shared duties were seen as something which entitled people to certain usufructuary rights in common property, was rational in a social context where all the parties involved had an equal economic position and, hence, an equal capacity to contribute to the community. In a situation where some members were willing and able to invest more and do more, the whole idea of a community became problematic, as the letter from the Noraskog *bergmästare* shows. When Swedish mining communities became economically differentiated, they had serious difficulty remaining in existence.

A general law concerning the right to prospect for and exploit minerals did not exist in Russia before the establishment of large-scale enterprises in the Urals around 1700. Minerals were indeed exploited before that, to some extent in the Ural region, but mainly in the regions of Tula and Olonets. But the principle of royal prerogative to ores was codified and made explicit through the ukase of Peter the Great, issued in 1719 and known as *Berg-Privilege*. According to this document, the state was looked upon as the owner proper of all the entrails of the earth, having the right to demand taxes.[63] This idea was combined with the principle of 'free mining', according to which anyone could apply for the right to exploit ore deposits.[64] The ukase stated that anyone could produce iron, provided that the state received its taxes[65] and that landowners were duly compensated.[66] It was explicitly stated that mining had hitherto been neglected because subjects had feared that their investments and exertions would be wasted because of a lack of legal security.[67] After a short interlude, when free mining was set aside, it was once again spelt out by the Empress Anna in 1739.

The *Bergregulation* of 1739 set out in greater detail how the state viewed ore prospectors, landowners, and the workforces of the ironworks. It strongly supported initiatives to search for new ore deposits. Foreigners were encouraged to come to Russia. Anyone who had found ore was to have a prior right over anyone else to exploit the deposit, and the capital invested in such enterprises

63. Peterson, *Peter the Great's*, 374ff.

64. See I. Dübeck, 'Europeisk privilegieret', in *Industri og bjergværksdrift*, for the interrelated concepts of *ius regale* and *Bergfreiheit*.

65. One tenth of the returns.

66. By a payment amounting to 1/32 of the returns.

67. Peterson, *Peter the Great's*, 376; Portal, *L'Oural*, 37ff.

was to be safeguarded even in cases where the exploiter had become indebted.[68] But at the same time, the decree contained passages which made this security of tenure conditional:

> If anyone finds an ore deposit.. he and his partners and their heirs will not be deprived of the works, *except in cases where the works is not properly taken care of and has been used with negligence.*[69]

> If a stranger, who does not belong to the company, should notice that the owners of the works *do not carry out mining* in their allotted district or neighbourhood *with due assiduity* … he can demand that the works be transferred to himself …[70]

Despite this, the principle of free mining, presupposing a royal prerogative to the entrails of the earth, was never fully accepted by landowners. The Stroganov family were particularly hostile to the state's claims. In the early eighteenth century, many complaints were registered against them for attempting to exclude ore prospectors from their estates. Throughout the century, the landowners continued their silent resistance. Roger Portal has argued that the Russian state never really believed in the possibility of enforcing free mining in the Urals. Rather, the formulations in the ukase of 1719 should be regarded as exhortations to the local landowners to start mining themselves, in order not to be troubled by ambitious prospectors.[71]

Approximately 90 per cent of the Ural lands belonged to the state, and the problem of private landowners refusing to accept free mining would thus appear to be a marginal problem. However, as many state-owned works were sold off to the aristocracy in the 1750s, the state lost control of mining activities. The hostility towards and neglect of free mining now became a major problem for ore prospectors. In this group, some iron-making peasants remained, despite severe competition from large-scale industry. De Hennin had consented to let them sell ore to the ironworks, but this of course presupposed access to ore deposits.[72]

The privatisation also severely influenced the rest of the local peasantry, since the new owners were seldom willing to accept and respect the customary rights which had previously governed access to land and forests. In the 1760s the ascribed peasants of a

68. *Bergregulation* 1739, §§ 1,3,4. Bergskollegii Arkiv, RA.
69. Ibid., § 3. Author's translation and italics.
70. Ibid., § 7. Author's translation and italics.
71. Portal, *L'Oural*, 43, 69.
72. Ibid., 69f. See also chapter 2.

village near the Sysert ironworks, which had recently been transferred to a private owner, complained that they were forbidden to fetch wood to meet their basic needs. Nor could they use the meadows freely.[73]

These complaints have to be understood against the backdrop of the traditional views on entitlement prevailing in eastern Russia. As Siberia was colonised, immigrating households took possession of land either by virtue of a state permit (the state being regarded as owner proper of all land) or by simple seizure. As time wore on, several households united as a *volost'* community in order to protect themselves against enemies, to help each other with demanding tasks (like tree felling) and to assist members in need. Subsequently, some *volost'* communities would split up into village communities, which could also involve a limited redistribution of land. However, such practices never seem to have affected the core areas of each household's land, merely the peripheral and uncultivated parts. Apparently, long usage and invested labour were seen as important in determining the allocation of resources. Consequently, households often retained their ancestors' land. In spite of the fact that the state claimed property in all Siberian land, and despite the fact that communities exerted considerable influence over land use, individual households maintained a kind of hereditary usufructuary rights which encompassed all types of land: arable, meadows and woods.[74]

The right to use woods to meet basic needs was also respected in Peter the Great's Great Ukase concerning Forests (1723). Wood was a material which had traditionally had a wide range of uses in Russian life, and it became even more coveted as early industry started to evolve. The introduction of large-scale enterprises in the Urals at the beginning of the eighteenth century put great pressure on the forest resources of this region. In the middle of the nineteenth century, two thirds of the area remained covered by woods, but it has been estimated that more than 70 per cent were forests in the early seventeenth century.[75] To avoid deforestation, new legislation and administrative procedures were devised. Forests were classified as either 'defended' or 'forbidden', and the ukase of 1723 stipulated that all woods were to be surveyed and their limits better known. In

73. Portal, *L'Oural*, 288.

74. Channon, 'Regional Variation', 67, 69f, 74.

75. Ia. E. Vodarskii, 'Zemel'nye ugod'ia i zemlevladenie na Urale v XVII – pervoi polovine XIX v.', in *Krest'ianstvo Urala v epokhu feodalizma*, Sverdlovsk 1988, 53; *Rossiia. Polnoe geograficheskoe opisanie nashego otechestva* t.5, 1914, 237.

principle, all use of forests was regarded as belonging to the state, but in reality factories and peasants could be granted the right to take trees to meet their 'essential needs'. Selling was prohibited, just as in Sweden.[76]

Thus, the use of meadows and woods by households, united loosely in *volost'* or village communities, formed part of their local customary rights, and such uses were also accepted by the state as long as they were intended to meet basic peasant needs. They would not be favoured by local ironmasters, though. Meadows provided grazing for horses, which were extremely important in transporting goods to and from the ironworks. Forests provided wood for making charcoal, which was also crucial to iron production. Ironmasters had several reasons for attempting to monopolise the use of such resources, just as they wanted to monopolise the mines.

In 1782, the stubborn resistance to free mining won the day, as this principle was finally and definitely abandoned by the Empress Catherine II.[77] Henceforward, the entrails of the earth were regarded as part of the landowner's property, which he could dispose of as he wished. Entrepreneurs wishing to search for minerals could no longer do so, since that would be deemed an encroachment. Likewise, the state had divested itself of the right to ensure that resources in private hands were used for the common good. In accordance with more liberal and anti-cameralist ideals, it was now taken for granted that private owners would take good care of their property, indeed, that they would even take better care of it if they were left completely in peace by the state administration. Catherine left more to the initiative and judgement of individuals.[78] This was a radical break with tradition, much more radical than the one that occurred in Sweden in 1723, where some conditions were retained in the mining ordinance.

Then again, the Russian state did not really divest itself of all the rights it had previously enjoyed on the basis of the royal prerogative, i.e. the right to ensure that ore deposits were not allowed to lie fallow. It retained many of these rights, but they were now enforced more clandestinely, under the veil of the state's own property rights. The *state* now also had a right to do anything it wanted with its land. But the Russian state did not choose to

76. D. Eeckaute, 'La législation des forêts au XVIIIe siècle', *Cahiers du monde russe et soviétique*, 1968, vol. IX.
77. Portal, *L'Oural*, 359.
78. Raeff, *Well-Ordered*, 233.

increase its direct involvement with the iron industry. Instead, in 1797, a new kind of land tenure was devised, *posessionnye*, which meant that private owners of works could use state land on certain conditions. Security of tenure presupposed that they did not decrease production or shut down operations without prior consent from the state. The owner *had* to use the land (and attached labour) to produce iron; three years of inactivity could mean that he lost his *posessionnye* rights. On the other hand, he could not increase production at will, since the state had now discovered that uncontrolled production threatened the forests. In principle, such rules had existed even before 1797, but like free mining, they had never been enforced to the letter.[79] Now, free mining was abolished but, in return, the conditional character of land holding was strictly emphasised. Only ironworks situated on patrimonial estates were free from these restrictions.

Thus, the introduction of *posessionnye* tenure was partly a means of ensuring that resources (mainly forests) were not over-exploited, partly a means of making sure that resources (mainly ores) were not allowed to lie fallow. An ironworks situated on *votchina* lands could be converted into a grain-growing farm at any time, although no examples of this are known; ironworks on *posessionnye* land were bound to produce iron and nothing else. Thus, without having to be very actively involved itself, the state managed to retain control of the iron industry. Ironworks owners who had their own patrimonial land were granted security from interference by the state or by ore prospectors. With regard to owners who held state land (the majority), the state retained its previous right of laying down conditions and making sure that ore deposits were duly worked. The royal prerogative was a coin from the past, now neatly exchanged for the currency of the nineteenth century – a more unlimited bundle of property rights – but the state hardly lost all its influence by this transaction. Nor did ironworks owners lose out.

The losers were the small-scale peasant cooperatives that had somehow managed to retain a hold for most of the eighteenth century, in spite of repeated prohibitions and increased competition from large-scale units.[80] The exact social position of such

79. Portal, *L'Oural*, 47ff, 85.

80. As early as 1717, the Siberian governor M. P. Gagarin had prohibited all domestic peasant production, and in 1723 this was confirmed by the mining authorities. A. V. Chernoukhov, *Istoriia medeplavil'noi promyshlennosti v Rossii XVII-XIX vv.*, Sverdlovsk 1988, 40.

cooperatives still remains somewhat obscure; in some cases they seem to have been free producers, selling ore from 'their' mine to whomever they liked. In other cases, they were a kind of hired labourers, extracting ore at the request of ironworks owners.[81] In all events, they were all seriously affected by the manifesto of 1782, since their mining activities rested on traditional use rights, not on land ownership.

However, the precarious position of small-scale cooperatives was not merely the effect of a more determined policy on the part of Catherine II. More important was the fact that the entire Ural region was now almost totally covered by the huge mining latifundia known as *gornozavodskiie okruga* or mining districts. In 1861, there were 52 such latifundia in the Urals, comprising 154 works of different kinds, including some cottage industries. Twenty-four belonged to the state, 52 were patrimonial property, and 78 were held on a *posessionnye* basis. With most of the ore deposits inside the boundaries of these vast estates, there was little room left for peasant cooperatives.

Despite the legal differences, all the mining districts had much in common. They were vast; each one was a closed technological cycle; their owners aimed to attain self-sufficiency in foodstuffs; the owners had an exclusive right to all the ore, woods and labour within the borders of their realms.[82] The mining district was modelled on the landed patrimonial estate of the Russian nobility (*votchina*). The core labour force consisted of the owner's serfs, who worked as skilled labourers but who at the same time, from a legal point of view, were estate peasants. It was almost incidental that iron was being produced in these places; it might just as well have been grain or pork.

With the establishment of the mining district as a means of organising metal production, the iron industries of the Urals and of Bergslagen became (even) more different than before. The Swedish system favoured many different types of economic actors

81. In 1720 in the Kungur *uezd*, there were still small water-driven works in peasant possession or owned by townspeople (GASO, F. 40, op. 1, d. 4a, 119–122). On the other hand, in 1724 peasant ore contractors supplied 6,863 tons (419,000 *pud*) of iron ore to the state works at Uktus.

82. See e.g. V. V. Adamov, 'Ob original'nom stroe i nekotorykh osobennostiakh razvitiia gornozavodskoi promyshlennosti Urala', in *Voprosy istorii kapitalisticheskoi Rossii: problema mnogoukladnosti,* Sverdlovsk 1972, 225–256; T. K. Gus'kova, 'Okruzhnaia sistema kak forma organizatsii ural'skoi gornozavodskoi promyshlennosti', in *Sotsial'naia i proizvodstvennaia organizatsiia metallurgicheskoi promyshlennosti Urala v XVIII–XIX vv. V pechati.* Forthcoming.

(ironmasters, *bergsmän*, charcoal-producing peasants, merchants, forgemen), all of whom were endowed with rights and duties. The Russian system mainly favoured one economic actor, the ironmaster in whose hands wide-ranging powers were laid. And just as property rights were spread among different social actors in Sweden, but monopolised and exclusive in Russia, the spatial setting of the iron industry in the two countries also differed. In the Urals, each mining district was a self-supporting, closed 'space' which did not really interact with other districts, but only with the market on which iron was sold. Within this space, hierarchy and subordination, rather than peer interaction, prevailed. In Bergslagen, many different economic actors sold and bought their products. It was an open 'space', a network supervised by the state but still mobile, characterised by the relatively free participation of several social strata.

Conclusions

Today, it is often assumed that a specific form of institutional arrangement, private property in the liberal sense, is especially beneficial from an economic point of view. If such 'full' property rights are granted to private individuals, it is believed that they will act in a way which promotes economic development in society at large. With security of tenure and a right to the capital that accrues from property (two important aspects of ownership in the liberal sense), for example, the owner will be willing to improve and invest, to ameliorate his property. Thus, 'full' private property is regarded as the most efficient incentive among members of society, making them take good care of what is theirs, and stimulating economic growth in general.[83] Conversely, it is sometimes assumed that the absence of this type of private property will make people refrain from working.[84] Against the backdrop of the institutional environment of early modern iron-making in (first) Sweden and (then) Russia, it will be possible to discuss these assumptions in greater detail.

83. F. S. Cohen, 'Dialogue on Private Property', *Rutgers Law Review* 9, 1957, 357–387. See also G. Rydeberg, *Skatteköpen i Örebro län 1701–1809*, Uppsala 1985, 9.

84. 'If everything belongs to everyone, the probability of anything being mine is very small. This means that nobody will work. One way to overcome this inefficiency is to bring in private property rights.' V. Gligorov, 'Justice and Privatisation', *Communist Economies and Economic Transformation*, vol. 4, no. 1, 1992, 47.

Do economic actors always try to maximise profits? Will they only work hard and spend their money on new investments if they feel sure that their labour and capital will return to them in an enhanced form, as profits? Not according to Chaianov, and those who have followed his lead. Taking as our examples Swedish peasants in general and *bergsmän* in particular, it should be obvious that their actions were not generally guided by a wish to maximise profits, even if there are exceptions to this rule. Their main goal was to retain their land and to be able to support their households.[85] For such a person, a 'full' bundle of property rights did not provide any extra incentive. What mattered to him was one specific right: security of tenure. Security of tenure so that the household could continue its sometimes precarious existence; security of tenure so that all the labour invested would not be spent in vain. In view of this, it is hardly surprising that security of tenure was accorded such importance in the discourse of property in early modern Sweden.

However, security of tenure was not granted unconditionally. It was only guaranteed *provided that* work was carried out continuously, communal duties were fulfilled and taxes were paid. Through these provisos, and through the demands and orders issued through the mining courts and the *bergmästare*, the state 'infiltrated' the economy of peasant households and steered them in a certain direction: towards maintaining production of pig-iron, and even increasing it. These devices were hardly required to prevent peasants from idling along, but they were regarded as necessary to make them concentrate on the tasks that the Board of Mines wanted them to carry out. However, such provisos also nurtured traditional conceptions about property rights and entitlement. If security of tenure presupposed the performance of duties towards state and community, the logic could easily be reversed. The fulfilment of duties then became the basis for property claims. Emphasis on the conditional character of property could be transformed into support for customary rights to use resources in a certain way. There was a close interaction between state views on entitlement, which aimed for the common good, and 'popular' views, aiming for the good of each household. At the intersection of these axes, the conception of property was created.

To use the expression of Marc Raeff, the Swedish state *co-opted* the *bergsman* communities in its efforts to ameliorate iron production and economic life in general. It did so partly by using the

85. M. Sjöberg, *Järn och jord. Bergsmän på 1700-talet*, Stockholm 1993.

mining courts as channels for communication, partly by manipulating the property rights with which *bergsmän* were endowed, making the much-coveted security of tenure conditional. This was nothing unique: 'ordinary' Swedish peasants were also subject to similar restrictions, since they were threatened with loss of 'their' *hemman* if they failed to pay taxes for three years. Nor could they use their land in a way which might endanger tax revenue.

Since unconditional private property was rare in pre-nineteenth century Europe, the impression is easily conveyed that in those days states did not show any interest in fostering economic growth. As we have seen here, and as is apparent from the works of Marc Raeff for example, this is totally wrong. Early modern states were fervently interested in making the economy work as well as possible. But they had *other methods* of providing security and encouraging investment among economic actors. The method most commonly recommended by seventeenth and eighteenth century cameralists was to use privileges. By granting specific privileges, the state forbade the rest of the world to infringe upon the economic activities of the privileged,[86] and security was granted to the economic actor. Privileges were widely used in Swedish iron production. Access to ore was granted through privileges, since ore deposits were regarded as a royal prerogative (until 1723). Permission to produce a certain amount of iron and to purchase charcoal was also given in the form of a privilege.

Privileges had one great advantage. They could establish a kind of property right to tangible *or* intangible assets (such as the right to buy or the right to sell), without making this right perpetual. Privileges could be withdrawn. In this way, they conferred security on economic actors such as the owners of ironworks, but at the same time they retained influence and power for the state. Thus, all the institutional arrangements discussed here had a conditional character, albeit to varying degrees. The state retained control, so as to prevent abuse of valuable resources. The Swedish state did not (yet) consider that individual economic actors could be trusted to use resources in a way which benefited the common good.

But the Swedish state *did* regard both ironworks owners and *bergsmän*/peasants as economic actors, worthy to cooperate with. The state did not wish either of these groups to outrival the other;[87] instead, they were to be linked together in a well-functioning

86. Dübeck, 'Europeisk privilegieret', 30f.

87. See E. F. Heckscher, 'Betydelsen av vår historiska brukspolitik', in *Ekonomi och historia*, Stockholm 1922.

production network. To make such a social organisation of production possible, it was regarded as absolutely necessary to give legal protection (albeit of different kinds) to all the parties involved.

Turning to Russia, we find one apparent similarity with Sweden. Throughout the period studied, the state secured a degree of influence over how minerals were used by emphasising certain conditions. First, before 1782, this was achieved by means of royal prerogative, combined with mining freedom for everyone; then, the goal was achieved through massive state ownership and through the conditional character of the rights with which *posessionnye* land was held. These conditions were essentially the same all the time: they said that minerals had to be used productively and that they could not lie fallow. As for the forests, efforts were made to safeguard them throughout the eighteenth century, but apparently restrictions were not enforced until the early nineteenth century.

There is also one glaring difference. The institutional arrangements in Russia show that the Russian state only reckoned with *one* economic actor: the owners of the large-scale ironworks. They were the ones that were to be given incentives and security of tenure. This was clear in the days of Peter the Great, when small-scale iron production was made punishable by death, and it was made clear once again with the manifesto of 1782 and with the gradual appearance of mining districts, covering most of the Ural area. Through these arrangements, the Russian state actively prevented the relatively peaceful interaction between bar iron-producing ironworks and the peasantry, selling pig-iron or charcoal, which was so typical of the Swedish experience and which was apparently capable of surviving even in the Urals for most of the eighteenth century.

Through networking, large-scale industry could maintain contact with domestic and/or cooperative production, but this method was finally shunned in the Urals at the close of the eighteenth century. Instead of a diversity of economic actors interacting through networks, the Russian solution favoured large, compact and self-sufficient units, held together by massive property rights vested in a single actor: the owner of the works. Within the borders of his mining district, he would be well-nigh omnipotent.

The only group which endangered the absolute power of the Russian ironmasters consisted of the ascribed state peasants, who had an ambivalent position. As was repeatedly confirmed by state officials, they were not serfs and were not to be treated as such. On the other hand, their actual situation deteriorated during the eighteenth century, culminating in the unrest of the 1750s and 1760s.[88] At this juncture, the Russian state had to decide which way to proceed with regard to these peasants: to regulate the use ironworks made of them, or to leave everything as before. Catherine II had plans to choose the first alternative; she even planned to acknowledge them as an estate in their own right.[89] However, these efforts were stillborn. The only tangible result for state peasants was that the ascription system was abolished (around 1800).

In his analysis of how the 'well-ordered police state' was achieved in Russia and the Germanies, Marc Raeff has underlined that Russia lacked pre-existing corporate bodies (such as clearly defined and acknowledged estates) in the early eighteenth century. The absence of such bodies, which the German states co-opted in their efforts to create order, forced Peter to enforce his various reforms entirely from above, with very little cooperation from what might be termed civil society. Catherine II, by contrast, realised that reforms had to reach the local level and that the state had to rely on the assistance and participation of civil society.[90] She was interested in ameliorating the conditions of the state peasants, apparently perceiving them as at least possible partners.

Her interest was never followed through, though. The late eighteenth-century Russian state showed a wavering attitude to the state peasantry. With regard to the iron industry, it was deemed better not to let state peasants (charcoal-producing or ore-selling) form independent parts of the production network, but to concentrate power entirely in the hands of ironworks owners. This resulted in the institutional arrangement known as the mining district. The solution was feasible partly because of the ownership structure in the Urals, with the state as the principal owner and with a growing serf population. In part, it was possible because ascribed peasants could now be replaced by a group which had hitherto been too small: wage labourers.

88. See chapter 4.

89. Raeff, *Well-Ordered*, 232 (free and state peasants were given greater freedom to trade), 235 (education was to be made available to state peasants), 243 (plan to give state peasants estate-like status). State peasants were also given a seat in the Legislative Commission of 1767.

90. Raeff, *Well-Ordered*, 237.

Chapter Eight

KNOWLEDGE

Its Transfer and Reproduction in Occupations

Rolf Torstendahl with Ludmila Dashkevich and Sergei Ustiantsev

Knowledge is a wide concept and we use it in its widest sense here, making no distinction between knowledge and skill. The limitations we make are imposed by the character of this chapter. Primarily, we will limit our attention to such knowledge as is relevant to mining and iron-making. Further, our perspective is limited to the forms in which knowledge was transmitted and the relationship of these forms to the state. Some forms were encouraged by the state, others were left outside its influence. The main discussion of Russian and Swedish forms of state influence in the field so defined is divided into two parts, one dealing with lines of communication for the transfer of knowledge from man to man, the other with institutional arrangements for the education and training of craftsmen, salaried employees and civil servants occupied in the mining and engineering trades or their administration.

Active Promotion of Industry and Business

At the time when Sweden was beginning to become a dominant iron producer in Europe – the time of Colbert and Louis XIV – there was no questioning of active state support for industry and business. When Sweden was superseded by Russia in the field of iron production in the eighteenth century, state promotion of industry was still unquestioned. On the contrary, every politician of stature realised that power politics called for financial resources, and these had to be gathered together by every means available. The underlying assumption was that what was good for society was also good for the state. The financial resources of industry and business were an asset for the state. The two countries met these requirements of politics in different ways, however.

Of course, in many respects the two states acted along the same lines in encouraging their iron industries. Here we will follow certain strands of industrial policy connected with the promotion of technical knowledge and the improvement of technology in the field of iron-making. We will leave aside the finer nuances of the economic reasoning of the times. Only one thing should be underlined here, and that is the absence of any ideological dividing line between state obligations and society before the late eighteenth century. The state not only levied taxes and duties on different industries, such as iron-making, but also played an active part in actual production and in the organisation of the trade in general. There may be some doubt as to whether these activities were always or even mostly profitable in a total calculus of investments and costs but, and this is also important, the economic calculation was not made in capitalistic terms.

We will mainly confine ourselves here to questions of formal and informal channels for the transfer of knowledge in the field of iron-making. Adjoining fields will be taken into account only when this is regarded as relevant to our primary concern.

Man-to-Man Transfer of Knowledge

Sweden

The Swedish *bergsmän* who established local cooperatives to produce iron were subjected to state control as early as the sixteenth century. This was hardly effective, and what was more important was that the state started to compete with peasant production through its own ironworks, although the idea of control was not abandoned. In the seventeenth century its form was settled through the establishment of the Board of Mines in 1637 (with a short prehistory beginning in 1630, and thrice renamed). This control of peasant production of iron was intended as an instrument to maintain quality at the different local works. Quality control was, of course, significant for domestic purposes, such as the production of shotguns and other weapons, which was organised by the state.[1] Most of all it, however, was regarded as important for the export of bar iron, which rose to prominence in the seventeenth century. Export duties were levied, and the state thus had a direct interest in the amount of iron which was exported.

The interest of the state was focused on the volume and value of the iron trade. The knowledge and skill of the producers was only indirectly affected, but soon both state officials and local producers became aware of the relationship between quality and market prices. This is one, but probably not the only, reason for the specialisation of duties that took place in local cooperatives of mining peasants (see chapter 2). Some men turned to furnace loading, blowing and tapping, others to the proper handling of the hammers and the refining of pig-iron into bar iron. A knowledge of the factors determining quality grew, and this was closely related to the skill exercised in handling iron-making equipment. The bulk of this knowledge was transferred from man to man.

In Sweden, the normal route by which knowledge came to be transferred in ironworks and peasant cooperatives was from father to son or some other close relative (such as a son-in-law). Journeymen were often related to a master of the trade, and the whole trade was organised in guild-like forms. Kin relations were not enough to maintain a knowledgeable labour force. Harsh legislation was introduced in the seventeenth century to bind workmen, masters as well as journeymen and apprentices, to their ironworks.[2] Not only were the titles of the different levels of workmen taken from guild organisations, but the forms of examination also came to be strict and of a guild-like character. Aldermen supervised the level of knowledge of journeymen and masters and inspected specimens of work.[3] Local variations and differences between periods can be observed. Göran Rydén has shown that a strengthening of the kinship ties among forgemen accompanied an improvement in the quality of iron from some works in

1. For one specific manufacturing unit, see A. Florén, *Disciplinering och konflikt. Den sociala organiseringen av arbetet. Jäders bruk 1640–1750* (*Studia hist. Upsaliensia*, vol. 147), Uppsala 1987, 11, 71. See also the dissertation on state interests in weapon manufacturing by S. Klingnéus, *Bönder blir vapensmeder. Protoindustriell utveckling i Närke under 1600- och 1700-talen (Studia hist. Upsaliensia* vol. 181), Uppsala 1997.

2. Florén, *Disciplinering och konflikt*, 69–72.

3. This overview of traditional man-to-man knowledge transfer is partly based on information from Anders Florén relating to his current research on Nora and Linde *bergslager*.

Gästrikland, the easternmost part of the Bergslagen region, in the first half of the nineteenth century.[4]

In Russia, man-to-man transfer of knowledge never had the same character, at least not in the Ural region. There, the absence of guild-like arrangements for the organisation of manpower may explain why kinship never played a major role in the transfer of positions from one generation to the next. The contrast between traditional Swedish and Russian forms of recruitment of the most important knowledgeable strata of workmen is striking.

In Sweden, the state was the active party in the relationship between state and iron producers. However, a complication arose in the seventeenth century. As one means of encouraging iron production, the state had invited rich individuals in general and certain foreigners in particular – notably a wealthy merchant from the Netherlands, Louis de Geer – to establish themselves as owners of ironworks, with a core workforce of competent immigrant craftsmen. This had proved a very successful policy. During the seventeenth and the first half of the eighteenth century, private ironworks owners grew in importance, and in the middle of the eighteenth century they produced around half the total amount of iron and considerably more of all the bar iron manufactured. They also became a power factor in the shaping of iron-making policy.

The wealthy merchants who formed the first generation of private owners of ironworks in Sweden had special links with the government. That this was the case with Louis de Geer himself is hardly surprising, as he was enormously rich and ennobled and had been specifically invited to take care of the iron industry. Several others, however, were tied up with de Geer's interests through family bonds. This 'cartel' could not be maintained in the same form during the latter part of the seventeenth and the first half of the eighteenth century. Instead the idea of an organisation was born.

The early capitalists in the Swedish iron industry found that they had common interests as opposed to those of peasant producers, who worked in cooperative collectivities, and the state, which continued to have certain limited interests in production but was mainly interested in obtaining revenue from the iron trade. The latter interest continued to be connected with quality control. After a lengthy discussion, in 1744 the private ironworks owners formed an organisation called the Ironmasters' Association, with

4. G. Rydén, 'Iron production and the household as a production unit in nineteenth-century Sweden', *Continuity and Change*, vol. 10, 1995, 69–104.

an executive known as the 'Iron Office' (*Jernkontoret*), which within a couple of years had become efficient and influential. The name of this executive is, in fact, the label under which the whole organisation is known.[5] (The official translation of the name of both the organisation itself and its executive is 'The Ironmasters' Association'. Here, however, we prefer to refer to the executive body as the Iron Office, as this corresponds closely to the Swedish term.) The Ironmasters' Association and the Iron Office were created to put pressure on government and state bodies on behalf of capitalist interests in iron working. They soon became more involved than an ordinary lobbying organisation. The Iron Office was equipped with expertise, not a very large staff, but some of the very best experts on iron chemistry in the country. It also tried to maintain its level of expertise in these matters. The members of the organisation could consult the staff of the Iron Office in order to improve their methods and results.

It is important to remember that the eighteenth century was a period of intense experimentation in general, and also that Swedish science was among the best in Europe at precisely this time, with physicists, chemists, biologists and mathematicians of world repute. When it was repeatedly stated by the experts of the period around and after the middle of the eighteenth century that knowledge regarding iron ore and its chemistry needed to be improved and was not yet reliable, this also represented an appeal for further experiments and 'scientification'. A group of specialists was formed who had knowledge based on foundations other than those of the workmen in the ironworks.

The state was by no means inactive in relation to the new organisation of ironworks owners, but during this period the initiative lay mainly with the latter. The state also realised the need for specialists, and in 1751 a new category of officials was created, first represented by a single post. The intimate relationship between the Iron Office and the Board of Mining and the battle for the lead in taking new action is well illustrated by the history of this new type of public servant.[6] The new posts were regional – only four in mainland Sweden and one in Finland – and carried the title of 'senior master furnaceman' (*övermasmästare*) the Swedish title including the word 'master'. This was no coincidence. The new officials also had a

5. B. Boëthius and Å. Kromnow, *Jernkontorets historia*, vol. 1, Stockholm 1947, 117–220.

6. Ibid., 463–87. See also G. Rydén, 'Skill and technical change in the Swedish iron industry, 1750–1860', *Technology and Culture*, 1998.

vital function in the guild system by which tasks were divided at the furnaces and forges. Master forgemen and master furnacemen had to be appointed by the guild and, in combination with other authorities, the *övermasmästare* were responsible for such appointments.[7] Another new category of civil servants, 'senior forge inspectors', was also created. Their function was to supervise technological standards in the forges. Beside them there were 'forge aldermen' at the local level. The latter were an old creation, midway between the guild and state organisations. From early on they were included in the state system, acting as a kind of local inspector and they were members of the special mining courts. These are examples of the attempts that were made to combine a new rational organisation and the traditional system with guild-like forms. It should be noted that there was no consistent guild system in iron-making, where production plants were normally too expensive to permit masters to establish themselves as owners of the entire means of production. Location and ownership of land were other factors which prevented the guild system gaining a firm hold of iron-making in Sweden.

However, civil servants did take on a function in the guild-like system in the late eighteenth century. This dual role does not seem to have bothered those who were active at the time. Control was control, and it could be carried out in the interests of the trade as well as in the interests of the state. Both sides were, in their own way, interested in maintaining the quality of Swedish iron-making. Cooperation with the capitalist owners of ironworks was likewise natural. They were concerned with improving output and processes, and this was also important to the state. One of the duties of the new inspectors was to travel around to the peasant cooperatives and not only supervise them, but also suggest improvements and, in some cases, prescribe what had to be done, especially as regards increasing the scale of production, when experts unanimously found that very small-scale production yielded iron of uneven quality and was uneconomical in its use of charcoal.[8] The inspectors had no obligation and no right to inspect the ironworks owned by capitalists. However, the latter were interested in good advice, and voluntarily reimbursed travel expenses and gradually also paid more and more in consultancy fees.[9]

7. B. Boëthius and Å. Kromnow, *Jernkontorets historia*, vol. 3:1, Stockholm 1955, 1–10.
8. Ibid., 5–7, 14f.
9. Ibid., 8–9.

Thus the new category of civil servants in the Mining Service were not administrators in a traditional sense, but rather, in a sense expert inspectors and consultants. Their main duty was to encourage technical improvements.

These people were recruited from the same sphere – the small circle of scientists who had specialised in mineral chemistry or related subjects – as the experts directly employed on the staff of the Iron Office. No clear division between the two categories is possible for, in fact, the same individuals often changed from one position to the other and the Iron Office saw it as one of its main duties, in pursuing the interests of its members, to suggest suitable candidates for state employment. From the very creation of the new posts of 'senior master furnacemen' it was stipulated that the Iron Office should have a say in the appointment of these officials. The Board of Mines, however, soon disregarded this obligation, but the Iron Office took the opportunity to remind the state agency of its duties when it found a suitable case.[10] The interaction between them was thus not without friction. The state agency was a board with traditions and standing and its members were academically trained people, while the Ironmasters' Association consisted of people who had made money, some of them without a university education. Cultural differences as well as different roles and interests led them in different directions. At the same time, they were both becoming more and more dependent on the experts in chemistry and mechanical engineering, who were neither traditional bureaucrats nor capitalist moneymakers. Their asset was one kind of knowledge, the systematised and generalised knowledge which constituted science.

The Iron Office sponsored a wide range of publications in the field of iron-making. Sven Rinman was the author of a famous 'dictionary of mining' and several other books on forging and refinery works. The praise and the prizes he received from the Iron Office for his writings were handsome, but not undisputed. One of the wealthy ironmasters found that few of his colleagues bought Rinman's works and warned that these books could do 'incredible damage' in the hands of foreign competitors.[11] However, the Iron Office continued its publishing activities, and started a very early journal of mining in 1806, *Samlingar i Bergsvettenskapen*, which was discontinued in 1811. In 1817 'the Annals of the Iron Office' (*Jernkontorets annaler*) were started under the editorship of

10. Ibid., 8.
11. Ibid., 48–62 (on Rinman's various writings).

Svedenstierna and, after him, other leading scientists in the field. Analysing the content of all the articles in these Annals over a long period, Göran Rydén has shown that the introduction of the Lancashire method meant much more than an adaptation of old technology to new forms, but rather a continuous series of steps towards an industrialised production process with a comprehensive view of raw materials, the technology to process them, and labour skills and costs. This new knowledge was consciously propagated through the Annals.[12]

Thus, in different ways, knowledge of mining and forging was disseminated through the activities of the Iron Office and the Swedish state in the late eighteenth and early nineteenth centuries.

Russia

Metallurgy in pre-industrial and proto-industrial civilisations is one of the most complicated material arts. The fine quality of iron and steel made by old masters who had no scientific conception of the structure of and specific laws of metamorphosis in metals amazes the metallurgists of a later industrial era. The skill, adroitness and strength of a master were fostered by many years' work on the shop floor. The traditional way of transferring production experience, from foreman to apprentice, was the most important one at Ural ironworks until the end of the nineteenth century. This is explained not only by the level of development of the productive forces, but also by the state of scientific and theoretical technology.

Apparently, all practical training consisted of a master and his apprentice working together. However, there are no sources describing the special methods used by Ural masters in the eighteenth and nineteenth centuries to transfer their production experience.

Almost all the achievements of the Ural ironworks are related somehow to an efficient combination of domestic and adopted technological experience. We would note that, while large ironworks were established in the early eighteenth century, they were only able to use the skills and knowledge accumulated by the peasant industry of the seventeenth century to a very limited extent. Peasants were not acquainted with the two-stage, blast furnace-conversion technology of iron production, for they obtained metal by the primitive bloomery-furnace method. In the early eighteenth century, a transfer of blast furnace and conversion technologies to

12. G. Rydén, 'Gustaf Ekman, Jernkontoret och lancashiresmidet. Ett inlägg i synen på teknisk utveckling', *Polhem. Tidskrift för teknikhistoria*, vol. 12, 1994, 132–164.

the Urals was a top-priority task. At that time, this meant first of all the introduction of masters acquainted with production via blast furnaces and blooms.

It is known that forty-three masters were moved in 1700 to 1701 from ironworks located near Moscow to the Urals to build and work at the first blast furnace ironworks in the region, the Kamenskii works and those in Nev'iansk. Masters of blast furnaces, blooms and charcoal-burning, foundry workers, constructors of dams and others were among them.[13] This group formed a very small core of builders and key workers at the ironworks mentioned, where they carried out the most complicated jobs and began to train local workers. Foreign specialists also played a significant role in the transfer of technology to the Urals. Blowers for blast furnaces and fineries for the two ironworks were built by a foreign master, Ivan Yanver. He is also said to have taught his successors. Three English masters erected a blast furnace and cannon factory for the Kamenskii ironworks[14] at the same time.

A large party of masters from the Olonets ironworks' arrived in the Urals in 1722. The group included Russians and foreigners – mainly Saxons.[15] Workers in highly qualified occupations now appeared at Ural ironworks: copper-founders, masters of wire work, blast furnace operators and specialists in white tin-plate production.

A tradition of using skilled workers of foreign origin or Russians from other parts of the country in order to guide the use of new inventions was maintained in the Urals in the second half of the eighteenth and in the nineteenth century. A few examples may be mentioned: An English mechanical engineer Joseph Hill, had organised a rolling mill at Chermoz ironworks in the 1780s (sheet iron had previously been obtained using a flat hammer);[16] Tagil's ironworks preferred the services of a Russian master, who assimilated the experience of sheet iron rolling gained in central Russia at the Batashov enterprise.[17] The puddling method was introduced in the Urals by the English puddling masters Samuel and

13. B. B. Kafengauz, *Istoriia khoziaistva Demidovykh v XVIII-XIX vv.*, Moscow-Leningrad 1949, 65.

14. D. Kashintsev, *Istoriia metallurgii Urala*, Moscow-Leningrad 1939, 49; S. G. Strumilin, *Istoriia chërnoi metallurgii v SSSR*, vol. 1, Moscow 1954, 146, 151. The English masters are called 'Levenfait, Jarton and Pankerst' in the Russian sources.

15. M. A. Pavlov, 'Predislovie', in V. Hennin, *Opisanie ural'skikh i sibirskikh zavodov 1735 g.*, Moscow 1937, 28–29.

16. A. G. Kozlov, *Tvortsy nauki i tekhniki na Urale. XVII - nachalo XX vekov*, Sverdlovsk 1981, 28.

17. TsGADA, F.1267. Op.1. D.313. L.388.

John Penn and Bernard Allender. The division of labour between the Penns is remarkable: the first constructed puddling and welding furnaces and the second trained Russian workers.[18] The Franche-Comté method of iron production which appeared in the Urals simultaneously with puddling in the late 1830s to 1840s is connected with the French Grandmontagne brothers. In addition, several Ural masters learnt this technology in central Russia at Evreinov's ironworks. Having returned home they taught the Franche-Comté technology to their comrades the same way they did it themselves – 'by preliminary training of bloom masters'.[19] And finally a large group of Swedish masters helped Ural workers to learn the Lancashire bloom method from the 1870s to the 1880s.

The list could go on because practically all the technical innovations introduced at Ural ironworks in the eighteenth and nineteenth centuries were somehow connected with specialists invited from abroad. The production experience of local masters constantly improved and often made their help more valuable. Sometimes, however, they opposed the introduction of new methods and the management of the ironworks had to take steps to overcome their opposition. A well-known Russian engineer, A. S. Iarstov, noted in 1798: 'Old masters are not used to any innovations and they do not like them, which is why innovations are often unsuccessful, however useful they may be.'[20]

Apart from the constant enrichment of technological experience with the help of invited masters, another way of accumulating experience existed in the Urals: a slow, gradual, empirical adaptation of already known technologies, equipment and manual methods of working to the local peculiarities of a particular ironworks. This is especially obvious with regard to technologies which were used for a long time, for example, the German bloom technology. As has been mentioned, some foreign masters near Moscow applied it in the early eighteenth century and it was common at all the Ural ironworks of that time. About ninety varieties of this old German bloom method existed in the Urals even before the nineteenth century. Practically every ironworks had its own variety.[21]

18. P. A. Olyshev, 'O pudlingovom proizvodstve v Kamsko-Votkinskom zavode', *Gornyi zhurnal*, 1843, vol. 6, 326.

19. TsGIA, F.44. Op.2. D.963. L.24 /ob./.

20. D. Lesenko, 'Materialy dlia istorii gornogo promysla v Rossii', *Gornyi zhurnal*, 1883, vol. 11, 287.

21. A. G. Kozlov, 'Ob osobennostiakh razvitiia tekhniki na kazënnykh zavodakh Urala, rubezh XVIII–XIX vekov', *Voprosy istorii Urala*, Sverdlovsk 1964, vol. 5, 13.

The situation was the same in blast furnace production. Up to the middle of the nineteenth century, all Ural blast furnaces were practically similar, but little by little all the ironworks changed the proportions and sizes of their furnaces. Blast furnaces even differed between ironworks belonging to the same owner. The manager of the Nizhnii Tagil ironworks, E. Nite, wrote in 1871: 'As regards the remaking of blast furnaces and hearths from earlier times, it is the norm for every ironworks to make hearths according to their previous sizes ... practically every one of our blast furnaces has its own size and shape, according to the general design of the furnace.'[22]

Thus in the late eighteenth and the nineteenth century we observe the adaptation of blast furnace and bloom technologies to the peculiarities of production at each individual ironworks, resulting in an increase in their efficiency.

There is no doubt that the masters were the main bearers of technological experience in metallurgical production in the eighteenth and nineteenth centuries although this does not mean that they were also the 'owners' of their skills, in the sense that they had a right to decide how they were to be used. An example from a Ural ironworks proves that the opposite was true: in most cases the administration and owners of ironworks directed the use of productive skills.

The task of foreign masters who were invited to the Urals was obviously to introduce new technologies. The duty to train local workers was laid down by officials at the ironworks and was explicitly stated in contracts. A master could decline the candidature of an apprentice only if he was unsuitable for the occupation. When invited specialists were liberally paid, this was usually not a payment for their labour, but a fee for their knowledge and skills.

Russian masters, including those hired voluntarily, were also forced by the administration to train apprentices. In the work of Vilhelm de Hennin – a handbook on labour and production organisation at all Ural ironworks in the eighteenth century – it was expressly stated: 'If any masters are not good at iron forging they must be sent by the foreman to skilled masters working at hearths to master iron forging, hearth construction, hammer and bellows work, and skilled masters must train them....'[23]

We could note that the owners of ironworks first trained their own serfs,[24] with the obvious aim of finally determining the right to use production experience in keeping with their own interests.

22. GASO, F.643. Op.2. D.348. LL.337–337 /ob./.
23. Hennin, *Opisanie*, 198.
24. A. S. Cherkasova, *Masterovye i rabotnye liudi Urala v XVIII veke*, Moscow 1985, 95.

Knowledge Transfer through Institutional Education

When the *École polytechnique* was created in the wake of the great French Revolution, it was intended as a means for the new French state to keep track of the diverse tasks and duties of the state and to give them a new foundation in scientific reasoning. This reasoning had as its core the application of mathematics in all branches of knowledge. The mathematical tradition was brought into the École polytechnique with its founder, the geometrician, mathematician and politician Gaspard Monge, and it was never given up.

This is often observed, but it is less often related to the fact that the École polytechnique was intended as a preparation for certain 'schools of application', one of these being the School of Mines (*École de Mines)* which already existed. This college – still the most prestigious of the continuations from the École polytechnique – was intended as a preparation for state service. After the general education of the École polytechnique and the special education provided at the École de Mines, the student was expected to be equipped with the knowledge needed for a position in the mining administration.[25]

Other educational institutions for mining came into existence in Europe in the late eighteenth century. One of the most important was the Academy of Mining (*Bergakademie*) in Freiberg in Saxony, Germany, situated in the middle of an old mining region. The creation of the Academy in 1765 was followed by a series of reforms of the institute during the following century aimed at raising its standards and making its teaching correspond to its name. Another institute was created in Clausthal in the Harz in 1775 and reformed in the nineteenth century to bring it the status of *Bergakademie* in 1864, corresponding to the level of Freiberg. Within Austria-Hungary an academy was created in Schemnitz in Hungary in 1770.[26]

Something new was emerging, of that there can be no doubt – the idea of improving mining through educational establishments had gained a foothold in European thought. As yet, the aim was not to promote profitability by controlling the invention of new processes and methods of production, but there certainly was a link with profitability. There were two main ingredients in this

25. T. Shinn, *L'École Polytechnique 1794–1914*, Paris: Presses de la Fondation nationale des sciences politiques, 1980, 89–94.

26. On the *Bergakademien*, see further *Das akademische Deutschland*, vol. 1, 603–620.

conception, good administration and a good knowledge of existing methods and processes.

It is possible to discern three steps in the development of educational thought in technology in general and in mining and iron-making in particular. The first step was to concentrate on the proper transfer of the local knowledge base from master to journeyman. This was developed into a conscious policy of importation of skilled workers and knowledgeable masters. The second step was to generalise different local traditions into one body of thought and to instil what were deemed to be the best parts of this generalised knowledge in the minds of beginners in the trade. This step required some institutional arrangements for the transfer of knowledge. The third step involved a systematic effort to improve methods and processes on a more or less theoretical bases with the help of standardised practical applications, i.e. laboratory work. Invention was the practical goal, and new explanations and an expansion of the theoretical base became the scientific goal.

The first step was the predominant mode of technological knowledge transfer up to the late eighteenth century. This was the traditional mode of working in a guild craftsmanship, and it remained the basis for the transfer of all practical knowledge in Britain, including 'professional knowledge' if this is taken as something more theoretical than handicrafts.

In the late decades of the eighteenth century we encounter frequent efforts to take the second step in a more systematic way. This step called for educational institutions, since a generalised study of methods and processes could not be achieved through practice, but required theoretical discussions of practical things. The early schools of mining were all representatives of this type of institution.

The third step was not taken until the late nineteenth century, a combination of theoretical schooling and controlled practice being the fundamental condition for achieving it. Many experimenters had been trying to invent new processes, products and methods from the eighteenth century onwards. Most of them kept their experimentation completely free from theory. They tried different materials, methods or ingredients in order to check their effects. The new type of experimentation that evolved in the late nineteenth century was dependent on a fundamental theoretical knowledge, combined with opportunities to experiment. This is not to explain away the arbitrary and accidental elements of invention, but it is quite obvious that the combination of theoretical schooling and practical work had a steering effect (sometimes, certainly, to the detriment of creativity).

Russia

Early efforts to create educational institutions for mining and engineering workers The establishment of large metallurgical and mining production units which applied complex technical methods (hydraulic engineering, lifts for ore and water, blowers) immediately created new areas for the transfer of technical experience. Mining officers, engineers and technical managers at Russian ironworks had to maintain existing technical devices and to introduce new ones. This led to a new conception: the idea of developing mining and metallurgy by creating special mining schools. Russia was not among the last countries to adopt and implement this idea.

The first school of mining appeared in 1716, attached to the Petrovskii ironworks in Karelia. In the first half of the eighteenth century the centre of mining education moved to the Urals, which became the primary metallurgical region in Russia (seventy-one ironworks had been built in the Urals by 1750). The creation of schools attached to the Ural ironworks is associated with the name of an outstanding statesman, the chief manager of the state mines, Vasilii Nikitich Tatishchev. In 1721 to 1722 he opened a number of elementary schools for workers´ families and, thanks to his initiative, the Ministry approved a project which provides schools for practically all the Ural state ironworks in the mid-1730s. However, these schools did not last long. By 1742, most of them had already been closed because maintenance costs could not be met.

The schools created by Tatishchev provided pupils with an elementary education. In foreign-language schools pupils were taught a similar curriculum. When these schools were opened, Tatishchev wanted to see a combination of general education with vocational training. Special attention was paid to the study of technical drawing, for drawing, according to the French education minister Durois, was 'the common language of industry, which ought to be clear to every worker' (Figs. 8.1 and 8.2).

From 1737 to 1742, after Tatishchev´s departure, vocational training of pupils was provided only in Ekaterinburg though not in the philological school there. In the late 1740s these lessons were dropped altogether. Ironworks managers preferred to train only a limited number of pupils in crafts, according to production needs. More than thirty decisions concerning the education of

FIGURE 8.1 Smithies (Dalsland 1758): Here the Swedish Master Forgeman Transferred His Knowledge to Journeymen

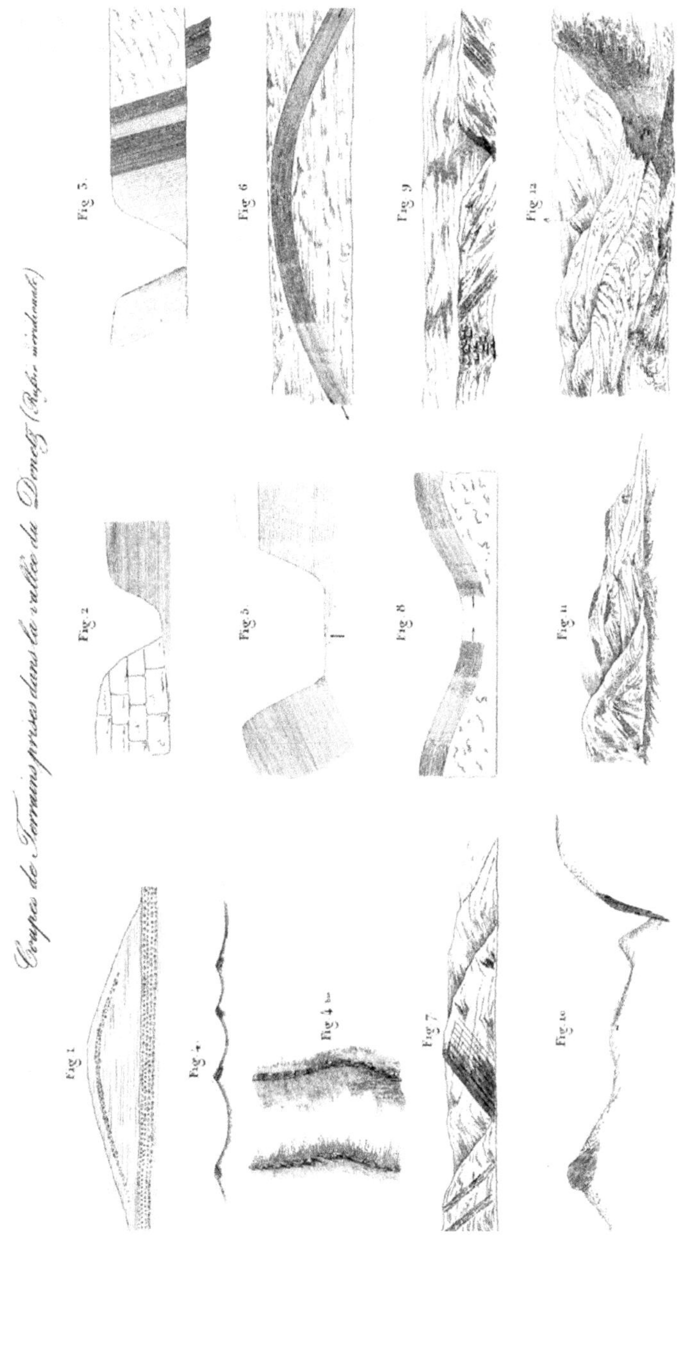

FIGURE 8.2 Drawings of the Geology of the Donetz Valley, Russia, 1839

Ekaterinburg pupils were recorded at the Ural Office from 1735 to 1750. This training was individual and the pupils were attached to masters as apprentices. Temporary use of schoolboys to help masters, mine surveyors and land surveyors thus played a remarkable role in vocational training at Ural ironworks in the first half of the eighteenth century.[27]

The St Petersburg Institute The growing complexity of mining technology, metallurgical production, thermal and hydraulic engineering and mechanics led to the idea of creating a high-level, specialised mining school which would provide future specialists with better and theoretically summarised knowledge, obtained empirically through production. In 1771 the Bashkir miner Ismail Tasimov and some friends sent a petition to the Mining Office, in which they proposed the founding of a mining school. Its running costs would be paid from 'the sum by which they profit from their own ore; namely one *polushka* from each *pud* of ore'.[28] The Mining Office informed the Senate about this petition. After a special Senate report of 21 October 1773 a ukase was signed concerning the foundation of the Mining School in St Petersburg, one of the oldest mining schools in the world. Its charter stipulated the creation of separate classes devoted to mine surveying, chemistry, mechanics, mineralogy, physics and drawing. Students in all classes were required to gain practical skills as well as theoretical knowledge. The chemistry class was to visit smelting furnaces and ore-washing plants specially constructed for study purposes and students in the mechanics class were to make models of machines. A special training mine was constructed for the practice of mining and mine surveying.[29]

Students from Moscow University were the first to study at the Mining School. Its first lecturers were academics and professors of the Russian Academy of Sciences and of the university, well-known scholars and mining specialists: the Corresponding Member of the Russian and Swedish Academies of Sciences A. M. Karam'shev, the academics V. M. Severgin, V. I. Viskovatov and others.[30] In

27. On the vocational training of pupils in the eighteenth century, see: L. A. Dashkevich and A. M. Safronova, 'Gornaia shkola v Rossii (XVIII - pervaia polovina XIX vv.)', in *Metallurgicheskie zavody i krest'ianstvo: Problemy sotsial'noi organizatsii promyshlennosti Rossii i Shvetsii v ranneindustrial'nyi period*, in the series *Sbornik nauchnykh trudov*, Ekaterinburg 1992, 87–94.

28. V. A. Frolov, 'Leningradskii Ordena Lenina Gornyi Institut (K 170-letiiu so dnia osnovaniia)', *Gornyi zhurnal*, 1944, no. 7, 4.

29. PSZ-1 T.XIX. no. 14048.

30. T. S. Dubrava, *Leningradskii gornyi institut*, Leningrad 1957, 12.

1806 this school was accorded equal status with the universities. The Mining School (later the Mining Institute) became one of the best university-level institutes in the country and a centre of technological thought. Scientific societies were initiated there and special theoretical publications were created. In 1817 the all-Russian mineralogical society, involving alumni of the Mining Corps, was organised. In 1825 the scientific committee of the Corps launched the 'Mining Journal'. The graduates of the St Petersburg Institute were not only managers in mining and metallurgical production, but also excellent scholars in geology, metallurgy, and thermal and hydraulic engineering.

The full teaching course at the Mining Cadet Corps (this had been the name of the Mining School since 1804) took eight years. In the first year students were given a general gymnasium education. The academic course for mining officers included descriptive and analytical geometry, differential and integral calculus, mechanics, physics, organic and inorganic chemistry, mining sciences and sciences related to ironworks.[31] Further, future production managers were instructed in mining law and bookkeeping. In their vacations, students trained in geodesy and mine surveying, mineralogical and geognostic prospecting and watched the production processes at ironworks located near the capital. Graduates from the Corps stayed in Petersburg for a further year to perfect their knowledge in laboratories and at ironworks. Then, according to the Charter of 1804, they were to enter mining enterprises as probationers for a period of two years. Students were to 'accustom themselves to all practical skills'[32] at ironworks and mines. Every half-year a student was to submit descriptions of mining and ironworks production with their own remarks, for consideration by the scientific council of the Institute. Only after this was he entitled to the rank of mining officer.

The reform of 1861 which abolished serfdom in Russia also led to a change in the status of higher educational institutions: they became open and unprivileged in terms of law. Some formerly unprivileged schools were accorded the status of higher educational institutions, e.g. the Petersburg Technological Institute and the Moscow Technical School. From the 1860s to the 1880s new high-level technical schools were established, e.g. the Technological Institutes in Riga, Khar'kov and Tomsk. The greatest changes

31. 'O publichnom ispytanii vospitannikov Gornogo Korpusa v 1825 godu', *Gornyi zhurnal*, 1825, no. 2, 207–211.

32. PSZ-1. vol. 28, no. 21233.

in technical education took place during the period of the industrial boom in the 1890s. From 1892 to 1902 polytechnic schools were opened in Warsaw, Kiev, Petersburg and Tomsk, along with the Engineering School in Moscow and the Mining School in Ekaterinoslav. By the beginning of the twentieth century there were thirteen engineering institutes in Russia in all.

Establishing a school system for engineering and mining Public opinion played a significant role in 'pushing' the Government to recognise the problems surrounding higher technical education. At that time, the problem of technical education was much discussed in the press by scientific societies and government institutions. Many zemstvos and town dumas in remote regions proposed their own plans to organise local technical institutes. In the 1890s, the authorities in Ekaterinburg and Ekaterinoslav sent in an application for permission to organise an Institute of Mines. The Government did not permit an Institute of Mines to be established in Ekaterinburg, but the Ural Mining School acquired a new status. In 1904 it became a secondary technical educational institution.

Engineering education was only developed within the state system of schooling during the period of the reforms. As regards the private sector, there was still no high-level technical school at the beginning of the twentieth century. Researchers attribute this not only to financial and organisational difficulties, but also to the attitudes of the domestic bourgeoisie.[33] The question of who was to subsidise technical educational institutions – the government or businessmen – was a matter of opinion.

In the reform period, several projects to organise mining schools at the private works for the training of senior foremen and technicians were developed. D. Dashkov, the owner of the works in Blagoveshchensk, wrote in a note presented to the Committee of Ural Owners of Ironworks that if hired labour could be freely used there was a need for qualified senior foremen who could be charged with the organisation and control of the work. He was opposed to state educational institutions, because their courses were too academic and encyclopaedic. Dashkov wrote that educated and learned people graduated from these institutions, but that these graduates had little practical understanding. Dashkov suggested that ironworks owners could organise

33. A. E. Ivanov, 'Vysshaia shkola kontsa XIX-nachala XX vv.: sostav i iuridicheskii status uchebnykh zavedenii', in *Istoriograficheskie i istoricheskie problemy russkoi kul'tury*, Moscow 1982, 189.

their own schools. The note included a draft set of school rules. The course was supposed to take five years.[34] The proposal stipulated that graduates would have to work ten years for the owner who subsidised their education. N. Vsevolozhskii proposed setting up a model workshop for a four to five-year training course for workers and a theoretical and practical school.[35] None of these projects came to fruition.

The middle and the lowest group of technical managers at state ironworks (supervisors, specialists' assistants and foremen) began to be given special education in the second half of the nineteenth century in the Urals. In 1847 the so-called district schools, one in the centre of every state mining district, were opened. Graduates from the lowest ironworks schools were trained at district schools for junior and middle technical and administrative posts. They studied ores and minerals, construction of mines and mechanisms of mining economics. Students received weekly training in workshops from skilled master joiners, metalworkers, smiths and other craftsmen.

In 1853 the Ural mining school was opened. Here special disciplines were studied, such as mineralogy, geodesy, mining skills, metallurgy, assay skills, and some geodesy and mine surveying skills. Practical training was conducted in the workshops of the local mechanical engineering factory, at the Verkh-Isetsk ironworks and in the Ural Chemical Laboratory.[36] Mechanical engineers for mining works were trained at the technical school of mining, which from 1834 was attached to the St Petersburg Technical Institute. Special attention was paid to practical education. Special workshops, cabinets and laboratories existed where collections of ores, fluxes and metallurgical products were exhibited. In the mornings students usually had lectures and after midday they trained in workshops. The theoretical and practical training of assay specialists was carried out at an assay school in St Petersburg.

The curriculum of most of the elementary schools connected to private ironworks did not differ from that of state schools. However, works owners also founded schools with a curriculum similar to that of the technical secondary schools. Thus the Viiskaia school at the Demidov works was created in 1806 as a continuation of the Nizhnii Tagil arithmetical school. There, general and

34. S. Ia. Bugaeva, 'Tekhnicheskaia intelligentsiia v gornozavodskoi promyshlennosti Urala (poreformennyi period)', Diss. kand. ist. nauk, Moscow 1979, 163.
35. Ibid.
36. A. A. Ryvkin, *Gornotekhnicheskoe obrazovanie na Urale*, Moscow 1956.

Russian history, geography, geometry, rhetoric, statistics of Russia and the German, French and English languages were taught. Special disciplines were also taught at the Chermoz school founded by the Lazarev family, at the Pozhevsk school founded by the Vsevolozhkii family, and at the Verkh-Isetsk school founded by the Iakovlev family. In addition, some Ural ironworks owners created schools in the capital. A school of agriculture and mining founded by the Stroganovs in Petersburg was well known. This school gave its students not only a wide knowledge of special disciplines equal to that provided at technical secondary schools (mining skills, metallurgy, forest economics, architecture), but a good general training as well.[37] Professors from the Mining Institute lectured on some of the courses. In Petersburg the Demidovs established a boarding school for the instruction of ironworks and estate administrators, which had a general curriculum similar to those of the district schools. The 'home school' of the Vsevolzhskiis provided its students with both general and specialised knowledge in mining skills, topography, forestry and architecture.[38]

The main method of training masters and workers at private ironworks in the Urals in the nineteenth century remained apprenticeships in the workshops. Owners tried to organise craft classes at the ironworks schools, but they were not widespread. It is known that in 1839 a craft class was established at the Dobriansk ironworks, which belonged to the Stroganovs.[39] An attempt to create a craft school was made in 1847 on the estate of the Vsevolzhskiis.[40]

The situation did not change in the second half of the nineteenth century. The reform of 1861, which abolished serfdom, put an end to traditional patriarchal relations between owners and the mining population. Consequently, vocational and special schools organised for the children of serfs and state artisans declined. In 1860 the last students graduated from the mining technical school and assay school in Petersburg. From 1879, all the mining schools set up by the ironworks and all the district schools were to be administered by the Ministry of Public Education. Their links with the ironworks were totally severed and all disciplines of a technical and practical character, geared to specific occupations, were excluded from the curriculum. The only remaining school for technical engineers and office workers sponsored by private ironworks was Nizhnii-Tagil's

37. TsGADA, F.1278. Op.1. D.480. L.1–12.
38. Kozlov, 'Ob osobennostiakh', 13.
39. GAPO, F.R-790. Op.1. D.2308. L.55.
40. GAPO, F.176.Op.1. D.692. L.13–14.

non-classical secondary school. By the end of the nineteenth century two technical schools had been opened in the Urals. In 1877 a technical school was opened in Kungur by Councillor of State Gubkin for the training of engineering masters, and in 1884 a mining school was established in Turinsk. In 1896 the Nizhnii Tagil technical school was transformed into a mining school.

Thus, in contrast to Sweden, Russia's most advanced technical education was organised in forms similar to those of the École polytechnique and the École de Mines. The engineers from the St Petersburg Institute were expected to enter state service as officers of one of the corps or as administrative civil servants. As in Sweden, the state in Russia organised the education of mining engineers at all levels, but one difference was the endeavour by private-sector organisations and individual owners of works to establish schooling for their own personnel. It seems that this endeavour in Russia sprang mainly from the patriarchal relations which were the outcome of serfdom, and that such relations were less prominent after its abolition in 1861.

Public discussion on engineering and mining education In the second half of the nineteenth century, the Russian technical society raised the question of special training for skilled workers or lower technical employees:

> From those middle ranks we demand neither the unskilled labour of a worker nor the knowledge and abilities of an administrator, nor the learning of a scientist, but we do demand the attentive and honest performance of a job which is not difficult ... We have had to replace them with either unskilled workers ... or people with higher education.[41]

At that time, however it was considered that no special education for foremen might be necessary. Workshop training was regarded as a cheaper and more rational way to learn. In the late 1860s the expediency of the Ural mining school was called into question. When officials from the Ministry of Mining visited the school they criticised its unwarranted theoretical teaching.

The discussion on specialist knowledge or broad encyclopaedic knowledge continued in Russia well into the twentieth century, and ironworks owners were unable to see how vast numbers of mining engineers could be used. As late as 1900, P. P. Boklevskii

41. *Zapiski Russkogo tekhnicheskogo obshchestva*, 1879, vol. 6, 65.

opposed the foundation of the Mining Institute in Ekaterinburg with the argument that an additional twelve ironworks every year would be necessary to provide jobs for all the engineers it produced.[42] Administrators preferred cheap training of practical specialists. Up to the end of the nineteenth century, the education of mining engineers at institutes covered the whole of the technical field. An engineer, G. Romanovskii, wrote to the Commission for the revision of the Mining Charter:

> A Russian mining engineer up to this moment ought to be a good specialist in chemistry, metallurgy, geognosy, geology, palaeontology, mechanics, architecture and ores … It is not physically possible to know specially practically everything concerning mining.[43]

The desire to see engineers receiving an education of a universal character arose from the fact that they went on to hold a wide variety of administrative posts in small and large industry.

In the late nineteenth century, the question of small industry which could be guided not by engineers but by technicians, whose training was much cheaper, was discussed. In 1884 the Ministry of Public Education proposed a general 'normal plan' of industrial education. This plan envisaged special schools, forming a system of general and special education for five qualification categories: engineers, technicians (the closest assistants of engineers), commercially educated managers, masters and workers. It was not fully implemented. In the Urals, only one of the technical schools – Krasnoufimsk – was reorganised in 1889 as an industrial school with a mining department. In 1896 this department was moved to the Perm' technical school.

The overwhelming majority of technical supervisory employees at the middle and lowest levels (masters, foremen, supervisors) at state-owned and private ironworks in the Urals received their practical training at the works themselves. This was considered to be the most profitable and rational method.

In the early twentieth century, certain efforts were made to create technical schools in all the mining regions, but the ironworks owners did not support the idea.[44] Another plan to link technical and mining schools received formal support from works owners, but the Ministry of Mining did not act on this proposal.

42. *Ural'skoe gornoe obozrenie*, 1900, no. 4, 4.
43. Bugaeva, 'Tekhnicheskaia intelligentsiia', 66.
44. *Materialy k voprosu o nizshikh gornotekhnicheskikh uchebnykh zavedeniiakh na Urale*, St Petersburg 1905.

Sweden

Early efforts to create educational institutions for mining and engineering workers In Sweden most ironworks were small, especially when compared with those in the Urals. The relatively large Swedish works were situated outside of the Bergslagen region which we are primarily considering in this comparison. Only a few works, for instance the Walloon works in Uppland, were of a size which made it natural to connect special teaching institutions to them. There are examples of such schools, e.g. those in Forsmark, which were elementary schools for workers´ children and provided a specific education. This type of school was similar to other private schools of the late eighteenth and early nineteenth century, the others often being connected with large-scale farming. The education provided at these schools was the normal elementary education. Up to the late nineteenth century there were no private schools specialising in the training of different categories of workmen for ironworks.

Institutionalised vocational training did not become a widespread phenomenon in Sweden in the eighteenth century. Normally training in crafts was obtained by joining a master's workshop as an apprentice. This is also true of those crafts which gradually became industrialised and specialised, such as iron-making, with its various specialised occupations. The guild and semi-guild system of occupational organisation certainly prevented the creation of educational institutions for vocational training.

In the eighteenth century several efforts were made to systematise training for the experts needed for the ironworks and mines. Christoffer Polhem (1661–1751), a Swedish engineer with a wide range of inventions relating to pumps and lifts for mines to his name, provided an incentive for systematic training in this field. In a report to the Board of Mines, Polhem proposed that a *Laboratorium mechanicum* be organised, in which both training and experiments could take place, but war and poor finances made public funding insecure and liable to fail altogether. Polhem's own models formed the core of a Chamber of Models, which was at times used for purposes of instruction. The Chamber became an important source of knowledge about the mechanical devices which were vital to all mining, and in 1798 a Mechanical School was founded, relying on the collections of the Chamber and subordinated to the Academy of Fine Arts. The Board of Mines was repeatedly asked by individuals interested in iron-making and mining to develop these activities further, but nothing was done and the Mechanical School was finally, in 1813, transferred to the responsibilities of the newly founded Academy for Agriculture. By then, the link with mining and iron-making had already been lost.[45]

The creation of the school of mining in Falun One of the lines dividing political life and activity in parliament and government in Sweden throughout the nineteenth century was the distinction between 'common' and 'restricted' (*allmän* and *enskild*), not to be confused with the 'public sector' and 'private sector' of later times. What belonged to the 'common interest' could, and sometimes should, be sponsored by the state, and taxes might be used for its promotion. What fell within the bounds of a 'restricted interest' was not to profit from state intervention, legal or financial.[46]

In order to overcome the difficulty constituted by this dividing line, the mining official (*bergmästare*) A. Pihl, in a report on the future constitution of *Stora Kopparbergs bergslag* in 1808, proposed the establishment of an educational institution for mining and iron-making in Falun, arguing that such a school would be useful and of a common character.[47] In 1812 a senior clergyman, Isak Mellgren, brought a bill before Parliament calling for a 'practical school of mining' to be instituted in Falun. He appealed to the paradigm of the École polytechnique and pointed to the needs of the Board of Mines, but also mentioned what had been done by the Iron Office to disseminate 'practical knowledge'. When it came to finance Mellgren proposed making a semi-public fund the main source of income for the institute. This proposal gave Parliament a perfect excuse for not reaching a position on the subject.[48]

In 1817 a special commission was appointed to plan a school of mining. In their report, the commissioners argued that the main aim of such a school was to 'educate able civil servants'.[49] The Board did not contradict the commissioners on this point.

45. P. Henriques, *Skildringar ur Kungl. Tekniska Högskolans historia*, vol. 1, Stockholm 1917, 55–69.

46. S. Kilander, *Den nya staten och den gamla* (*Studia hist. Upsaliensia*, vol. 164), Uppsala 1991, esp. 23–34.

47. Abstract of Pihl's report enclosed with a communication from the Board of Mines to the Government, 18 May 1819.

48. Parliamentary papers 1812: Minutes of Estate of Clergy, vol. 1, 339–347; Standing committee ABEU, Addendum vol. 2, 1156–1158; cf. vol. 4, 2388–2390.

49. Report of commission set up to plan a school of mining, 12 April 1817, enclosed with report of Board of Mines to Government, 18 May 1819.

During the parliamentary session of 1817 to 1818, efforts were made to bridge the gap between common and restricted interests in the case of a school of mining,[50] but the standing committee on taxation, in a long argument, refused any use of taxes for the purpose. The committee also responded to the proposals of the 1817 commission, and argued that a scientific school of academic standing was beyond the intentions of earlier parliamentary debates and the committee indicated that the universities already offered an appropriate education for the needs of the civil service.[51]

A solution was found in 1818. A renowned chemist, J. G. Gahn, owned a house in Falun with extensive mineral collections. After his death in 1818 his house and collections were bought by the state to form the core of a practical school, for which the Ironmasters' Association would have financial responsibility. After some initial difficulties, this school was opened in 1822.[52]

The School of Mining in Falun was a small institution. For a long time the number of students was limited to twenty. Its importance was due to the fact that during the period from 1822 to 1838, its first one and a half decades, it was led by a man of standing in chemistry, Nils Gabriel Sefström, who was a successful student of Berzelius. His laboratory proved costly for the Iron Office, but his teaching was regarded as useful and he was forgiven. Other teachers at the school were competent scientists and even if none of them attained Sefström's fame, the main difficulty was the location of the institute and the fact that many teachers, though not Sefström, preferred to live in Stockholm and visit the institute only at times and then give concentrated courses.

In 1830 the Falun school was followed by another School of Mining, in Filipstad. The latter became a lower-level counterpart of the school in Falun, and also sometimes a stepping-stone to a further education in Falun. They were regarded as related, though a system of education in mining, in the sense of having different stages of schools in a definite relationship to each other, had never been discussed.

50. Parliament 1817/18, Bg, vol. 1, 643 f (Mitander); Nobility: vol. 2, 357 (Löfvenskjöld).

51. Parliament 1817/18, Addendum vol. 4:2, no. 194 (897–901).

52. See R. Torstendahl, *Teknologins nytta* (*Studia hist. Upsaliensia*, vol. 66), Uppsala 1975, 18f. Further, see 'Promemoria ang. Bergsskolan' in 'Handlingar rör. bergsskolan i Falun', Bergskollegii arkiv, RA.

Iron-making as an aspect of technology The commission for a school of mining of 1817 had already emphasised what were called the 'theories of practical science' relating to mining and iron-making. The Board of Mines concurred with this opinion and argued that both scientific theory and practice must be the objectives of an education and that they must accompany one another in achieving a common goal. Observations based on experimentation must form the foundations of such practical sciences, and they could be continued almost *ad infinitum*, which made practice necessary. The practical goal, on the other hand, could only be attained if practice was incessantly led and guided by scientific principles.[53]

The School of Mining in Falun met a new challenge in the middle of the 1840s, when its former principal, Joachim Åkerman, publicly demanded that the school should be merged with the Institute of Technology in Stockholm. The advantages of its location in Falun were well known. The assumed benefits of moving to Stockholm were related to the scientific studies of the teaching staff, given possible links with other institutes there and their personnel. The Ironmasters' Association gave priority to the practical education of the students for the mining and iron-making industry (*bergshanteringen*). In this respect, Falun had precedence.[54]

Less than fifteen years later, the Ironmasters' Association had revised its opinion. This was no total reversal, but on one point the change was fundamental. Through its board, the Society wrote to the Government and proposed a merger. Only the theoretical, scientific, part of the education provided by the School was to be moved to the Institute of Technology. This was 'surely not the most important part' of the education of future mining engineers.[55]

To a considerable extent this proposal from the Ironmasters' Association had its origins in a plan submitted to the association by Victor Eggertz, the then principal of the School of Mining. He strongly emphasised the position that had earlier been advanced by the association, namely that 'only in connection with well-ordered practical teaching' was it possible to give satisfactory and fruitful instruction

53. Board of Mines to Government, 18 May 1819, enclosures: Report of commission set up to plan a school of mining, 12 April 1817; Communication of Board of Mines to Government, 16 April 1817.

54. The Ironmasters' Association to Government, 6 June 1847, in Konseljakt Civildepartementet, 16 July 1847.

55. The Ironmasters' Association to Government, 24 Aug. 1859, in Konseljakt Civildepartementet, 24 Feb. 1860, no. 18.

in the metallurgy of iron, and this was a conviction which he had developed as both a student and a teacher at the School.[56]

It might seem that the most important obstacle to the merger of the School of Mining with the Institute of Technology was removed when the Ironmasters' Association gave up its objections and no longer threatened economic sanctions in the event of a merger of the two institutes. However, a broad body of opinion was not prepared to accept the change immediately.

The Government proposed in a bill to Parliament in 1860 that funds should be provided in order to make possible the move of the School of Mining from Falun to Stockholm and its merger with the Institute of Technology. As soon as the bill was referred to the standing committee, some of the representatives of mining and iron-making among the Estate of Burghers emphasised purely practical training as the real purpose of the School of Mining.[57] The standing committee on taxation attempted a compromise. The School of Mining had a fundamentally practical purpose, while the Institute of Technology was mainly a theoretical institute.[58] Parliament accepted this report of the committee. However, the question was not buried. The parliamentary decision contained a call for a further inquiry.

The merger with the Institute of Technology The board of the Institute of Technology was not quite happy with the contrast that had been suggested during the parliamentary session of 1860. It could not accept the distinction that had been made between the School of Mining and the Institute. Its aims and its teaching were as practical as those of the School.[59] After another fruitless effort to effect a merger at the parliamentary session of 1862 to 1863, the Government returned to Parliament once more with the same demand.

In the additional material presented to the session of Parliament held in 1865 to 1866, Professor J. S. Bagge, one of the teaching staff in Falun, stated very clearly that the teaching at the School of Mining must be practical. 'By this I mean not only that the application of scientific theories is taught, but also that the very execution of the operations is taught, which is important both

56. Eggertz to the Ironmasters' Association 28 Dec. 1858, printed in V. Eggertz, *Utlåtanden rörande Falu bergsskolas förening med Teknologiska institutet*, Falun 1863, 4–8.

57. See Torstendahl, *Teknologins nytta*, 113f.

58. Ibid., 114.

59. The Board of the Institute of Technology to Government, 20 Jan. 1863, in Konseljakt Civildepartementet, 22 March 1867.

for the understanding of the working of the theory in the trade, and for future supervision of others in the practice of this trade.'[60]

Bagge also believed that the Institute of Technology could never be practical in this sense. The aim of this institute was to teach 'preparatory' knowledge in mathematics, physics, chemistry and technology for occupations for which these sciences provided the necessary basic knowledge.[61]

The Board of Commerce did not accept Bagge's view. It argued that mining technology could be studied as a continuation of the two years of basic education at the Institute of Technology and that it could be completely integrated there, except as regards practical experience in ironworks and mines.[62] The President of the Board dissented and in the following discussion in Parliament all kinds of arguments were put forward. One nobleman, von Schulzenheim, argued that those who 'nowadays' sought admission to the School of Mining were mainly young owners of ironworks or their salaried employees. They already knew the practice of the trade. They wanted theoretical knowledge, theoretical evidence and explanations for phenomena with which they were well acquainted.[63]

Conclusions

The basic fact common to both Russia and Sweden as regards the transfer of practical knowledge relating to mining and iron-making is that personal, man-to-man training was gradually replaced or complemented with formalised, institutional education. The older system, only exchanged for educational institutions to a very limited extent before 1800, was based on different social principles in Sweden and Russia. In Russia, the works owner played a central role in the selection of trainees and forms of training for the young workmen. This was possible to a large extent because workers were often his serfs or heavily dependent on him in other ways. In Sweden kinship was the normal selection principle for

60. J. S. Bagge, *Tankar rörande beskaffenheten af ett bergsundervisningsväsende i Sverige*, Falun 1865, 6. Cf. Torstendahl, *Teknologins nytta*, 117.

61. Ibid.

62. The Board of Mines to Government, 4 Oct. 1865, in Konseljakt Civildepartementet, 22 March 1867; cf. Torstendahl, *Teknologins nytta*, 117f.

63. Parliament 1865/66, Nobility: vol. 4, 240–244, 261f (Schulzenheim); cf. Torstendahl, *Teknologins nytta*, 122.

training in highly skilled iron-making crafts. This was possible because guild or semi-guild situations were dominant in all crafts in the iron-making industry and also in very small workplaces.

It may be noted that the change towards institutional training that took place mainly in the nineteenth century seems to have been ideologically much more important in Sweden, where heated debates raged for decades over the proper way of teaching mining and iron-making, than in Russia, where the very size of the country made it impossible to regard an institute in St Petersburg as extremely important for iron-making in the Urals. However, this dissimilarity should not be exaggerated. The leading industrialists in the Ural region influenced the education provided in Petersburg through specialised institutes and scholarships, just as the industrialists of Bergslagen influenced education in Stockholm through their votes in the *Riksdag* and at the Iron Office. At all events, one important difference was that Russian industrialists primarily tried to improve standards of knowledge among their dependent workers, in the first instance serfs who served in all positions up to the abolition of serfdom in 1861. Swedish industrialists were eager to gain knowledge for their sons and relatives in the first instance.

The channels used had to be different in countries with systems of government as different as those of Sweden and Russia. However, it should be noted that pressure groups emerged in both countries, though only at a much later stage in Russia. In Sweden, the Iron Office was an interest organisation and, at the same time, an organ of state policy, and the ironworks owners used it to put pressure on the government in educational policy. The congresses of ironworks owners in Russia were a much later creation and played a less prominent role in shaping educational policy and as a link with the state.

The main difference in educational policy as regards training for iron-making is that in Russia this training remained primarily a private duty, even though the state took several early initiatives for the education of workers' families. Theirs was not a specialised education (Tatishchev) and this type of initiative, by private owners and public authorities, had few counterparts in Sweden. Even though the state assumed responsibility for some major institutions like the Petersburg Institute of Mining, it was far less active than could be expected, given that it was both a key interested party in the business of iron-making and made profits from the flourishing private exports of iron. It would seem that the Swedish state, without any vested interests as a producer and obtaining its only revenue from export duties, played a greater role in the system of knowledge transfer. One factor explaining this difference may be the dominance of domestic demand for Russian iron while exports tended to remain a dominant motive in all the Swedish debates on skill and training from the early eighteenth century onwards. Proprietors and state officials alike saw the education of the workforce as a primary condition for good and profitable exports.

Another difference is no doubt that different proprietary attitudes to knowledge prevailed in the two societies. While Swedish forgemen and furnacemen used, in the seventeenth and eighteenth centuries, to build their relative independence in the production process on their exclusive knowledge, in Russia owners and their works managers tended to control the workforce more closely. In the nineteenth century the growth of schools of mining technology entailed a sort of expropriation of knowledge. The alliance formed in Russia at that time between the owners of large ironworks and school organisers was similar to the Swedish situation, though it will seem that Swedish mining engineers were more closely connected with other professional groups, too.

Though the differences in the training of ironworkers are striking in some respects, this chapter must end on a note of similarity. The importation of technological innovations through invited foreigners, or specialists sent abroad to gain practical experience or engage in straightforward industrial espionage was for a very long time common to both Russia and Sweden. They used it in relation to each other, but above all to introduce any new inventions and processes that public authorities and private entrepreneurs got wind of in any European country. Locally, the man-to-man transfer of knowledge was the norm in both countries for centuries. Also, in both countries, institutional education of technical specialists took shape in the late eighteenth and early nineteenth centuries. Both countries were well in the middle of the field, or even among the front runners in European developments in this respect. It was hardly the fault of the advanced parts of these systems that both Sweden and Russia fell back in the rank order of iron-producing countries in the nineteenth century.

Chapter Nine

IRON-MAKING SOCIETIES

The Development of the Iron Industry in Sweden and Russia, 1600–1900

Anders Florén

Iron-Making Societies, the title of this volume, can be understood in two different ways. During the eighteenth century the two regions under study, the Urals in Russia and Bergslagen in Sweden, were the principal areas in which iron was 'made' to supply the European market. Given our approach to regions and regional development, as outlined in chapter 1, which views them as functional entities made up by human interaction, rather than applying a purely geographical concept of physical and natural regional borders, it also seems relevant to assert that it was iron that 'made' these societies. To a large extent it was the socio-spatial network of iron production that formed the everyday lives of people living in Bergslagen and the Urals. However, this forging of new regional relationships did not take place in a social vacuum. It interacted with pre-existing social as well as political institutions in Swedish and Russian society.

Comparative studies always involve an interplay of similarities and differences. As was outlined in the introduction our study has been of a rather complex kind striving to look at both structures and processes in two regions over an extended period of time loosely defined as 'early industrial', i.e. from the seventeenth century until the second half of the nineteenth century. Our starting-point was the assumption that iron was produced in both Russia and Sweden using largely the same techniques, but that the industries were integrated into different social and political landscapes.

The economies of both Russia and Sweden were based on agriculture. In order to understand how the iron industry fitted into its agricultural setting, it is necessary to grasp the founding principles of rural life. It is vital to note the differences between the two regions. Serfdom, in the legal sense of the word, did not exist in Sweden, while it was a dominant feature of the rural world in Russia. It is furthermore important to state that the situation in Siberia and the Urals challenges such a simplistic view of Russia. In this region, the majority of land was owned by the Crown and its cultivators were never subjected to legal unfreedom as serfs. This makes the situation in Bergslagen and the Urals more similar than might have been expected.

Another general difference between the two countries is the extent to which the collectivity of the village community influenced the daily lives of peasants. In Russia the power of the villages was crucial, and even if land was not redistributed by the *obshchina* in the Urals, its position was nevertheless vital. Even if villages also existed in Sweden, they played a minor role since both agrarian production and taxation were mainly based on the household of the individual peasant.

Central government administration expanded in both Russia and Sweden during the period studied. The chronology of these processes, however, shows striking differences. While a proto-bureaucratic administration was hammered out in Sweden as early as the first half of the seventeenth century the same movement is perceivable in Russia a century later, i.e. during the reign of Peter the Great. The two processes were related. Russia's desire to Westernise, and its aim of competing with Sweden as the superpower of the Baltic region, led to efforts to copy the Swedish administrative model. However, the copy differed from the original in certain important ways. The most striking difference was that while the Swedish state succeeded in its aim of building up a network structure, in which central institutions were connected with regional and local ones, the Tsarist state was modernised at the central level, but generally lacked the important links downwards. In chapter 7, Marc Raeff's concept of co-optation has been used to accentuate this difference. While the Swedish state integrated the traditional arenas of local administration in its network, these bodies as a rule remained untouched in Russia.

In both countries, the state showed a keen interest in the iron industry, which is quite natural bearing in mind both the economic and the military value of the trade. It is also possible to perceive similarities in the way these interests were mirrored in policy making. We can thus see a development from ownership to administration in both regions, although it followed quite different chronologies. In Sweden the state-owned ironworks were made over to private owners during the first half of the seventeenth century. At the same time a solid administration was built up to take care of the taxation and administration of the trade. During the latter part of the eighteenth century this administrative network was set in motion primarily in order to safeguard forest resources, by restricting the total output of the industry. The foundation during this period of the Ironmasters' Association, which to begin with was closely connected to the Board of Mines, meant that the administrative structure was further developed. During the first half of the nineteenth century the old system of restrictions withered away and the leading role in the industry switched, in accordance with the new and more liberal trends in bourgeois society, from the Board of Mines, i.e. the state, to the Ironmasters' Association.

In the Urals the take-off of large-scale iron industry began later than in Sweden and during the first period of expansion the role of the Tsarist state as owner was even more pronounced than that of the Swedish state two centuries earlier. Following the general pattern of developments under Tsar Peter, the prototype of industrial administration was found abroad, namely in Sweden. The general failure of the Russian reforms to link the central and local levels was also apparent in the case of the Russian Board of Mines, which mostly came to serve as a tool for supervising and organising the state-owned plants. During the latter part of the eighteenth century the Ural ironworks were sold off to private ironmasters. In Sweden moves towards privatisation had been accompanied by an elaboration of the mining administration, a correlation that was not at hand in Russia, although there are traces of initiatives to re-establish the Board of Mines during the last decades of the century. The absence of a smoothly functioning administration led the Tsarist regime to choose another means of promoting the expansion of the industry, namely a feudal distribution of the Crown's landed estates. Possession was conditional, as the recipient of such *posessionnye* rights had to use the land and its inhabitants to supply his ironworks. In this way the national interest in a flourishing iron trade was safeguarded, but the policy also led to a regionalisation of the industry, dividing it into closed and independent mining districts (*gornozavodskiie okruga)* (Figs. 9.1 and 9.2).

FIGURE 9.1 **A Mining District in Blagodats, the Urals, 1870s**

FIGURE 9.2 **The Mines at Blagodats, 1870s: The Stage for Early Large-Scale Iron Production in Russia**

The Network of Iron Production

In the introductory chapter it was argued that studies of early industrial development should be based on the concept of production as a socio-spatial network. This enables us to study the entire sequence of production, leaving out neither those parts of it that were centralised in proto-factories nor its purely agrarian elements, for example organised as putting-out systems. However suitable to our aim of discussing the production of iron in relation to other early industrial activities though, the network concept retains something of a bias towards functionalistic assumptions, focusing rather on spatial and horizontal relationships. In order to understand not only the social structure of iron production and its reproduction, but also its development, it is necessary to pay attention to the way power relations were established between the different stages of the production process and also to such relationships within the different parts of the actual network.

The network concept is based on a prior understanding of the different tasks that necessarily had to be performed in order to fulfil the aim of the network, i.e. the production of a certain item. The foundation of the network can thus be said to be technical, and iron production to consist of four different parts, namely:

(1) Mining
(2) Charcoal-burning
(3) Melting
(4) Forging

Spatially and socially these four steps can be tied together in different ways. Depending on the structure of the network, interaction with other production networks, e.g. agriculture or transportation, may be necessary for its proper functioning.

When we speak about the iron production network in such general terms it becomes obvious, as was discussed in chapter 2, that there existed not one, but actually two technically as well as socially different networks in the two countries. Iron-making based on the so-called direct method coexisted with production in blast furnaces and forges. An older view of a straightforward development from direct to indirect iron-making must thus be questioned. In both countries these two networks were regionally separate. In Sweden this partitioning of the industry was a logical outcome of the fact that iron-making using the direct method was dependent on iron from bogs and lakes, while the blast furnaces used ironstone. As the central iron-production region in Sweden, Bergslagen, had developed round the ore deposits, the direct method was mainly practised outside that region. In Russia direct iron production remained in the western parts of the country as well as in Tula, south of Moscow.

Direct, or low-technology, iron production in Sweden was organised on the basis of the peasant household, which produced iron and turned it mainly into tools, nails etc. The markets for these products were different from those wares originating from the network based on indirect methods, which were aimed at the world market for bar iron. This distinction between the two networks becomes somewhat blurred when we turn to Russia. Bar iron both from the western part of the country and from Tula was actually sold on the British market. In these regions of Russia we can discern a productive network in which petty producers of blooms were integrated as suppliers to forges owned by ironmasters. It thus seems relevant to speak of a network consisting of a group of peasants and a group of merchant capitalists who interacted through the market. In the Urals, where such a small-scale industry had previously flourished, it was more or less swallowed during eighteenth century by the expansion of the large-scale industry. Smiths, petty producers of blooms and suppliers of iron ore were transformed into furnace and forge workers.

Taking a closer look at the networks that developed in the Urals and Bergslagen, we find similarities, but also striking differences. Both of them were directed towards the international market for bar iron, but broadly speaking that is as far as the parallelism goes. In general terms, the network in Bergslagen was socially multiform, including a variety of interacting social groups who were associated in different ways with the production of iron. Unlike their counterparts in the Urals, Swedish petty producers of iron had, by cooperative means, been able both to establish themselves in the mining part of the chain of production as well as producers of pig-iron and, during the sixteenth century, also of iron bars. At the beginning of the seventeenth century iron production in Sweden was thus dominated by networks consisting of cooperating petty producers, known as *bergsmän*. When the merchant capitalist entered the arena he strove to monopolise the sectors of the network closest to the market, i.e. the forges. The network thus assumed its socially multiform character, with wage labourers being employed in the forges, while pig-iron was bought from the *bergsmän*. The market thus functioned as an intermediary between the two groups. Its dynamism was limited, both as a consequence of the regulatory framework imposed by the state and because the *bergsmän* were often linked to the merchant capitalists by economic constraints, bonds institutionalised as *Kauf-* or *Verlagsystem*. Although the network thus included actors from different social classes, it is nevertheless, evident that those actors were not viewed as equals. The dominant ideological discourse implied, rather that the merchant capitalists ought to have the power to direct activities in the other parts of the network.

The social multiformity of the network was an important factor behind the decentralised landscape of the Swedish industry. Pig-iron was produced by *bergsmän* in one region and then transported to forges in a neighbouring area. From the viewpoint of the authorities, this was a desirable situation, since it helped to protect the forest resources as well as keeping down the prices of raw materials for the production of iron bars. As a result of the power relations existing within the network a clear-cut nodal relationship was carved out, in which the forge and the merchant capitalist played a central role.

In the Urals we find a socially more homogeneous network. The central position of the ironmaster (during the first part of the eighteenth century, often the state) was more marked and the network was held together by a variety of feudal devices. Even though they were paid, the workers at the furnaces and forges held the legal status of serfs. The social homogeneity of the industry left its imprint on its spatial arrangements. Furnaces and forges thus often, but not always, formed rather impressive industrial complexes.

These archetypal structures of the industries in Bergslagen and the Urals, however, were not left unchanged during the period under study. Around the middle of the eighteenth century about half the total amount of Swedish pig-iron was produced in furnaces owned by ironmasters. A century later the *bergsmän* had in most areas lost both their role as producers of pig-iron and their property rights in the mines. The network had thus become more homogeneous and its nodal character was more clearly underlined. This process also meant that the ironworks became less dependent on the market for their raw materials: ore and pig-iron were, as it were, produced within the realm of the enterprise. On a very general level the network in Bergslagen thus became more like the Ural one. A vital difference, however, was that while the Swedish network was dominated during the early nineteenth century by wage labour, feudal characteristics lingered on in the Urals and were indeed strengthened by the establishment of the mining districts.

So much for the network made up of what have previously been called the *industrial part* of the line of production, which, in terms of the size of the labour force needed only accounted for a tiny fraction of the total population of workers. The majority were employed in producing charcoal. The question of how these workers were attached to the network of iron production points directly to the problem of how to integrate an industrial activity in a rural world.

As long as the *bergsmän* in Sweden, and the agrarian petty producers of iron in Russia, divided their hours of labour between agricultural work, charcoal-burning and iron-making, the acquisition of charcoal never really posed a problem. The problem arose with the development of large-scale iron production directed by social groups outside the ranks of the actual producers. In both the Urals and Bergslagen the situation was exacerbated by the absence of a substantial group of pauperised rural dwellers, suitable for employment as wage labour. This meant that 'ordinary peasants' had to be linked to the network. As was discussed in chapters 4 and 5, three main alternatives were open to the ironmasters when it came to establishing such a link, namely:

(1) To use feudal exaction
(2) To establish commercial ties
(3) To deprive the peasants of their material resources and thus make them ready to become wage labourers

In Sweden all three strategies were used. Feudal exaction, i.e. duties imposed on the tenants on the landed estates of the ironworks and the transfer of taxation rights from the Crown to the ironmasters, could never fully keep up with demand at the ironworks. Charcoal thus had to be purchased on the market, which was highly regulated by the state in order to control prices and to safeguard an adequate supply for each ironworks. The establishment of the charcoal districts has thus been seen as an institutional device to force the Swedish freeholding peasantry to sell their charcoal to certain ironworks at a fixed price.

From the middle of the eighteenth century the ironmasters became more eager to expand their forest resources. In a move that is to be regarded in the light of the third strategy, they established agricultural holdings (crofts) of such a restricted size that the crofters had to work for the ironworks in order to make a living. This strategy was facilitated by the fact that a process of differentiation among

Swedish peasants gave rise to a rural substratum of landless people. It also meant that the ironworks´ tendency towards self-sufficiency and withdrawal from the markets for raw materials, underlined above, became even more marked. During the first part of the nineteenth century charcoal was produced, to a higher degree than previously, on the landed estates of the ironworks.

In the Urals, feudal exaction dominated. Not, however, in the sense that the majority of the peasants were made serfs under the rule of the ironworks. The number of serfs in the Urals did, though, expand during the eighteenth century, a development closely connected with the expansion of the iron industry. As was the case with the tenants of the Swedish ironworks, there were too few serfs to support the industry. In a manner similar to the way tax obligations to the Swedish Crown were transformed into duties to the ironworks, peasants in the Urals were ascribed to work for the industry. A major difference, however, was that while taxes in Bergslagen were paid by the individual household, in Russia the *obshchina* was collectively responsible. Furthermore, taxes in the Urals were collected as labour rent, while in Bergslagen they were paid in fixed quantities of charcoal. This meant that the ironmasters in the former region were fairly free to use the forced labour as they saw fit. During the latter part of the eighteenth century ascription made way for a system of *posessionnye* serfs, belonging not to the ironmaster but to the industrial establishment.

Several similarities can be seen in the way rural inhabitants were involved in iron production in Bergslagen and the Urals during the eighteenth century. In both regions feudal means were used and the state interfered to facilitate production by transferring taxes to the ironmasters and by regulating the trade in charcoal. However, while regulation of the market was the most important measure used in Bergslagen, the system of ascription was its counterpart in the Urals. One reason for this difference was that regulation of the market required an administrative network that was not at hand in Russia. The antithesis also reflected the differences in social structure or perhaps rather in social attitudes in the two countries. Ascription is closely associated with a feudal vision of power, asserting the sovereign's right to use the labour power of his subjects. This contrast between feudal and capitalist means of networking became clearer during the nineteenth century, when charcoal-burning in Bergslagen was often performed by crofters, working as wage labourers for the ironworks, while feudal property rights over land and people were strengthened in the Ural mining districts.

The different ways of linking the agrarian labour force to the 'industrial' production of iron thus accentuate the general features of the social networks in the two regions. In Bergslagen the structure seems even more multifaceted when we take into account the feudal means of acquiring charcoal and transport. In the Urals, on the other hand, the mode of charcoal production rather moderates the picture of an industry entirely built upon the social concept of serfdom. In both regions development of the social relations in the 'agrarian' parts of the chain of production points in the same direction as that seen in the industrial parts of the chain: within the realm of the ironworks, production became more closed and self-sufficient. In Bergslagen, this development meant a homogenisation of social relations in the form of free wage labour, while in the Urals feudal bonds became more tightly bound.

The Organisation of Work

The general feature of the social networks of the two regions left a clear imprint on the actual organisation of work. On the purely technical side of iron production there were no striking differences, either in the furnaces and forges or in the production of charcoal. What differed was how the work was actually organised.

The *bergsmän* used cooperative means to bear the costs of constructing and maintaining mines and furnaces. Mining was also often organised collectively, while the ore extracted was distributed among the part-owners who individually, using the labour force provided by their households, produced pig-iron and further refined products. Following the discussion of Sabel and Zeitlin in chapter 3, their mode of production was characterised as an alternative to large-scale production. The fact that in the seventeenth century the *bergsmän* became attached to the network as producers of pig-iron influenced the way they organised their work. As the ironmasters and the state strove to secure a steady flow of high-quality pig-iron from the furnaces to the forges, the organisation of the furnaces changed. A master furnaceman, not the *bergsmän*, was to be in charge of production, according to the Board of Mines. Up to the nineteenth century, however, the *bergsmän* or members of their households served as hands at the furnaces. At blast furnaces run directly by the ironmasters, work was performed by wage labourers; leaving the formal organisation aside, however, work was ordered in roughly the same way as at the *bergsmän*'s plants. A trend towards further specialisation of the furnace labour force began during the first half of the nineteenth century, interacting with more intense technical progress.

In a similar way as at the furnaces, work in the forges was put in the hands of specially trained masters during the seventeenth century, and crews were formed around them. Examples from the first half of the nineteenth century show that production went on night and day, using an informal shift system which demanded a high degree of coordination within the crew. Probably the same arrangement was at hand in earlier times, although the small size of the work crews in several forges indicates that many of them lay idle during the night.

The master forgeman was responsible for production, and requested raw materials as well as delivered the bar iron to the ironmaster or his servants. To underline the autonomous position of the masters in the workshop, we have characterised this way of organising work as a 'centralised putting-out system'. Though rather contradictory on an analytical level, this concept also implies that the ironmaster's influence over production was not pursued at the shop-floor level. From the viewpoint of the ironmaster, the forge was a 'black box' whose input and output he carefully controlled. From the seventeenth and the first half of the eighteenth century we also have several examples of even more far-reaching responsibilities for the master forgeman, who acted as an entrepreneur, himself purchasing raw materials and hiring hands. As the markets for both charcoal and pig-iron became more firmly regulated, such a position became difficult to maintain. However, it is quite evident that even during the nineteenth century master forgemen had a say in determining the composition of work crews. During this period, however, the forges became larger and new techniques were applied which increased production and also specialisation among the workforce. On the whole, the Swedish iron industry had by then assumed a more industrial shape.

In contrast to the scattered landscape of Swedish iron production, the works in the Urals often formed industrial complexes comprising several furnaces and forges (Figs. 9.3 and 9.4). Two furnaces were often built together, one of them serving as a reserve. While the forges in Bergslagen generally incorporated two hearths and one hammer, we find forges in the Urals consisting of as many as twenty hearths. Consequently, the number

FIGURE 9.3 Ironworks in the Urals Forming Industrial Complexes, Early Twentieth Century

FIGURE 9.4 **Industrial Landscape, the Urals, Early Twentieth Century**

of workers was much higher and their specialisation was also more far-reaching. We thus find a work organisation that seems more industrial, a picture that is underlined by the fact that direct control was exerted in the workshop. Supervisors and technical experts were thus part of the labour force and played the role of the people in charge, which the master forgemen held in Bergslagen.

As in Sweden, operations were supposed to be continuous. However, the shift system was not informal, but formalised meaning that entire crews relieved each other at the hearth. This points to the more highly developed organisation in the Ural industry, requiring a certain degree of coordination not only among the members of each crew, but also between crews, i.e. at the level of the workplace. The relatively modern features of the Ural system, however, are balanced by the fact that the workers were considered as serfs and thus legally bound to either the ironworks or their owners.

In both the Urals and Bergslagen, workers were categorised by titles drawn from the artisan's vocabulary: masters, journeymen, etc. In Bergslagen, but hardly in the Urals, the position of the master forgemen also greatly resembled the status of the master artisans of urban crafts. In the former region this parallel was further nourished by the establishment of the forge courts, which resembled the artisan guilds in both form and functions. It was thus fellow masters and the regional authorities of the Board of Mines, not the ironmaster, who examined the worker's skill as a master forgeman and fined him if he violated the regulations of his trade. Using Marc Raeff´s concept of co-optation, we here have an illustrative example of the Swedish state using traditional organisations, or creating new ones with a rather archaic design, to administer and regulate the economy of the local community. When work began to be organised along more industrial lines and the artisan workers became proletarians, the forge courts withered away. As the guild tradition did not exist in Russia and its principles of corporate responsibility would have counteracted the power technology in place at the Ural ironworks, where the master forgemen were given a less influential position, no such institutional framework was ever built up.

The organisation of work in the rural parts of the iron industry bore the same hallmarks as in its industrial parts. In Bergslagen, charcoal-burning and transportation were mainly carried out by peasant households, who supplied the charcoal as rent or sold it to the ironworks on a regulated market. Whatever the connection was between the producer of charcoal and the ironworks, a small-scale organisation based on the rural household prevailed in Sweden. In the Urals, on the other hand, charcoal-burning was organised as 'gang work' under command of a group of foremen. Unlike in Sweden, charcoal-burning was broken down into specialised tasks, carried out by distinct categories of workers. During the eighteenth century ascribed peasants were used for woodcutting, but not for the technically more demanding tasks involved in constructing the kiln or producing charcoal.

In Bergslagen, charcoal-burning as well as mining and pig-iron production among the *bergsmän* were essentially household-based tasks. Among other things, this meant that the female share of the labour force in mining, transport and charcoal-burning was considerable, in contrast to the Urals. Studies have also shown that households were substantially larger than those of other rural groups in Sweden. It is likewise clear that forgemen´s households played an important role when work crews were put together. Relationships within the crew were not simply professional, but also based on kinship.

The fact that households served as a foundation for iron production implies that, in occupational terms, they were fairly homogeneous. As was discussed in chapter 6, the structure of households differed between Sweden and Russia, as households in the latter country were more often of an extended type, comprising several generations and relatives. This meant that the households associated with the Ural ironworks often consisted of a varied blend of professions related to the trade. An interesting trend can be discerned, however, in that during the nineteenth century forgemen and skilled workers separated themselves from these households and established nuclear families. We thus find a multiform household structure, in which unskilled workers remained attached to the traditional form of the extended household, while skilled workers began to form smaller units. Bearing in mind the organisational structure of work at the Ural furnaces and forges, it cannot be asserted that this development corresponded to a shift towards household-based work crews. Perhaps it served more of an ideological or rather a cultural function, as a way for the skilled workers to demonstrate their more important role in the production process, compared with other groups in the community. The status accorded by their occupation thus formed the basis for their self esteem and gave them their societal standing.

Dynamics of the Network

The aim of our comparison has been to study similarities and differences in both the structures and the transformation of the iron industries of the Urals and Bergslagen. As regards the dynamics of the production network, we can single out internal and external factors promoting change. As both networks were directed towards the market for bar iron, changing conditions on the international market ought to be seen as an important external factor. It is nevertheless important to stress that signals from the market were filtered through the social composition of the network, and responses were made to these filtered signals. It was in the social context of the network that the actors decided what action it was possible and necessary to take.

Ural iron began to be exported during the first half of the eighteenth century. The Swedish ironmasters and also the state instantly perceived the new threat from the East. The more detailed regulation of the internal raw materials market and the restrictions on the total output of bar iron should be seen, at least partly, as answers to the new competitive situation on the British market. The aim of the Swedish state was to avoid internal competition for raw materials, in order to keep down the price of Swedish bar iron on the market. In a way this policy can be seen as counterproductive, since the halt to the expansion of iron-making cleared the way for Russian iron. Nevertheless, an important goal was achieved as the Swedish industry did not decline.

During the late eighteenth century, the position of both Swedish and Russian iron was threatened by British puddled iron. In this context, however, it is important to underline that the market consisted of different segments rather than constituting one homogeneous arena. Swedish iron had a strong position as a raw material for the steelworks, where puddled iron could not be used. British iron thus competed primarily with the Russian brands. Consequently, the strategies chosen in Sweden and Russia to respond to this threat differed. While the Swedish ironmasters strove to concentrate their works and to increase productivity, the Russian answer seems to have been the establishment of the mining districts. The Russians thus tried to lower production costs by strengthening the bonds of unfreedom, while Swedish ironmasters sought to make production more efficient and to enhance the quality of their products. The Russians lost this contest and their iron came to play a less important role on the world market. This decline, however, was partly offset by the subsequent expansion of their internal market.

Social constraints were embedded in the networks of both regions. This led to conflicts which played an important role in changing the structure of these networks. The fiercest conflicts were fought out between ironmasters and rural groups. The coercive nature of peasant work in the Urals explains the uprisings there, just as well as the collective foundations of rural life account for the scale of the rebellions. Peasants protested in Bergslagen too, but their complaints never assumed such violent forms as in the Urals. One reason why they chose other means of protest was their wider scope to use the legal and political arenas, another was probably that they lacked the organisational means which the village community offered their counterparts in the Urals.

However, a similarity between the rural protests in Bergslagen and the Urals was that they both sprang from the dual character of the workers´ economy, giving them a choice between different types of occupation. If they could feed their families with the products and earnings from their agriculture, peasants in neither of the two regions were economically forced to work for the iron industry. In Sweden, negative attitudes to industrial work came to dominate in the mid-eighteenth century, as the state and the ironmasters strove to keep charcoal prices down so that the industry would not lose its position on the world market, at the same time as grain prices were rising. It thus became more profitable for peasants to specialise as farmers. Broadly the same pattern of action is visible among the *bergsmän*: as the prices paid for pig-iron stagnated, they developed their agrarian occupations. In the Urals we perceive the same tendencies in the southern parts of the region.

It seems possible to explain vital changes in the network structure in terms of the effects of these waves of resistance and protest. That ironmasters in both the Urals and Bergslagen endeavoured to gain a firmer control of the agrarian workforce seems logical. Their strategies differed, however, in the light of what was viewed as possible and acceptable in their different social settings. The Ural ironmasters strengthened their property rights in a feudal context, while their counterparts in Bergslagen built up comparable complexes (of rather modest dimensions though, compared with the Ural establishments), based on capitalist ownership and wage labour.

As a consequence of the household economy forming the kernel of resistance, the protests were most ferocious in the rural parts of the chain of production. In the Urals, the skilled workers did not follow under the banners of the ascribed peasants, but stayed and defended their workplaces against attack. We know very little at all about workers' conflicts within the walls of the forges and furnaces of the Urals, but are better acquainted with the situation in Bergslagen. Here, the social organisation of work has been characterised as a centralised putting-out system. As generally occurs where such modes of production are applied, this system generated conflicts between workers and putters-out over the amount and quality of the raw materials and over the end-product. In the Swedish iron industry, however, the protests never led to widespread social unrest. Embezzlement by workers was a well-known fact in early industrial iron-making in Bergslagen. Such actions could have been devastating for the ironmasters since they increased the cost of production, but they may also have had beneficial effects as an incentive for workers to develop their skills. Cheating can thus be productive in the long run. The dearth of information about resistance among workers in the Urals may be due to neglect on the part of historians, but it also seems evident that the mode of work in Bergslagen and also the workers' self esteem as artisans gave them far greater scope both to present demands to the ironmasters and to benefit personally from production.

The skills of workers were the most important force of production in all early industrial activities, and iron production was no exception to this. The social organisation of the Swedish forges seems to have reflected this fact more clearly, giving the master forgeman a high degree of autonomy. Knowledge was passed on in an informal way from master to journeymen, who were often members of the same household. By contrast, in the Urals knowledge was regarded as the property of the ironmaster. It is possible that such a view of the ownership of knowledge facilitated technological change in the industry during the eighteenth century. As was discussed in chapter 8, the early appearance of formal education of mining engineers in Russia should be seen against such a background, underlining the need for training of foremen and technical experts to equip them to perform their tasks in the forge. In Bergslagen, the technological take-off began later, in response to the new situation on the world market, i.e. at the beginning of the nineteenth century. It had become more than evident to the Swedish ironmasters that they had to grasp the mysteries of production, to light up the shadows inside the black box, in order to develop technology and make production more efficient. The expansion of formal training of mining engineers in Sweden should be seen in this perspective.

Technical change is thus to be seen as an answer to specific problems, rather than as a transforming force in its own right. These problems could be of a varying nature, social as well as market-related. In both the Urals and Bergslagen, technicians tried to overcome the problems posed by peasant charcoal-burners by industrialising charcoal production or reducing the consumption of charcoal in furnaces and forges. In Bergslagen, the threat from British puddled iron was countered by technological means. During the seventeenth and eighteenth centuries the Board of Mines had acted as an institutional bearer of new techniques, in the nineteenth century the Ironmasters´ Association took over this role. The existence of such an institutional framework for technical change certainly facilitated industrial development in Sweden. In the Urals, on the other hand, it was the responsibility of the individual ironmaster to promote technological change, which he often did with the assistance of foreign experts.

Proto-industrial Iron-Making

In the introductory chapter, the studies in this volume were related to a proto-industrial approach. At this stage, it seems fairly evident that early industrial iron-making shares several features which in earlier research have been interpreted as tokens of a proto-industrial evolution:

- Iron was produced for the world market and the structure of this market influenced the development of the trade.
- Specialisation in iron-making in certain regions led to a regional division of labour, as foodstuffs were imported from neighbouring areas.
- In important parts of the network, production was based on peasant households. This influenced the size of these households and also (in Bergslagen) made women an important part of the labour force. In other parts of the network households played an important, although not a dominant role, e.g. in the Swedish forges.
- The manner in which the social organisation of work developed shows striking similarities with the development

that some researchers have described as proto-industrial, i.e. from the *Kauf*- and putting-out systems to the 'mature' industrial system.

- Capital and knowledge accumulated during the early industrial period were used in order to industrialise production.

This list of similarities does not mean that we insist on characterising the iron industry as proto-industrial, but merely reflects our desire, which we share with several other scholars, to widen the concept of early industry to comprise sectors other than textiles. We argue that it is more fruitful to use the network concept as a theoretical tool to study the dynamic features of an industry, than to concentrate the analysis only on those parts with special characteristics, such as household production. As parts of a chain of production, the different elements interacted. The demand for cheap raw materials, in order to keep the price of bar iron down on the international markets, partly explains the Swedish peasants' unwillingness to burn charcoal in the mid-eighteenth century. The rise in grain prices during the same period further elucidates the peasants' choice of strategy. Such problems with the agrarian labour force in both Bergslagen and the Urals forced the ironmasters to reorganise the social foundations of the production network. Industrial activities no longer fitted into their agrarian surroundings in the way they had once done, and as a consequence the networks were reconstructed. In Sweden this reconstruction meant a leap forward towards industrialisation proper, while in Russia the mining districts were stigmatised by feudal property relations.

Glossary

Explanations in quotes are direct quotations from *Land Commune and Peasant Community in Russia: Communal Forms in Imperial and Early Soviet Society* (ed. R. Bartlett, Macmillan 1990).

allmoge	(Swedish) rural population
backstugusittare	(Swedish) person occupying a very small cottage, partly below ground level, with practically no land attached to it
Bergamt, pl. Bergämte	(German term) office of mining administration
bergmästare	(Swedish) state official in charge of a certain geographical district
bergshantering	Swedish term denoting the entire production process, from mining to forging
Berg(s)hauptman	(originally German term); Russian state official responsible for supervising mining and metal production
Bergskollegium	Swedish Board of Mines
bergslag	community of *bergsmän*; region in which metal production took place
bergsman, pl. -män	Swedish peasants in charge of pig-iron production
bergsmanshemman	farmstead intended for pig-iron production; it was a prerogative of *bergsmän* to possess such farmsteads
bonde, pl. bönder	Swedish peasant or farmer
borgerlig näring	old Swedish expression, meaning economic activities regarded as fitting for people of wealth and/or people belonging to the burgher estate
bruk	traditional Swedish metalworks
butsmide	variant of the German forging method
byggningabalk	part of the Swedish legal code dealing with the internal life of the hamlet/village

chetvert'	measure: grain 209.91 litres; liquids 3.08 litres; 'six bushels'
desiatskii, pl. desiatskie	'tensman, decurion, low-level peasant official'
desiatina, pl. desiatiny	'measure of area, = 2400 sq. sazheni, 1.09 hectares'
dvor, pl. -y	'household, family unit, peasant farm; courtyard'
duma	council, representative assembly
frälsebönder	Swedish peasants who were tenants of the nobility
gornozavodskii okrug, pl. gornozavodskie okruga	mining district
gruvlag	community of those who had a mine in common
guberniia, pl. gubernii	'province (principal Imperial geographical administrative division)'
guliashchii	ambulating person
hemman	Swedish farmstead, the principal unit of taxation in early modern Sweden
husbehov	domestic needs, essential needs
hushåll	household
hyttelag	community of those who had a blast furnace in common
hyttfogde	furnace bailiff
izba, pl. izby	peasant's log hut
jordabalk	part of the Swedish legal code dealing with landed estates (inheritance, purchase, etc.)
koksmide	variant of the German forging method
kolare	charcoal-burner; a person who produces charcoal on a wage basis
krest'ianin, pl. krest'iane	peasant
kronobönder	Swedish peasants who were tenants of the state/the Crown
kureni	place where charcoal was produced
kustar', pl. kustari	craftsman, 'worker in cottage industry'
köpkol	charcoal bought from freeholders
landbo, pl. landbor	Swedish tenant
legofolk	domestic servants
masmästare	master furnaceman
mir, pl. -y	'group of peasants with some powers of self-government, based on a village; commune, community'
mulltimmerhytta	timber-clad blast furnace
nepremennye rabotniki	permanent workers; replaced ascribed peasants c. 1800

obshchina, pl. obshchiny	'community, commune, collective'
okrug, pl. -a	'administrative division, region'
osmund	wrought iron cut into small pieces
ostrog, pl. -a	stockaded town
övermasmästare	senior master furnaceman
podzavodskie	literally: at or under an ironworks, here: serfs used in iron production
polushka, pl. polushki	a quarter-copeck piece
posessionnye	in eighteenth–first half of nineteenth century: a category of peasants belonging to private enterprises/works; also, conditional form of land tenure
prikaz, pl. -y	order, command; department, office
pripisnye	ascribed peasants
pud, pl. -y	pood, measure of weight: 16.38 kg
raskladchik	person in charge of apportioning work
rekognition	fee which Swedish ironmasters were liable to pay in return for the right to use state forests
Riksdag	Diet; early modern Swedish parliament
sazhen', pl. sazheni	'measure of length: 2.134 m'
selo, pl. sëla	village with a church
skattebönder	Swedish peasants who possessed land of their own
skattefrälse	*skattebönder* who were to pay their land taxes to a nobleman instead of to the state
sloboda, pl. slobody	a large village or settlement (often inhabited by free, non-serf peasants)
soslovie, pl. sosloviia	'social estate, estate of the realm'
statare	farm worker paid partly in kind
stig, pl. stigar	20 hectolitres
torpare	crofter
uezd, pl. -y	'district, sub-division of guberniia, q.v.'
ustavshchik	technical head of production
verst	measure of length: 1.067 kilometres
voevoda	governor of a province
volost', pl. volosti	'canton, township, unit of administration next above the village', small rural district
votchina	ancestral lands, patrimonial estate
votchinnye	patrimonial serfs
zemstvo, pl. zemstva	'system of rural local government introduced in 1864, finally implemented in 34 gubernii of European Russia', elective district council in pre-revolutionary Russia
zhuki	solid lumps of spongy iron

Notes on Contributors

Swedish Contributors

Maria Ågren is associate professor of history at the University of Uppsala. Her research deals with the social differentiation among peasants due to debt during the seventeenth and eighteenth centuries, and, generally, ownership, property rights, and legal discourse. Her publications include *Jord och gäld. Social skiktning och rättslig konflikt i södra Dalarna ca 1650–1850* (diss. 1992), 'Land and Debt: On the Process of Social Differentiation in Rural Sweden, circa 1750–1850' in *Rural History* (1994), 'Labour and Property: Questions at the Intersection of Intellectual History and Social History' in *Societies Made Up of History* (1996), and *Att hävda sin rätt. Synen på jordägandet i 1600-talets Sverige, speglad i institutet urminnes hävd* (1997).

Anders Florén is associate professor of history at the University of Uppsala. His research focuses on the social organisation of work in pre-industrial society, nationally and internationally. Among his publications are *Disciplinering och konflikt. Den sociala organiseringen av arbetet: Jäders bruk 1640–1750* (diss. 1987), 'Social organization of work and labour conflicts in proto-industrial iron production in Sweden, Belgium, and Russia' in *International Review of Social History* (1994) and (with G. Rydén) 'A Journey into the Market Society: A Swedish Pre-industrial Spy in the Middle of the Eighteenth Century' in *Societies Made Up of History* (1996).

Maths Isacson is professor of economic history at the University of Uppsala. Among his fields of research are rural history, proto-industry, and modern labour history. Among his main publications are *Ekonomisk tillväxt och social differentiering 1680–1860. Bondeklassen i By socken, Kopparbergs län* (diss. 1979), and (with L. Magnusson) *Proto-industrialisation in Scandinavia: Craft Skills in the Industrial Revolution* (1987).

Göran Rydén holds a Ph.D. in economic history from the University of Uppsala. His main line of research is the nineteenth-century iron industry, focusing on the relation between labour and household. Among his publications are *Hammarlag och hushåll. Om relationen mellan smidesarbetet och smedshushållen vid Tore Petrés brukskomplex, 1830–1850* (diss. 1990), 'Iron production and the household as a production unit in nineteenth-century Sweden' in *Continuity and Change* (1995), and the above-mentioned article, written with A. Florén.

Maria Sjöberg holds a Ph.D. in history from the University of Stockholm. Her research deals with small-scale iron production within the framework of peasant households, and with inheritance strategies in proto-industrial areas. Her main publication is *Järn och jord. Bergsmän på 1700-talet* (diss. 1993).

Rolf Torstendahl is professor of history at the University of Uppsala. His fields of research cover among others historiography, historical social theory, professionalisation, and technology and social change. Among his many publications, *Bureaucratisation in North-Western Europe, 1880–1985: Domination and Governance* (1991) can be mentioned. He is a member of *Academia Europaea.*

Russian Contributors

All Russian contributors are from the Institute of History and Archaeology in Ekaterinburg.

Ludmila Dashkevich

Dmitri Gavrilov

Svetlana Golikova

Nina Minenko

Igor Poberezhnikov

Vladimir Shkerin

Anton Tomilov

Sergei Ustiantsev

Vladimir Zhelezkin

Bibliography

Unpublished Sources

Sweden

Riksarkivet, Stockholm
Bergskollegii Arkiv
Huvudarkivet
Bergverksrelationer Stora Kopparberg
Inkomna skrivelser och akter
Relationer och skrivelser angående utländska bergverk
Advokatfiskalskontorets arkiv
Riksdagshandlingar
Bergsdeputationen 1734
Konseljakter
Civildepartementet
Skrivelser till Kungl. Majt
från Bergskollegium
Jernkontoret, Stockholm
R. Angersteins reseberättelser
Landsarkivet, Uppsala
Bergmästarens i Nora och Linde bergslager arkiv
Bergmästarens i Gävleborg, Stockholm och Uppsala arkiv
Bergstinget i Österbergslagen
Gruv- och hammarsmedsting vid Bispberg
Gruvting vid Säter
Uppsala universitetsbibliotek
Das Nord- und Ostliche Theil von Europa und Asia. Map, Stockholm 1730
Gävle kommunarkiv, Gävle
Tolvfors arkiv

Russia

Gosudarstvennyi Arkhiv Sverdlovskoi Oblasti
F. 24
F. 40
F. 643
Gosudarstvennyi Arkhiv Permskoi Oblasti
F.R.-790
F.176

Tsentral'nyi Gosudarstvennyi Arkhiv Drevnikh Aktov
F. 1267
F. 1278
Tsentral'nyi Gosudarstvennyi Istoricheskii Arkhiv
F. 44

Published Sources

Sweden

Kongl. Stadgar, Förordningar, Brev och Resolutioner från år 1528 intill 1701 angående Justitiae och Executionsährender, ed. J. Schmedemann, Stockholm 1706.
Kongl. Stadgar, Förordningar, Privilegier och Resolutioner angående Justitien och Hushållningen wid Bergwerken och Bruken, vol. I–III, Stockholm 1736–1797.
Konung Gustaf den förstes registratur, Stockholm 1891, 1910.
Riksdagshandlingar 1818–1860: Allmänna berednings- och ekonomiutskottets handlingar; Borgarståndets protokoll; Ridderskapets och adelns protokoll.
Tilas, D. 'Curriculum vitae I–II, 1712–1757', in *Historiska Handlingar* 38:1, Stockholm 1966.
Utdrag utur alle ifrån den 7 decemb. 1718 utkomne publique handlingar, placater, förordningar, resolutioner och publicationer, ed. R. G. Modée, Stockholm 1746.

Russia

Polnoe sobraine zakonov (Code of law II), vols 19, 20,22, 28, 29.
Svod zakonov Rossiiskoi imperii (Code of the Russian Empire) St Petersburg 1857, vol. 8.

Literature

Adamov, V. V. 'Ob original'nom stroe i nekotorykh osobennostiakh razvitiia gornozavodskoi promyshlennosti Urala', *Voprosy istorii kapitalisticheskoi Rossii: problema mnogoukladnosti*, Sverdlovsk 1972.
———. 'Ob original'nom stroe Urala', in *Nauchnaia sessiia, posviashchënnaia problemam mnogoukladnosti*, Moscow 1969.
Adamson, R. 'Borrowing and adaptation of the British technology by the Swedish iron industry in the early nineteenth century', in K. Bruland, ed., *Technology transfer and Scandinavian industrialisation*, Oxford 1991.
———. *De svenska järnbrukens storleksutveckling och avsättningsinriktning 1796–1860*, Göteborg 1963.
———. *Järnavsättning och bruksfinansiering 1800–1860*, Göteborg 1966.
Ågren, M. *Att hävda sin rätt. Synen på jordägandet i 1600-talets Sverige, speglad i institutet urminnes hävd.* Stockholm 1997.
———.'Att lösa ekonomiska tvister - domstolarnas främsta sysselsättning på 1700-talet?', *Svensk Historisk tidskrift* 1988:4.
———. *Jord och gäld. Social skiktning och rättslig konflikt i södra Dalarna ca 1650–1850*, Studia Historica Upsaliensia 166, Uppsala 1992.
———. 'Land and Debt. On the Process of Social Differentiation in Rural Sweden, circa 1750–1850' *Rural History* 1994:5(1).
———. 'Nordvikens bergsmän och deras syn på rättten', *Husby-Rocken* 1993, 1994.

Agricola, G. *De Re Metallica*, translated by H.C. Hoover and L.H. Hoover, London 1912.
Alefirenko, P. 'Ekonomicheskie vzgliady V.N. Tatishcheva', *Voprosy Istorii*, 1948:12.
Aleksandrov, V. A. *Sel'skaia obshchina v Rossii (XVII–nachalo XIX v.)*, Moscow 1976.
Alfërov, N. S. *Zodchie starogo Urala. I polovina XIX veka*, Sverdlovsk 1960.
Almquist, Joh. A. *Bergskollegium och Bergslagsstaterna 1637–1857*, Stockholm 1909.
Althoff, G. *Verwandte, Freunde und Getreue. Zum politischen Stellenwert der Gruppenbildungen im früheren Mittelalter*, Darmstadt 1990.
Amussen, S. *An Ordered Society. Gender and Class in Early Modern England*, Oxford 1988.
Andersson, I. *Uddeholms Historia till 1914*, Stockholm 1960.
Andrushchenko, A. I. *Krest'ianskaia voina 1773–75 gg. na Iaike, v Priural'e, na Urale i v Sibiri*, Moscow 1969.
Antipov, A. I. 'Kharakter rudonosnosti i sovremennoe polozhenie gornogo, t.e. rudnogo, dela na Urale', *Gornyi zhurnal*, 1860.
Arpi, G. *Den svenska järnhanteringens träkolsförsörjning, 1830–1950*, Jernkontorets Bergshistoriska Skriftserie 14, Stockholm 1951 .
Aston, T. H. and Philpin, C. G. E, eds, *The Brenner Debate: Agrarian Class Structure and Economic Development in Pre-industrial Europe*, Cambridge 1985.
Attman, A. *Fagerstabrukens historia. Adertonhundratalet*, Uppsala 1958.
———. *Svenskt järn och stål 1800–1914*, Jernkontorets Bergshistoriska Skriftserie 21, Stockholm 1986.
Bagge, J. S. *Tankar rörande beskaffenheten af ett bergsundervisningsväsende i Sverige*, Falun 1865.
Bagger, H. 'Reformernas tidevarv - och reaktionens', in *Ryssland, ett annat Europa*, Stockholm 1995.
Baidin, V. I. 'Formirovanie burzhuazii v srede gosudarstvennogo krest'ianstva Srednego Urala vo vtoroi polovine XVIII v.', in *Krest'ianstvo Urala v epokhu feodalizma*, Sverdlovsk 1988.
Baron, S. H. 'The Transition from Feudalism to Capitalism in Russia: A Major Soviet Historical Controversy', *American Historical Review* 1972.
Bartlett, R. 'The Russian peasantry on the eve of the French revolution', *History of European Ideas*, 1990:12(3).
Bartlett, R., ed. *Land Commune and Peasant Community in Russia. Communal Forms in Imperial and Early Soviet Society*, London 1990.
Bengtsson, T. and Ohlsson, R. 'Sveriges befolkning - myter och verklighet', in B. Furuhagen, ed., *Äventyret Sverige. En ekonomisk och social historia*, Stockholm 1993.
Berg , M. *The Age of Manufactures 1700–1820*, London 1985.
———. 'Markets, Trade and European Manufacture', in M. Berg, ed., *Markets and Manufacture in Early Industrial Europe*, London, New York 1991.
———. 'Revisions and revolutions: Technology and productivity change in manufacture in eighteenth-century England', in P. Mathias and J. A. Davies, eds, *Innovations and Technology in Europe*, Oxford 1991.
Bergensköld, E. 'Nora Bergslag vid 1784 års slut', in J. Johansson, ed., *Om Noraskog vol. 1*, Stockholm 1875.
Berger, S., Bohlin, M. and Forsberg, G. *Bergslagsprojektet. ERU-rapport 6*, Stockholm 1981.
Berkner, L. K. 'The stem family and the developmental cycle of the peasant household: An eighteenth-century Austrian exemple', *American Historical Review* 1972:77.
Bhaduri, A. *The Economic Structure of Backward Agriculture*, London, New York 1983.
The Birth of a Consumer Society, eds N. McKendrick, J. Brewer and J. H. Plumb, London 1982.
Björkenstam, N. *Västeuropeisk järnframställning under medeltiden*, Stockholm 1991.

Bladh, G. *Finnskogens landskap och människor under fyra sekler - en studie av natur och samhälle i förändring*, Göteborg 1995.
Blickle, P. *Landschaften im alten Reich. Die Staatliche Funktion des gemeinen Mannes in Oberdeutschland*, München 1973.
Blum, J. *Lord and Peasant in Russia from the Ninth to the Nineteenth Century*, Princeton 1961.
Boëthius, B. *Gruvornas, hyttornas och hamrarnas folk. Bergshanteringens arbetare från medeltiden till gustavianska tiden*, Stockholm 1951.
———. 'Ramnäs bruk under inflationen på 1700-talet. Studier i Jakob Tersmdens enskilda räkenskapsbok', *Svensk Historisk tidskrift* 1954:1.
———. *Ur de stora skogarnas historia*, Stockholm 1917.
Boëthius, B. and Kromnow, Å. *Jernkontorets historia*, vols 1–3, Stockholm 1947–1968.
Bonde, C. and Fleming, E. 'Ett principbetänkande om bergshantering från mitten på 1600-talet. Christer Bonde och Erik Flemings relation till drottning Kristina år 1652', *Jernkontorets Annaler*, Stockholm 1901.
Braudel, F. *Civilization and Capitalism: Fifteenth–Eighteenth Century. Vol. 2: The Wheels of Commerce*, London 1982.
———. *Marknadernas Spel. Civilisationer och kapitalism 1400–1800*, vol. 2, Stockholm 1986.
Braun, R. 'The impact of cottage industry on an agrarian population' in D. Landes, ed., *The Rise of Capitalism*, New York 1966.
———. *Industrialization and Everyday Life*, Cambridge 1979 (first edition 1960).
Bredefeldt, R. *Tidigmoderna företagarstrategier. Järnbrukens ägar- och finansieringsförhållanden under 1600-talet*, Stockholm 1994.
Bubnov, E. N. *Russkoe dereviannoe zodchestvo Urala*, Moscow 1988.
Bugaeva, S. Ia. 'Tekhnicheskaia intelligentsiia v gornozavodskoi promyshlennosti Urala (poreformennyi period)', Diss. kand. ist. nauk, Moscow 1979.
Bursell, B. *Träskoadel. En etnologisk undersökning av lancashiresmedernas arbets- och levnadsförhållanden på Ramnäs bruk vid tiden kring sekelskiftet 1900*, Stockholm 1974.
Bäck, K. *Bondeopposition och bondeinflytande under frihetstiden*, Stockholm 1984.
Cantelaube, J. 'La société Pyrénéenne, La forge à la Catalane et Le Haut Fourneau (XVIIe - XIXe siècles)' in G. Magnusson, ed., *The Importance of Ironmaking. Technical Innovation and Social Change, vol. 1*. Stockholm 1995.
Cerman, M. 'Proto-industrialization in an urban environment: Vienna, 1750–1857', *Continuity and Change* 1993:8.
———. 'Proto-industrielle Entwicklung in Österreich', in M. Cerman and S. Ogilvie, eds, *Protoindustrialisierung in Europa*, Vienna 1994.
Chaianov, A. V. *A.V. Chayanov on the Theory of Peasant Economy*, Homewood, Illinois 1966.
Channon, J. 'Regional Variation in the Commune: The Case of Siberia', in R. Bartlett, ed., *Land Commune and Peasant Community. Communal Forms in Imperial and Early Soviet Society*, London 1990.
Cherkasova, A. S. 'Demograficheskaia kharakteristika masterovykh i rabotnykh liudei Zlatoustovskogo i Miasskogo zavodov Urala', in *Demograficheskie protsessy na Urale*, Sverdlovsk 1990.
———. 'Khoziain ili tvorets slavy Starogo sobolia', *Metallurg* 1993:8.
———. *Masterovye i rabotnye liudi Urala v XVIII veke*, Moscow 1985.
———. 'Sotsial'no-ekonomicheskie sviazi gornozavodskikh tsentrov i dereven' Urala v seredine XVIII v.', in *Derevnia i gorod Urala v epokhu feodalizma: problemy vzaimodeistviia*, Sverdlovsk 1986.
Chernoukhov, A. V. *Istoriia medeplavil'noi promyshlennosti v Rossii XVII–XIX vv.*, Sverdlovsk 1988.
Clarke, E. D. *Travels in Various Countries of Europe, Asia and Africa. Part the Fourth*, London 1810.

Clay, C. G. A. *Economic Expansion and Social Change: England 1500–1700*, Cambridge 1984.
Cohen, F. S. 'Dialogue on Private Property', *Rutgers Law Review* 1957:9.
Crisp, O. 'Labour and Industrialization in Russia', in *The Cambridge Economic History of Europe* vii:2, Cambridge 1978.
Czap, P. 'A Large Family: the Peasant's Greatest Wealth: Serf Households in Mishino, Russia, 1814–1858', in R. Wall, J. Robin and P. Laslett, eds, *Family Forms in Historic Europe*, Cambridge 1983.
———. 'Marriage and the Peasant Joint Family in the Era of Serfdom', in D. Ransel, ed., *The Family in Imperial Russia*, Urbana 1978.
———. 'The Perennial Multiple Family Household, Mishino, Russia 1782–1858', *Journal of Family History* 1982:7.
Dahlgren, E. W. 'Kolböndernas sammansvärjning vid Lövsta 1718', *Med hammare och fackla* II, Stockholm 1930.
Dankoviskii, 'Ob uglezhzhenii na Ural'skikh kazënnykh zavodakh v nastoiashchee vremia', *Gornyi zhurnal* 1858:1.
Das akademische Deutschland, vol. 1, ed. M. Doeberl, Berlin 1930.
Dashkevich, L. A. and Safronova, A. M. 'Gornaia shkola v Rossii (XVIII - pervaia polovina XIX vv.)', in *Metallurgicheskie zavody i krest'ianstvo: Problemy sotsial'noi organizatsii promyshlennosti Rossii i Shvetsii v ranneindustrial'nyi period*, in the series *Sbornik nauchnykh trudov*, Ekaterinburg 1992.
Dashkevich, L. A., Ustiantsev, S. V. and Tret'iakova, Z. F. 'Ob obiazannostiakh sluzhashchikh i zavodskikh liudei pri zavodskikh proizvodstvakh', in *Organizatsiia proizvodstva i truda v metallurgicheskoi promyshlennosti Urala XVIII–nachala XIX vv. Nauchnye doklady*, Sverdlovsk 1990.
Daunton, M. J. *Progress and Poverty. An Economic and Social History of Britain 1700–1850*, Oxford 1995.
Davies, R. S. W. and Pollard, S. 'The Iron Industry, 1750–1850', in C. Feinstein and S. Pollard, eds, *Studies in Capital Formation in the United Kingdom 1750–1920*, Oxford 1988.
De Vries, J. *The Dutch Rural Economy in the Golden Age*, London, New Haven 1974.
———. *The Economy of Europe in an Age of Crisis, 1600–1750*, Cambridge 1976.
DeHart, B. 'The Pugachev Revolt and its Effect on the Industrial Work-Force of the Urals: a Reconsideration' in R.P. Bartlett, A.G. Cross, and K Rasmussen, eds, *Russia and the World of the Eighteenth Century*, Columbus, Ohio 1986.
Dmitriev, A. V. *Osnovnye napravleniia razvitiia ekonomiki Urala v poreformennyi period 1861 - 1900 gg*. Sverdlovsk 1991.
Douhan, B. *Arbete, kapital och migration. Valloninvandringen till Sverige under 1600-talet*, Studia Historica Upsaliensia 140, Uppsala 1985.
Dübeck, I. 'Europeisk privilegieret' in *Industri og bjergværksdrift. Privilegering i Norden i det 18. århundrede*, Oslo, Bergen, Stavanger, Tromsø 1985.
Dubrava, T. S. *Leningradskii gornyi institut*, Leningrad 1957.
Eeckaute, D. 'La législation des forêts au XVIIIe siècle', *Cahiers du monde russe et soviétique* 1968:9.
Eggertz, V. *Utlåtanden rörande Falu bergsskolas förening med Teknologiska institutet*, Falun 1863.
Ehn, W. *Mötet mellan centralt och lokalt. Studier i uppländska byordningar*, Uppsala 1991.
Ericsson, B. 'Privilegiegivning till bergverk, järnbruk och exportsågar i komparativt perspektiv', in *Industri og bjergværksdrift. Privilegering i Norden i det 18. århundrede*, Oslo, Bergen, Stavanger, Tromsø 1985.
———. 'Privilegiegivningen till järnbruk och järnmanufaktur i Sverige under frihetstiden', in *Industri og bjergværksdrift. Privilegering i Norden i det 18. århundrede*, Oslo, Bergen, Stavanger, Tromsø 1985.
Eriksson, E. *Greksåsar. En bergslagsby*, Hembygdsföreningen Noraskogs skriftserie, no. 3, Nora 1963.

Esper, T. 'The condition of the serf workers in Russia's metallurgical industry, 1800–1861', *The Journal of Modern History* 1978:4.

———. 'The Incomes of Russian Serf Ironworkers in the Nineteenth Century', *Past & Present* 1981:93.

———.'Industrial Serfdom and Metallurgical Technology in 19th-Century Russia', *Technology and Culture* 1982:23(4).

Essemyr, M. *Bruksarbetarnas livsmedelskonsumtion. Forsmarks Bruk 1730–1880*, Uppsala 1989.

Essen, Iu. *Istoriia gornozalozhnykh SSSR*, vol. 1, Moscow 1926.

Evans, C. 'The Corporate Culture of the British Iron Industry 1650–1830', Paper presented at the Eleventh International Economic History Congress in Milan 1994.

Evans, C. and Rydén, G. 'Iron in Sweden and Britain 1600–1800: Interdependence and Difference', in G. Magnusson, ed., *The Importance of Ironmaking. Technical Innovation and Social Change*, Stockholm 1995.

———. 'Recruitment, kinship and the distribution of skill: Bar iron production in Britain and Sweden 1500–1860' in M. Berg and K. Bruland, eds, *Technological Revolutions in Europe 1760–1860*, Edward Elgar, Aldershot 1998.

Florén, A. *Disciplinering och konflikt. Den sociala organiseringen av arbetet. Jäders bruk 1640–1750*, Uppsala 1987.

———. *Genus och producentroll. Kvinnoarbete inom svensk bergshantering, exemplet Jäders bruk 1640–1840*, Opuscula Historica Upsaliensia 7, Uppsala 1991.

———. 'Järnproduktionens sociala mångfald. Brukspatroner, bergsmän och hammarsmeder i Nora och Linde bergslagar 1620–1750', in M. Essemyr, ed., *Bergshantering i Örebro län: en arkivinventering*. Forthcoming.

———. 'Klasskamp utan fackförening, om förindustriella arbetskonflikter', *Arkiv för studier i arbetarrörelsens historia* 1990:45.

———. 'Patterns of Crime, Protest and Conflict in the Nora and Linde Mining Region 1650–1720', Unpublished paper presented at the third meeting on iron-making in Russia and Sweden before the twentieth century, Ekaterinburg 1993.

———. 'Social Organization of Work and Labour Conflicts in Proto-Industrial Iron Production in Sweden, Belgium and Russia', *International Review of Social History*, Supplement, 1994:39.

———. 'Voyager, c'est apprendre! Le voyage de l'ingénieur suédois Reinhold Angerstein en 1755', *Patrimoine industrielle Wallonie-Bruxelles* 1996:33.

Florén, A. and Rydén, G. 'A journey into the market society. A Swedish preindustrial spy in the middle of the 18th century' in R. Björk and K. Molin, eds, *Societies made up of history. Essays in historiography, intellectual history, professionalisation, historical social theory & proto-industrialisation*, Uppsala 1996.

———. 'Social organisation of the Swedish bar iron production, 1600–1880', Unpublished paper.

———. *Arbete, hushåll och region. Tankar om industrialiseringsprocesser och den svenska järnhanteringen*, Uppsala Papers in Economic History, Research Report no. 29, Uppsala 1992.

Florén, A., Isacson, M., Rydén, G., and Ågren, M. 'Swedish Iron Before 1900', in G. Rydén and M. Ågren, eds, *Ironmaking in Sweden and Russia. A Survey of the Social Organisation of Iron Production before 1900*, Uppsala 1993.

Foster, G. M. 'Peasant Society and the Image of Limited Good', *American Anthropologist* 1965:67(2).

Freeze, G. L. 'The *Soslovie* (Estate) Paradigm and Russian Social History', *The American Historical Review* 1986:91(1).

Friberg, N. 'The growth of population and its economic-geographic background in a mining district in central Sweden 1650 - 1750', *Geografiska Annaler*, Stockholm 1956.

Frolov, V. A. 'Leningradskii Ordena Lenina Gornyi Institut (K 170-letiiu so dnia osnovaniia)', *Gornyi zhurnal* 1944:7.

From Supplication to Revolution. A Documentary Social History of Imperial Russia, ed. G.L. Freeze, New York, Oxford 1988.
Furuskog, J. *De värmländska järnbruken*, Filipstad 1924.
Gaier, C. 'Un artisanat spécialisé: l'armurerie liégeoise', in R. LeJeune and J. Stiennon, eds, *La Wallonie le pays et les hommes*, vol. IV, Bruxelles 1981.
Gaunt, D. 'Familj, hushåll och arbetsintensitet. En tolkning av demografiska variationer i 1600- och 1700-talens Sverige', *Scandia* 1976.
———. 'Pre-Industrial Economy and Population Structure', *Scandinavian Journal of History* 1977:2.
Gavrilov, D. V. *Rabochie Urala v period domonopolisticheskogo kapitalizma, 1861–1900*, Sverdlovsk 1985.
Geary, D. 'Protest and Strike. Recent Research on 'Collective Action' in England, Germany and France', *Historische Zeitschrift, Sonderheft 15, Arbeiter und Arbeiterbewegung im Vergleich*, ed., K. Tenfelde, München 1986.
Givens, R. D. 'To Measure and to Encroach: the Nobility and the Land Survey' in R.P. Bartlett, A.G. Cross, and K Rasmussen, eds, *Russia and the World of the Eighteenth Century*, Columbus, Ohio 1986.
Gligorov, V. 'Justice and Privatisation', *Communist Economies and Economic Transformation* 1992:4(1).
Gorlovskii, M. A. and Piatnitskii, A. N. *Iz istorii rabochego dvizheniia na Urale*, Sverdlovsk 1954.
Gornozavodskaia promyshlennost' Urala na rubezhe XVIII–XIX vv., Sverdlovsk 1956.
Gorovoi, F. S. *Padenie krepostnogo prava na gornykh zavodakh Urala*, Perm' 1961.
Granlund, J. 'Greksåsars Bergsmän i Hyttelag och Gruvlag', *Meddelanden från Föreningen Örebro Länsmuseum*, 1947.
Gregory, J. R. 'Image of Limited Good, or Expectation of Reciprocity?', *Current Anthropology* 1975:16(1).
Gromyko, M. M. *Traditsionnye normy povedeniia i formy obshcheniia russkikh krest'ian XIX v.*, Moscow 1986.
Gullickson, G. 'Love and power in the proto-industrial family' in M. Berg, ed., *Markets and Manufacture in Early Industrial Europe*, London, New York 1991.
Gurevitj, A. J. *Feodalismens uppkomst i Västeuropa*, Stockholm 1970.
Gus'kova, T. K. 'Okruzhnaia sistema kak forma organizatsii ural'skoi gornozavodskoi promyshlennosti', in *Sotsial'naia i proizvodstvennaia organizatsiia metallurgicheskoi promyshlennosti Urala v XVIII–XIX vv.* Forthcoming.
Haggrén, G. 'Från brukspatrons disciplineringsredskap till kronans förvaltningsorgan. Bergstingen i västra Nyland under stormaktstiden', *Historisk Tidskrift för Finland* 1994:1.
Hajnal, J. 'Two Kinds of Pre-industrial Household Formation System', in R. Wall, J. Robin and P. Laslett, eds, *Family Forms in Historic Europe*, Cambridge 1983.
Hansotte, G. *La clouterie liégeoise et la question ouvrière au XVIIIe siècle*, Bruxelles 1972.
Hareven, T. 'The History of the Family and the Complexity of Social Change', *The American Historical Review* 1991:96(1).
Harris, J. *The British Iron Industry 1700–1850*, London 1988.
Hasquin, H. *Une mutation, le pays de Charleroi aux XVIIe et XVIIIe siècles*, Bruxelles 1971.
Heckscher, E. F. 'Betydelsen av vår historiska brukspolitik' in *Ekonomi och historia*, Stockholm 1922.
———. *Svenskt arbete och liv. Från medeltiden till nutid*, Stockholm 1968 (first edition 1941).
———. *Sveriges ekonomiska historia från Gustav Vasa 1:1–2:2*, Stockholm, 1935–1949.
———. 'Un grand chapitre de l'histoire du fer: le monopole suédois', *Annales ESC* 1932:14, 15.
Hennin, V. De *Opisanie ural'skikh i sibirskikh zavodov 1735 g.*, published Moscow 1937.

Henriksson, H. 'Kvinnor i Gruvarbete', *Med Hammare och Fackla* 1994:33.
Henriques, P. *Skildringar ur Kungl. Tekniska Högskolans historia*, vol. 1, Stockholm 1917.
Hildebrand, K.-G. 'Foreign Markets for Swedish Iron in the 18th Century', *The Scandinavian Economic History Review* 1958:6(1).
———. *Fagerstabrukens historia. Del I, Sexton och sjuttonhundratalen*, Uppsala 1957.
———. *Swedish Iron in the Seventeenth and Eighteenth Centuries. Export Industry before the Industrialization*, Jernkontorets Bergshistoriska Skriftserie 29, Stockholm 1992.
———. *Svenskt Järn. Sexton- och Sjuttonhundratal. Exportindustri före Industrialismen*, Stockholm 1987.
Hoch, S. *Serfdom and Social Control in Russia. Petrovskoe, a Village in Tambov*, Chicago 1986.
———. 'Serfs in Imperial Russia: Demographic Insights', *Journal of Interdisciplinary History* 1982:13(2).
Holderness, B. A. 'Credit in a Rural Community, 1660–1800. Some Neglected Aspects', *Midland History* 1975:3(2).
Hudson, H. D. Jr. *The Rise of the Demidov Family and the Russian Iron Industry in the Eighteenth Century*, Newtonville, Mass 1986.
Hudson, P. *The Genesis of Industrial Capital*, Cambridge 1986.
———. 'Landholding and the organization of textile manufacture in Yorkshire rural townships c. 1660–1810' in M. Berg, ed., *Markets and Manufacture in Early Industrial Europe*, London, New York 1991.
———. 'The regional perspective' in P. Hudson, ed., *Regions and Industries in the Industrial Revolution in Britain*, Cambridge 1989.
Hyde, C. *Technological Change and the British Iron Industry 1700–1870*, Princeton 1977.
Hyenstrand, Å. 'Iron and Iron Economy in Sweden', in H. Clarke, ed., *Iron and Man in Prehistoric Sweden*, Stockholm 1979.
Högberg, S. *Utrikeshandel och sjöfart på 1700-talet*, Stockholm 1969.
Iarovoi, G. V. 'K voprosu o mnogoukladnosti ural'skoi gosudarstvennoi derevni nakanune reformy 1861 goda', *Voprosy istorii kapitalisticheskoi Rossii*, Sverdlovsk 1972.
———. *Regulirovanie krest'ianskogo otkhoda na gornye zavody Urala v doreformennyi period. Genezis i razvitie kapitalisticheskikh otnoshenii na Urale*, Sverdlovsk 1980.
Iggesunds Bruk 1685–1869, ed. G. Utterström, Iggesund 1985.
Ihrfors, C. G. *Om häradsallmänningar enligt svensk rättsutveckling*, Uppsala 1916.
Imperial Russia A Source Book, 1700–1917, ed., B. Dmytryshyn, Portland 1974.
Industrialization before Industrialization. Rural Industry in the Genesis of Capitalism, eds P. Kriedte, H. Medick and J. Schlumbohm, Cambridge 1981.
Industri og bjergværksdrift. Privilegering i Norden i det 18. århundrede, Oslo, Bergen, Stavanger, Tromsø 1985.
Ironmaking in Sweden and Russia. A Survey of the Social Organisation of Iron Production before 1900, eds G. Rydén and M. Ågren, Opuscula Historica Upsaliensia 12, Uppsala 1993.
Isacson, M. 'The decline of peasants' dependence on iron industry. 17th to 19th century' in *Metallurgical Works and Peasantry. Problems of Social Organization of Industry in Russia and Sweden in Early-Industrial Period*, Ekaterinburg 1992.
———. *Ekonomisk tillväxt och social differentiering 1680–1860, Bondeklassen i By socken, Kopparbergs län*, Uppsala Studies in Economic History 18, Uppsala 1979.
Isacson, M. and Magnusson, L. *Protoindustrialisation in Scandinavia*, Leamington Spa 1987.
Isaksson, O. *Vallonbruken i Uppland. Människor och miljöer*, Stockholm 1995.
Istoriia Urala s drevneishikh vremën do 1861 g., Moscow 1989.

Ivanov, A. E. 'Vysshaia shkola kontsa XIX–nachala XX vv.: sostav i iuridicheskii status uchebnykh zavedenii', in *Istoriograficheskie i istoricheskie problemy russkoi kul'tury*, Moscow 1982.
Jars, G. *Voyages métallurgiques*, Paris 1769.
Jonsson, U. 'Bruksdöden i Örebro, Västmanland och Kopparbergs län', *Bebyggelsehistorisk tidskrift* 1983:5.
Journal of the House of Commons, XXII, April 1737.
Jörberg, L. *A History of Prices in Sweden 1732–1914*, 2 vols, Lund 1972.
Kafengauz, B. B. *Istoriia khoziaistva Demidovykh v XVIII–XIX vv.*, Moscow-Leningrad 1949.
Karlsson, P.-A. *Järnbruken och ståndssamhället. Institutionell och attitydmässig konflikt under Sveriges tidiga industrialisering 1700–1770*, Jernkontorets Bergshistoriska Skriftserie 25, Stockholm 1990.
Karpenko, Z. G. *Gornaia metallurgicheskaia promyshlennost' Zapadnoi Sibiri v 1700–1860 gg.*, Novosibirsk 1963.
Kashintsev, D. A. *Istoriia metallurgii Urala*, Moscow, Leningrad 1939.
Kaufhold, K. H. 'Gewerbelandschaften in der frühen Neuzeit', in H. Pohl, ed., *Gewerbe- und Industrielandschaften vom Spätmittelalter bis ins 20. Jahrhundert*, Stuttgart 1986.
———. *Das Metallgewerbe der Grafschaft Mark im 18. und frühen 19. Jahrhundert*, Dortmund 1976.
Kellenbenz , H. 'Europäisches Eisen. Produktion - Verarbeitung - Handel, Vom Ende des Mittelalters bis ins 18. Jahrhundert', in H. Kellenbenz, ed., *Schwerpunkte der Eisengewinnung und Eisenverarbeitung in Europa 1500–1650*, Köln 1974.
Khanykov, Ia. 'Obozrenie rudnogo proizvodstva chastnykh Orenburgskikh zavodov v 1838 g.' *Materialy dlia statistiki Rossiiskoi imperii*, part 4, St Petersburg 1884.
Kilander, S. *Den nya staten och den gamla*, Studia Historica Upsaliensia 164, Uppsala 1991.
Klingnéus, S. *Bönder blir vapensmeder. Protoindustriell tillverkning i Närke under 1600-och 1700-talen*, Studia Historica Upsaliensia 181, Uppsala 1997.
Kondrashenkov, A. A. *Krest'iane Zaural'ia v XVII–XVIII vekakh*, part 2, Cheliabinsk 1969.
Kozlov, A. G. 'K voprosu o predposylkakh tekhnicheskogo perevorota v gornozavodskoi promyshlennosti na Urale', *Voprosy istorii Urala* 10, Sverdlovsk 1970.
———. 'Ob osobennostiakh razvitiia tekhniki na kazënnykh zavodakh Urala, rubezh XVIII- XIX vekov', *Voprosy istorii Urala* 5, Sverdlovsk 1964.
———. *Tvortsy nauki i tekhniki na Urale. XVII - nachalo XX vekov*, Sverdlovsk 1981.
Kriedte, P. 'The Origins, the Agrarian Context and the Conditions in the World Market', in P. Kriedte, H. Medick and J. Schlumbohm, eds, *Industrialization before Industrialization. Rural Industry in the Genesis of Capitalism*, Cambridge 1981.
Kriedte, P., Medick, H., and Schlumbohm, J. 'Sozialgeschichte in der Erweiterung - Proto-Industrialisierung in der Verengung? Demographie, Sozialstruktur, moderne Hausindustrie: ein Zwischenbilanz der Proto-Industrialisierungs-Forschung', *Geschichte und Gesellschaft* 1992, 1, 2.
———. *Industrialisierung vor der Industrialisierung. Gewerbliche Warenproduktion auf dem Land in der Formationsperiode des Kapitalismus*, Göttingen 1977.
———. *Industrialization before Industrialization. Rural Industry in the Genesis of Capitalism*, Cambridge 1981.
Krivonogov, V. Ia. 'Krest'ianskii podnevol'nyi i naëmnyi trud na vnezavodskikh operatsiiakh na Urale v kontse XVIII–pervoi polovine XIX v.,' *Voprosy agrarnoi istorii Urala*, Sverdlovsk 1975.
———. *Naëmnyi trud v gornozavodskoi promyshlennosti Urala v XVIII veke*, Sverdlovsk 1959.

———. *Naëmnyi trud v gornozavodskoi promyshlennosti Urala v doreformennyi period 1800–1860, Naëmnyi trud v gornozavodskoi promyshlennosti Urala v doreformennoe vremia,* Sverdlovsk 1964.
Kromnow, Å. 'Övermasmästarämbetet under 1700-talet', *Med Hammare och Fackla* 9, 10.
Krupianskaia, V. Iu. and Polishchuk, N. S. *Kul'tura i byt rabochikh gornozavodskogo Urala,* Moscow 1971.
Kula, W. *An Economic Theory of the Feudal System. Towards a Model of the Polish Economy 1500–1800,* London 1976.
Kuylenstierna , C. W. U. *Om rekognitionsskogar och under bruk skatteköpta hemman,* Lund 1916.
Kuzin, A. A. *Istoriia otkrytiia rudnykh mestorozhdenii v Rossii,* 1961.
Land Commune and Peasant Community in Russia. Communal Forms in Imperial and Early Soviet Society, ed. R. Bartlett, London 1990.
Larsson , A. 'Nora socken förr och nu', in *Från bergslag och bondebygd* 1982.
Larsson, M. *Arbete och lön vid Bredsjö bruk. En studie av olika lönepринciper och lönenivåer för olika yrkeskategorier vid Bredsjö Bruk 1828–1905,* Uppsala 1985.
Laslett, P. 'Family and Household as Work Group and Kin Group', in R. Wall, J. Robin and P. Laslett, eds, *Family Forms in Historic Europe,* Cambridge 1983.
———. 'Introduction: The History of the Family', in P. Laslett and R. Wall, eds, *Household and Family in Past Time,* Cambridge 1972.
Leboutte, R. *Reconversions de la main-d'oeuvre et transition démographique. Les bassins industriels en aval de Liège XVIIe–XXe siècles,* Paris 1988.
Lenin, V. I. *The Development of Capitalism in Russia,* Moscow 1974 (first edition 1899).
Lepëkhin, I. I. *Dnevnye zapiski puteshestviia po raznym provintsiiam Rossiiskogo gosudarstva,* St Petersburg 1802.
Lesenko, D. 'Materialy dlia istorii gornogo promysla v Rossii', *Gornyi zhurnal* 1883:11.
Lewin, M. 'The *Obshchina* and the Village' in R. Bartlett, ed., *Land Commune and Peasant Community in Russia. Communal Forms in Imperial and Early Soviet Society,* London 1990.
Levine, D. and Wrightson, K. *The Making of an Industrial Society, Whickham 1560–1765,* Oxford 1991.
Lewis, G. 'The constraints of a protoindustrial society on the development of heavy industry: the case of coalmining in the south-east of France, 1773–1791', in P. Mathias and J.A. Davies, eds, *Innovations and Technology in Europe,* Oxford 1991.
Lindegren, J. *Utskrivning och utsugning. Produktion och reproduktion i Bygdeå 1620–1640,* Uppsala 1980.
Lindqvist, S. *Technology on Trial. The Introduction of Steam Power Technology into Sweden 1715–1736,* Uppsala 1984.
Lindström, D. *Skrå, stad och stat. Stockholm, Malmö och Bergen ca. 1350–1622,* Uppsala 1991.
Lucassen, J. 'The other proletarians: Seasonal Labourers, Mercenaries and Miners', *International Review of Social History* Supplement, 1994:39.
Lundh, C. 'Households and Families in Pre-industrial Sweden', *Continuity and Change* 1995:10(1).
Lushinkova, N. M. 'K voprosu o klassovoi bor'be na gornykh zavodakh Permskoi gubernii v 1801–1860 gg. (Uchastie lesorabochikh v massovykh vystupleniiakh)', *Voprosy istorii Urala,* Sverdlovsk 1975.
———. 'Vliianie toplivnoi bazy na tekhnicheskie usovershenstvovaniia v metallurgicheskoi promyshlennosti Urala', *Voprosy istorii Urala,* Collection 10, Sverdlovsk 1970.
Lynch, K. 'The Family and the History of Public life', *Journal of Interdisciplinary History* 1994:24(4).

Magnusson, G. 'Ironmaking in a Long Time Perspective. Some Important Questions', in G. Magnusson, ed., *The Importance of Ironmaking. Technical Innovation and Social Change.* Papers presented at the Norberg Conference on May 8–13, Stockholm 1995.

Markets and Manufacture in Early Industrial Europe, ed. M. Berg, London, New York 1991.

Materialy k voprosu o nizshikh gornotekhnicheskikh uchebnykh zavedeniiakh na Urale, St Petersburg 1905.

Mathias, P. 'Resources and Technology', in P. Mathias and J. A. Davies, eds, *Innovations and Technology in Europe*, Oxford 1991.

Mel'nikov, P. I. 'Dorozhnye na puti iz Tambovskoi gubernii v Sibir', *Otechestvennye zapiski*, vol. 19, no. 4, St Petersburg 1841.

Mendels, F. 'Proto-industrialization: the first phase of the industrialization process', *Journal of Economic History* 1972:32.

———. *Industrialization and Population Pressure in Eighteenth-Century Flanders*, New York 1981.

Metallurgical Works and Peasantry. Problems of Social Organization of Industry in Russia and Sweden in Early-Industrial Period, Ekaterinburg 1992.

Minenko, N. A. *Russkaia krest'ianskaia obshchina v zapadnoi Sibiri XVIII - pervaia polovina XIX* veka, Novosibirsk 1991.

Minenko, N. A. and Poberezhnikov, I. V. 'The interaction of industry and agricultural environment', Unpublished paper at the third meeting on iron-making in Russia and Sweden before the twentieth century, Ekaterinburg 1993.

Minenko, N. A. et al., 'Ural Iron Before the Industrial Revolution', in G. Rydén and M. Ågren, eds, *Ironmaking in Sweden and Russia. A Survey of the Social Organisation of Iron Production before 1900*, Opuscula Historica Upsaliensia 12, Uppsala 1993.

Mitchell, B. R. 'Statistical Appendix', in C. M. Cipolla, ed., *The Fontana Economic History of Europe*, 4(2) third edition, Glasgow 1976.

Mitterauer, M. 'Produktionsweise, Siedlungsstruktur und Sozialformen im Österreichischen Montanwesen des Mittelalters und der frühen Neuzeit', in M. Mitterauer, ed., *Österreichisches Montanwesen. Produktion, Verteilung, Sozialformen*, Vienna 1974.

Mitterauer, M. and Kagan, A. 'Russian and central European family structures: a comparative view', *Journal of Family History* 1982:7.

Mitterauer, M. and Sieder, R. *The European Family*, Oxford 1982.

Montelius, S. '1600–1815' , in Montelius, S., Utterström, G. and Söderlund, E. *Fagerstabrukens historia V. Arbetare och arbetarförhållanden*, Uppsala 1959.

———. 'Iggesunds Bruk 1685–1869', in G. Utterström, ed., *Iggesunds Bruks Historia 1685–1985*, Iggesund 1985.

———. 'Relationer mellan bruk och bönder i Dalarna. Den äldsta järnhanteringens tekniska och ekonomiska struktur', in *Från kulturdagarna i Bonäs Bygdegård 23–25 juni 1975*, Uppsala 1976.

———. *Säfsnäsbrukens arbetskraft och försörjning 1600–1865*, Geographica, Skrifter från Uppsala universitets Geografiska institution, Falun 1962.

Montelius, S., Utterström, G. and Söderlund, E. *Fagerstabrukens historia V. Arbetare och arbetarförhållanden*, Uppsala 1959.

Mozel', Kh. *Materialy dlia geografii i statistiki Rossii, sobrannye sluzhashchimi general'nogo shtaba: Permskaia guberniia*, vol. 2, St Petersburg 1864.

Mukhin, V. V. 'Rabochaia sila na predpriiatiiakh ural'skoi gornozavodskoi votchiny Vsevolozhskikh v pervoi polovine XIX veka', in *Iz istorii rabochego klassa i krest'ianstva Permskogo kraia*, Perm' 1965.

Myrdal, J. and Söderberg, J. *Kontinuitetens dynamik. Agrar ekonomi i 1500-talets Sverige*, Stockholm 1991.

Myska, M. 'Pre-industrial iron-making in the Czech lands: the labour force and production relations circa 1500 - circa 1840, *Past & Present* 1979.

Nechaeva, M. Iu. and Golikova, S. V. 'Households of the Zlatoustovsky okrug residents on the turn of the XVIII–XIX centuries', in *Metallurgical Works and Peasantry. Problems of Social Organization of Industry in Russia and Sweden in Early-Industrial Period*, Ekaterinburg 1992.

Neeson, J. M. *Commoners: Common Rights, Enclosure and Social Change in England, 1700–1820*, Cambridge 1993.

Nelson, H. 'En bergslagsbygd, en historisk-geografisk överblick', *Ymer* 1913.

Nisser, M. 'Forsmark - ett av vallonbruken kring Dannemora gruvor', in *Forsmark och vallonjärnet*, Forsmarks kraftgrupp 1987.

Nogin, K.-I. 'K istorii uglezhzheniia na Urale', *Ural'skii tekhnik* 1918:10–12.

Noraskogs Arkiv: bergshistoriska samlingar och anteckningar, ed. J. Johansson, Stockholm 1889–1928.

Norberg , P. *Engelsbergs Bruk. En historisk framställning*, Stockholm 1934.

———. *Avesta under kopparbrukets tid*, del 1, Avesta 1956.

———. *Forna tiders järnbruk i Norr- och Västerbotten*, Stockholm 1957.

Nordström, O. *Relationer mellan bruk och omland i östra Småland 1750–1900*, Lund 1952.

Norman, H. 'Den utomäktenskapliga fruksamhetens lokala variationer. En fallstudie från Örebro län', in *Historieforskning på nya vägar. Studier tillägnade Sten Carlsson 14.12 1977*, Uppsala 1977.

'O publichnom ispytanii vospitannikov Gornogo Korpusa v 1825 godu', *Gornyi zhurnal* 1825:2.

Odén, B. 'Komparation över tid. Till de långa linjernas metodologi' in *Den kritiske analyse. Festskrift til Ottar Dahl på 70-årsdagen den 5. januar 1994*, Oslo 1994.

Ogilvie, S. 'Proto-industrialization in Europe', *Continuity and Change* 1993:8(2).

———. 'Soziale Institutionen und Proto-industrialisierung', in M. Cerman and S. Ogilvie, eds, *Protoindustrialisierung in Europa*, Vienna 1994.

Olyshev, P. A. 'O pudlingovom proizvodstve v Kamsko-Votkinskom zavode', *Gornyi zhurnal* 1843:6.

Omberg, T. *Bergsmän i hyttelag. Bergsmansnäringens utveckling i Linde och Ramsberg under en 100-årsperiod från mitten av 1700-talet*, Stockholm 1992.

Om Noraskog, äldre och nyare anteckningar, ed. J. Johansson, Stockholm 1875–1884.

Ordbok över svenska språket utg. av Svenska Akademien, Lund 1826– (present day).

Orlov, A. S. *Volneniia na Urale v seredine XVIII veka. K voprosu o formirovanii proletariata v Rossii*, Moscow 1979.

Oskolkov, G.I. 'Vnedrenie ruchnykh pil na kazënnykh ural'skikh zavodakh v 30-50-e gody XIX veka', *Voprosy istorii Urala*, Collection 5, Sverdlovsk 1964.

Österberg, E. *Kolonisation och kriser. Bebyggelse, skattetryck, odling och agrarstruktur i västra Värmland ca 1300–1600*, Lund 1977.

Pallas, P. S. *Puteshestvie po raznym mestam Rossiiskogo gosudarstva*, Part 2, Book 1, St Petersburg 1786.

Pamiatniki sibirskoi istorii XVIII v, Tome 1, St Petersburg 1882.

Pavlenko, N. I. *Razvitie metallurgicheskoi promyshlennosti Rossii v pervoi polovine XVIII v.: Promyshlennaia politika i upravlenie*, Moscow 1953.

Pavlov, M. A. 'Predislovie', in V. De Hennin, *Opisanie ural'skikh i sibirskikh zavodov 1735 g.*, Moscow 1937.

Perechod ot feodalizma k kapitalizmu v Rossii: Materialy Vsesoiuznoi diskussii, Moscow 1969.

Peterson, C. *Peter the Great's Administrative and Judicial Reforms: Swedish Antecedents and the Process of Reception*, Stockholm 1979.

———. ''En god ämbetsman är bättre än en god lag...'. Frågan om justitiekanslern som en allmogens besvärsinstans i klagomål över kronobetjänternas ämbetsutövning' in Y. Blomstedt et. al., *Administrasjon i Norden på 1700-talet*, Oslo 1985.

Pettersson, I.-M. *Norbergs Bergslag*, Stockholm 1994.

Pettersson, J.-E. *Från kris till kris. Den svenska stålindustrins omvandling under 1920- och 1970-talen*, Stockholm 1988.

Pettersson, T. J.-E. 'Blästan och blåsningen', in *Lima and Transtrand. Ur två socknars historia. Del 1. Myrjärn och smide,* Stockholm 1982.

———. 'Myrjärnshanteringens storhetstid och nedgång', in *Lima och Transtrand. Ur två socknars historia. Del 1. Myrjärn och smide,* Stockholm 1982.

Pipes, R. *Russia under the Old Regime,* Harmondsworth 1977.

Pliushchevskii, B. G. 'Gosudarstvennye krest'iane Viatskoi gubernii v 1-i polovine i seredine XIX v.', *Sbornik Udmurtskogo NII.* no. 17.

———. 'Imushchestvennoe i sotsial'noe rassloenie krest'ianstva gubernii Bol'shogo Urala v 1-i polovine XIX veka', in *Problemy sotsial'no-ekonomicheskogo razvitiia derevni Srednego Povolzh'ia v period feodalizma,* Kazan' 1986.

Poberezhnikov, I. V. *Obshchestvenno-politicheskie vzgliady russkikh krest'ian Sibiri v period pozdnego feodalizma,* Novosibirsk 1989.

Pohl, H. 'Einführung', in H. Pohl, ed., *Gewerbe- und Industrielandschaften vom Spätmittelalter bis ins 20. Jahrhundert,* Stuttgart 1986.

Pohl, H., ed. *Gewerbe- und Industrielandschaften vom Spätmittelalter bis ins 20. Jahrhundert,* Stuttgart 1986.

Polianskii, F. Ia. *Ekonomicheskii stroi manufaktury v Rossii xviii veka,* Moscow 1956.

Pollard, S. 'Regional and inter-regional economic development in Europe in the eighteenth and nineteenth centuries', in *Debates and Controversies in Economic History. Eleventh International Economic History Congress,* Milan 1994.

———. 'Regional markets and national development' in M. Berg, ed., *Markets and Manufacture in Early Industrial Europe,* London, New York 1991.

Popov, N. S. *Khoziaistvennoe opisanie Permskoi gubernii,* Perm' 1804.

Portal, R. 'The industrialisation of Russia', in *The Cambridge Economic History of Europe* vol. vi:2, Cambridge 1978.

———.'Manufactures et classes sociales en Russie au xviiie siècle, *Revue Historique* 1949.

———. *L'Oural au XVIIIe siècle. Étude d'Histoire Économique et Sociale,* Paris 1950.

———. *The Slavs,* Studies in World History, London 1969.

Poulsen, B. '"Alle myne rent', bondekredit i 15-1600-tallet', *Dansk Historisk tidsskrift* 1990.

Prak, M. 'Regions in Early Modern Europe', in *Debates and Controversies in Economic History. Eleventh International Economic History Congress,* Milan 1994.

Preobrazhenskii, A. A. 'Rabotnye liudi na Urale v XVII–nachale XVIII veka. K voprosu o rynke rabochei sily', in *Iz istorii rabochego klassa Urala,* 1961.

———. *Ural i Zapadnaia Sibir' v kontse XVI–nachale XVIII veka,* Moscow 1972.

Promyshlennost' i rabochie Urala v period kapitalizma (1861–1917 gg.), Sverdlovsk 1991.

Protoindustrialisierung in Europa, eds M. Cerman and S. Ogilvie, Vienna 1994.

Putilova, I. V. *Kazënnye zavody Urala v period perechoda ot krepostnichestva k kapitalizmu,* Krasnoiarsk 1986.

Putilova, M. V. 'O sostoianii toplivnoi bazy zavodov Ekaterinburgskogo gornogo okruga v 40-70-kh godakh XIX veka', *Voprosy istorii Urala,* Perm' 1966.

Rabochee dvizhenie v Rossii v XIX v., vol. 1, Moscow 1955.

Raeff, M. *The Well-Ordered Police State. Social and Institutional Change through Law in the Germanies and Russia, 1600–1800,* Yale 1983.

Ragozin, E. I. *Zhelezo i ugol'na Urale,* St Petersburg 1902.

Reddy, W. *The Rise of Market Culture. The Textile Trade and French Society, 1750–1900,* Cambridge 1984.

Ringmar, R. 'Bergsmän i Bergslag. En undersökning av bergsmansbruket i Ovansjö och Torsåkers socknar, Gävleborgs län', Unpubl. diss., Department of Economic History, Uppsala University, 1960.

Rinman, C. *Handbok uti den gröfre jern- och stålförädlingen,* Falun 1823.

Rinman, S. *Allmogesmidet i Dalarna 1764 års undersökningar,* ed. C. Sahlin, Stockholm 1936.

Rogov, N. 'Raznye svedeniia o lesakh Permskoi gubernii i o zagotovke glavnykh lesnykh materialov dlia gornykh zavodov i solianykh promyslov', Tome 1, *Gornyi zhurnal* 1877.

Rosander, G. 'Skogsarbetaren och hans verktyg', *Kungliga Skogs- och Lantbruksakademiens Tidskrift* 1988:127.
Rossiia. Polnoe geograficheskoe opisanie nashego otechestva, tome 5, 1914.
Rowlands, M. *Master and Men in the West Midland Metalware Trades before the Industrial Revolution*, Manchester 1975.
Rudolph, R. L. 'Agricultural structure and proto-industrialisation in Russia: Economic development with unfree labor', in *VIIIe Congrès international d' histoire Economique, Budapest 16–22 août 1982, section A2: La protoindustrialisation: théorie et réalité.*
———. 'Family structure and proto-industrialization in Russia', *Journal of Economic History* 1980.
Rule, J. *The Experience of Labour in Eighteenth-Century Industry*, London 1981.
Rydeberg, G. *Skatteköpen i Örebro län 1701–1809*, Uppsala 1985.
Rydén, G. 'Bergsmän, skog och tackjärn', *Svensk Historisk tidskrift* 1994.
———. 'Gustaf Ekman, Jernkontoret och lancashiresmidet - Ett inlägg i synen på teknisk utveckling', *Polhem, Tidskrift för teknikhistoria* 1994:2.
———. *Hammarlag och hushåll. Om relationen mellan smidesarbetet och smedshushållen vid Tore Petrés brukskomplex 1830–1850*, Stockholm 1991.
———. 'Iron production and the household as a production unit in nineteenth-century Sweden', *Continuity and Change* 1995:10.
———. 'Skill and technical change in the Swedish iron industry, 1750–1860', *Technology and Culture* 1998.
Rydén, G. and Ågren, M., eds, *Ironmaking in Sweden and Russia. A Survey of the Social Organisation of Iron Production before 1900*, Opuscula Historica Upsaliensia 12, Uppsala 1993.
Ryvkin, A. A. *Gornotekhnicheskoe obrazovanie na Urale*, Moscow 1956.
Rösener, W. *The Peasantry of Europe*, Oxford UK & Cambridge USA 1994.
Sabean, D. *Property, Production, and Family in Neckarhausen, 1700–1870*, Cambridge 1990.
Sabel, C. and Zeitlin, J. 'Historical Alternatives to Mass Production', *Past & Present* 1985.
Safronova, T. V. 'Zakonodatel'nye istochniki o formirovanii i polozhenii rabochei sily na metallurgicheskikh predpriiatiiakh Rossii i Urala vo vtoroi polovine XVIII veka', *Polozhenie i bor'ba trudiashchikhsia Urala XVIII–nachala XX v.*, Ekaterinburg 1987.
Schlumbohm, J. ''Proto-Industrialisierung' als forschungsstrategisches Konzept und als Epochenbegriff - eine Zwischenbilanz', in M. Cerman and S. Ogilvie, eds, *Protoindustrialisierung in Europa*, Vienna 1994.
Scott, J. C. *Weapons of the Weak. Everyday Forms of Peasant Resistance*, New Haven 1985.
Seccombe, W. *A Millennium of Family Change. Feudalism to Capitalism in Northwestern Europe*, London 1992.
———. *Weathering the Storm. Working-Class Families from the Industrial Revolution to the Fertility Decline*, London 1993.
Segalen, M. *Love and Power in the Peasant Family*, Oxford 1983.
Semëvskii, V. I. *Krest'iane v tsarstvovanie imperatritsy Ekateriny II*, St Petersburg 1901.
Serbina, K. N. *Krest'ianskaia zhelezodelatel'naia promyshlennost' Severo-Zapadnoi Rossii XVI - pervoi poloviny XIX v.*, Leningrad 1971.
———. *Krest'ianskaia zhelezodelatel'naia promyshlennost' Tsentral'noi Rossii XVI – pervoi poloviny XIX v.*, Leningrad 1978.
Servais, P. 'Les structures agraires du Limbourg et des pays d'Outre-Meuse du XVIIe au XIXe siècle', *Annales ESC* 1982:37.
Shinn, T. *L'École Polytechnique 1794–1914*, Paris: Presses de la Fondation nationale des sciences politiques, 1980.
Shkerin, V. A. 'Rebellious crowds in social conflicts at the Ural private works in the first half of the XIX century', in *Metallurgical Works and Peasantry. Problems*

of Social Organization of Industry in Russia and Sweden in Early-Industrial Period, Ekaterinburg 1992.
———. 'The process of adoption and social conflicts at the mining and ironworks in the Urals during the feudal period', Unpublished paper at the second meeting on iron-making in Russia and Sweden before the twentieth century, Uppsala 1992.
Sjöberg, M. 'Järn bryter bygd - en sanning med variation', *Scandia* 1992:1.
———. *Järn och jord. Bergsmän på 1700-talet*, Stockholm 1993.
Smetanin, S. I. 'Razlozhenie gornozavodskogo naseleniia i formirovanie rabochego klassa na Urale', *Voprosy istorii* 1982:2.
Smirnov, S. S. 'Sel'skoe khoziaistvo pripisnykh krest'ian gornozavodskogo Urala na rubezhe XVIII–XIX vv.', *Gosudarstvennye krest'iane Urala v epokhu feodalizma*, Ekaterinburg 1992.
Somers, M. 'The 'misteries' of property. Relationality, rural-industrialization, and community in Chartist narratives of political rights' in J. Brewer and S. Staves, eds, *Early Modern Conceptions of Property*, London, New York 1995.
Sommarin, E. *Bidrag till kännedom om arbetareförhållanden vid svenska bergverk och bruk i äldre tid till omkring 1720*, Lund 1908.
Sonenscher, M. 'Journeymen's migrations and workshop organization in eighteenth century England', in S. L. Kaplan and C. J. Koepp, eds, *Work in France. Representations, Meaning and Practice*, Cornell 1986.
———. *Work and Wages. Natural Law, Politics and the Eighteenth-Century French Trades*, Cambridge 1989.
Sotsial'no-ekonomicheskoe polozhenie kadrov gornozavodskoi promyshlennosti v doreformennyi period, Sverdlovsk 1989.
Spufford, M. *Contrasting Communities. English Villagers in the Sixteenth and Seventeenth Centuries*, Cambridge 1974.
Stromer, W. von 'Gewerbereviere und Protoindustrien in der Spätmittelalter und Frühneuzeit', in H. Pohl, *Gewerbe- und Industrielandscahften vom Spätmittelalter bis ins 20. Jahrhundert*, Stuttgart 1986.
Strumilin, S. G. *Chërnaia metallurgiia v Rossii i v SSSR. Tekhnicheskii progress za 100 let*, Moscow-Leningrad 1935.
———. *Istoriia chërnoi metallurgii v SSSR*, Moscow 1954.
Ström, E. T. *Hällefors och Grytthytte bergslag från år 1640 till omkring 1700*, Lund 1956.
Styles, J. 'Embezzlement, industry and the law in England 1500–1800', in M. Berg and M. Sonenscher, eds, *Manufacture in Town and Country Before the Factory*, Cambridge 1983.
Sundbärg, G. *Emigrationsutredningen V*, Stockholm 1910.
Sundin, J. 'Family Building in Paternalistic Proto-Industries: A Cohort Study from Nineteenth-Century Swedish Iron Foundries', *Journal of Family History* 1989:14(3).
Svensk handelsstatistik 1637–1737, eds E. F. Heckscher and B. Boëthius, Stockholm 1938.
Svod materialov po kustarnoi promyshlennosti v Rossii, St Petersburg 1874.
Tal'skaia, O. S. 'Iz istorii uglezhzheniia na ural'skikh zavodakh', *Trudy Sverdlovskogo oblastnogo kraevedcheskogo muzeia*, Sverdlovsk 1960.
———. 'Pamiatniki uchëta truda zavisimykh krest'ian v metallurgicheskoi promyshlennosti Rossii', *Istochnikovedenie i arkheografiia Sibiri*, Novosibirsk 1977.
Tamm, A. G. *Försök till besvarandet af den af herrar bruks-ägare i Wermland och Dahl år 1827 framställde prisfråga*, Karlstad 1831.
Tarkiainen, K. *'Vår gamble Arffiende Ryssen'. Synen på Ryssland i Sverige 1595–1621 och andra studier kring den svenska Rysslandsbilden från tidigare stormaktstid*, Uppsala 1974.
Thirsk, J. 'Industries in the Countryside', in *Essays in the Economic and Social History of Tudor and Stuart England in Honour of R.H. Tawney*, Cambridge 1961.
Tigerstedt, Ö. *Kavalkad. Fagerstabrukens historia*, Uppsala 1957.

Tomilov, A. 'Metallurgical trades of the Ural peasantry in the mining industry at the beginning of the 18th century.', in *Metallurgical Works and Peasantry. Problems of Social Organization of Industry in Russia and Sweden in Early-Industrial Period*, Ekaterinburg 1992.

Tomsinskii, S. M. 'Rassloenie ural'skoi derevni v sviazi s razvitiem manufaktury (pervaia polovina XVIII v.)', in *Ezhegodnik po agrarnoi istorii Vostochnoi Evropy*, Leningrad 1972.

Torstendahl, R. *Teknologins nytta*, Studia Historica Upsaliensia 66, Uppsala 1975.

Trempé, R. *Les Mineurs de Carmaux 1848–1914*, vol. I, Paris 1971.

Tret'iakov, A. 'Shadrinskii uezd Permskoi gubernii v sel'skokhoziaistvennom otnoshenii', *Zhurnal ministerstva gos. imushchestv.*, part 45, 1852.

Tylecote, R. F. *The Early History of Metallurgy in Europe*, New York 1987.

Ural'skoe gornoe obozrenie 1900:4.

Ustiantsev, S. V. and Logunov, E. V. *British Technological Experience and Ural Mining and Iron-Making Works of the XIX century*, Ekaterinburg 1992.

Utterström, G. '1815–1870', in S. Montelius, G. Utterström and E. Söderlund, *Fagerstabrukens historia V. Arbetare och arbetarförhållanden*, Uppsala 1959.

Vagina, P. A. 'Lichnoe khoziaistvo i rabochaia kvalifikatsiia masterovykh chastnykh zavodov Iuzhnogo Urala v poslednei chetverti XVIII v.', *Voprosy istorii Urala*, Collection 8, Sverdlovsk 1969.

Vanja, Chr. 'Mining Women in Early Modern European Society', in T. M. Safley and L. N. Rosenband, eds, *The Workplace Before the Factory*, Cornell 1993.

Verley, P. 'La révolution industrielle Anglaise, une révision (note critique)', *Annales ESC*, 1991.

Vershinin, E. V. 'Istochniki po istorii organizatsii pervogo zhelezodelatel'nogo predpriiatiia na Urale.', in *Istochniki po sotsial'no-ekonomicheskoi istorii Urala dooktiabr'skogo perioda*, Moscow 1992.

Vilkov, O. N. *Ocherki sotsial'no-ekonomicheskogo razvitiia Sibiri kontsa XVI–nachala XVIII v*, 1990.

Vodarskii, Ia. E. 'Zemel'nye ugod'ia i zemlevladenie na Urale v XVII – pervoi polovine XIX v.', in *Krest'ianstvo Urala v epokhu feodalizma*, Sverdlovsk 1988.

Watelet, H. *Une industrialisation sans développement. Le bassin de Mons et le charbonnage de Grand-Hornu du milieu de XVIIIe siècle au milieu de XIXe siècle*, Louvain 1980.

Weinhagen, A. 'Gruvor, hyttor och hamrar under 1600- och 1700-tal', in *Norberg genom 600 år. Studier i en gruvbygds historia*, Uppsala 1958.

———. *Norbergs Bergslag samt Gunnilbo och Ramnäs, till omkring 1820*, Lund 1947.

Winberg, C. *Folkökning och proletarisering. Kring den sociala strukturomvandlingen på Sveriges landsbygd under den agrara revolutionen*, Göteborg 1975.

———. *Grenverket, Studier rörande jord, släktskapssystem och ståndsprivilegier*, Stockholm 1985.

Wolf, E. *Peasants*, Englewood Cliffs 1966.

Wolf, P. and Mauro, F. L. 'L'âge de l'artisanat', in L.-H. Parias, ed., *Histoire Générale du Travail*, Paris 1977.

Woronoff, D. *Histoire de l'industrie en France du xvie siècle à nos jours*, Paris 1994.

———. *L'industrie sidérurgique en France pendant la révolution et l'empire*, Paris 1984.

Wrigley, E. A. *Continuity, Change and Change. The Character of the Industrial Revolution in England*, Cambridge 1988.

Zaozërskaia, E. I. *U istokov krupnogo proizvodstva v russkoi promyshlennosti XVI–XVII vekov*, Moscow 1970.

Zapiski Russkogo tekhnicheskogo obshchestva, vol. 6, 1879.

Zlotnikov, M. F. *Pervoe opisanie ural'skikh i sibirskikh zavodov*, Moscow 1937.

INDEX

www.ingramcontent.com/pod-product-compliance
Ingram Content Group UK Ltd.
Pitfield, Milton Keynes, MK11 3LW, UK
UKHW021823100926
8556IPUK00003B/131

9 781571 819550